Advances in Wireless Engineering

Volume I

Advances in Wireless Engineering Volume I

Edited by **Bernhard Ekman**

NY RESEARCH PRESS

New York

Published by NY Research Press,
23 West, 55th Street, Suite 816,
New York, NY 10019, USA
www.nyresearchpress.com

Advances in Wireless Engineering: Volume I
Edited by Bernhard Ekman

International Standard Book Number: 978-1-63238-039-5 (Hardback)

Printed in China.

Contents

Preface

The term wireless is self-explanatory. It refers to any form of communication, in which the transfer of information does not occur over a connected electrical conductor. Instead, electromagnetic waves carry the signal over most of the communication path. The range of the transmitted distance can be anywhere between a few meters to thousands of kilometers. The origin of wireless communication can be traced to the year 1880, when Alexander Graham Bell and Charles Sumner Tainter invented and patented the photophone, and had their first conversation through it.

The discipline of Wireless engineering deals with wireless technology components such as radio communications and radar. It also covers the usage of other techniques such as acoustic, infrared, and optical technologies. Though cellular networks have recently gained a lot of popularity, it was with the invention of the FM Radio in 1935, that wireless communications attracted the focus of both private and government sectors.

With the advent of technology, the scope of wireless technology has magnified. To simplify the experience for the user, new applications have been developed. For instance, Wi-Fi is a wireless local area network which enables computing devices to connect easily to the Internet. Similarly, Satellite communications are particularly important in domains of transportation, aviation and maritime.

Editor

1

Spectrally Efficient Multi-Carrier Modulation Using Gabor Transform

Shravan Sriram, Naveenkumar Vijayakumar, P. Aditya Kumar, Akash S. Shetty, V. P. Prasshanth, K. A. Narayanankutty

Department of Electronics and Communication Engineering, Amrita School of Engineering, Amrita Vishwa Vidyapeetham University, Coimbatore, India.

ABSTRACT

Non Orthogonal Frequency Division Multiplexing (NOFDM) systems make use of a transmission signal set which is not restricted to orthonormal bases unlike previous OFDM systems. The usage of non-orthogonal bases generally results in a trade-off between Bit Error Rate (BER) and receiver complexity. This paper studies the use of Gabor based on designing a Spectrally Efficient Multi-Carrier Modulation Scheme. Using Gabor Transform with a specific Gaussian envelope; we derive the expected BER-SNR performance. The spectral usage of such a NOFDM system when affected by a channel that imparts Additive White Gaussian Noise (AWGN) is estimated. We compare the obtained results with an OFDM system and observe that with comparable BER performance, this system gives a better spectral usage. The effect of window length on spectral usage is also analyzed.

Keywords: Non-Orthogonal Frequency Division Multiplexing (NOFDM); Spectrally Efficient Frequency Division Multiplexing (SEFDM); Gabor Transform; OFDM; Reisz Bases; Multi Carrier Modulation (MCM)

1. Introduction

With a growth in the number of users, there is an increase in the demand for spectrum. Research pertaining to the efficient use the available spectrum has produced numerous results over the past few years. This work came to fruition in the form of Orthogonal Frequency Division Multiplexing, which is a special class of Multiple Carrier Modulation that has emerged as a leading candidate for high data rate wireless communication. The main advantage of this is its ease of implementation that eliminates ICI at low receiver complexity.

Ahmed *et al.* in [1] introduced a Spectrally Efficient Frequency Division Multiplexing (SEFDM) system which is spectrally more efficient than OFDM. In this SEFDM system non-orthogonality was introduced by a parameter α, where $\Delta f = \alpha/T$ and $0 < \alpha < 1$. The above discussed system requires high frequency precision in order to reduce frequency offset effects. In addition, it will require different equipment for different values of frequency spacing. To overcome the limitations of this system a technique is discussed in [2,3] where, the transmitted symbol set was modified by zero padding which leads to higher frequency resolution on performing IDFT. The partial transmission of the IDFT results in spectrum compression.

By exploiting the circular conjugate symmetry property of Discrete Fourier Transform (DFT) better spectral efficiency & BER-SNR performance was achieved as discussed in [4]. All these systems make use of non-orthogonal bases for which an inverse doesn't exist. This results in greater receiver complexity. In this paper we develop an NOFDM system that deviates from the conventional system as discussed in [5] where $\Delta fT < 1$, by introducing a non-orthogonal basis set that possess a dual, which can be used in the receiver to retrieve the transmitted symbol, hence simplifying the receiver design. In order to understand the NOFDM system, a clear understanding of the working of an OFDM system is a prerequisite.

This paper is organized as follows: Section 2 gives an insight into the critical concepts behind the working of an OFDM system that are critical for a clear understanding of an NOFDM system. Section 3 begins with a brief introduction on NOFDM, and proceeds to discuss the significance of Gabor bases and their use in the implementation of an NOFDM system. A typical OFDM system has been considered as the baseline system. Section 4 gives the simulation results of the proposed NOFDM system in comparison with the baseline OFDM system. Conclusions are drawn from the simulation results that help in developing an NOFDM system with optimal parameters resulting in better spectral efficiency and BER-SNR performance.

2. OFDM

In an OFDM system, the symbols obtained after constellation mapping are modulated onto orthogonal signals of different frequencies corresponding to each of the subcarriers used [6].

$$S(n) = X_k * S_k(n); \quad (1)$$

where, X_k represents the symbols to be transmitted and $S_k(n)$ represents each subcarrier given by,

$$S_k(n) = \cos 2\pi kn/N - j\sin 2\pi kn/N \quad (2)$$

On imposing Hermitian symmetry to the symbols *i.e.* $X_{N-K} = X_k$ for a symbol stream of length $N/2$ we end up with a stream of length N where the symbols are symmetric about $X_{N/2}$. Because of the use of orthogonal carriers, this modulation of symbols can be simplified and represented as taking an IFFT of the stream of symbols. Thus the IFFT and FFT pair performs the operation of modulation and demodulation of the symbols onto orthogonal Sine and Cosine subcarriers (depicted in **Figure 1**).

Another concept of interest is the use of cyclic prefix for the diagonalization of the channel matrix. It is this diagonalization that enables channel partitioning. Cyclic prefix also helps in the removal of ISI. Thus the transmitter portion of the OFDM system can be visualized as depicted in **Figure 2**.

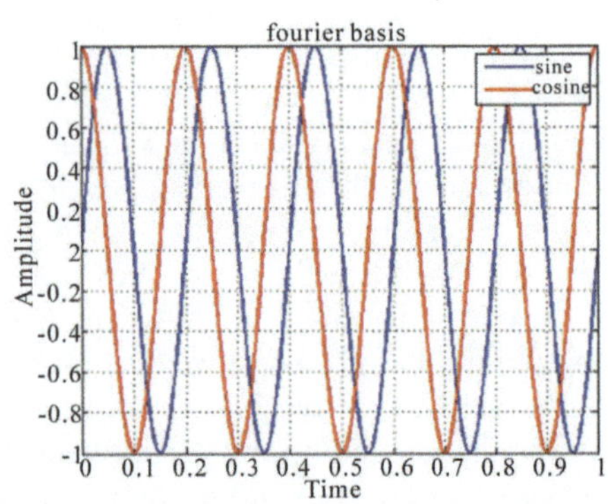

Figure 1. Fourier basis for specific channel.

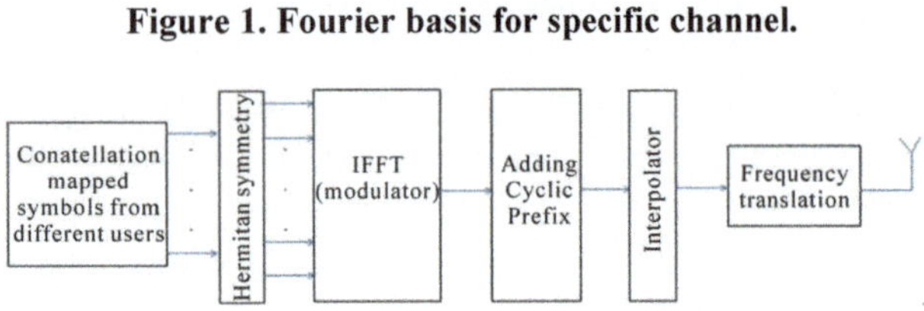

Figure 2. Transmitter of an OFDM system.

3. NOFDM

Non Orthogonal Frequency Division Multiplexing provides us with a major advantage of spectral efficiency, as the underlying pulse can be chosen with sharper frequency domain decay. Also the modulation scheme based on Reisz bases tends to be more robust against frequency selective fading [5]. The use of non-orthogonal basis functions (Riesz basis) is the key idea in this paper.

3.1. Gabor Transform

Gabor transform is a type of Short Time Fourier Transform (STFT). In STFT, the input symbols are first windowed in the time domain and then the Fourier transform operation is performed on the windowed symbol set. The windowing operation removes the orthogonality of the basis functions [7]. In Gabor transform a Gaussian window is used for windowing. It is given by,

$$w(n) = \left(\frac{1}{2\pi}\right)^{1/2} \cdot e^{-\sigma n^2/2} \quad (3)$$

Gabor transform is a two dimensional transform in which the response depends on the frequency of the signal and the time at which windowing is performed.

The basis functions of Gabor transform are non-orthogonal which means that the basis functions are not invertible. But, a unique property of a Gabor bases is that it is that there exists a dual basis [8] set such that the original signal can be recovered by a transformation operation which uses the dual basis set.

3.2. Implementation of NOFDM System

In the NOFDM system the loss of orthogonality calls for the use of non-orthogonal signals for mapping the symbols onto. Hence to induce non orthogonality we make use of a form of Gabor transform for modulation. Here, we go with a combination of Gaussian and hanning window as the final combination that performs windowing in the time domain. This window is given by,

$$w(n) = \left(\frac{1}{2}\right) \cdot \left(\frac{1}{2\pi}\right)^{\frac{1}{2}} \cdot \left(1 - \cos\left(\frac{2\pi n}{N-1}\right)\right) \cdot e^{-\frac{\sigma n^2}{2}} \quad (4)$$

We use Hanning window in order to provide finite support for the Gaussian window. We select this window approximately eight times the 3σ of the Gaussian window. This window and FFT combination can be approximated as performing the modified Gabor transform (since the window is not just Gaussian in nature) on the signal. This Gabor basis set (each basis corresponding to a specific carrier) obtained, used for modulating and demodulating the symbols from different users is as shown in **Figure 3**.

It was observed that for σ value of $2/\pi$ we obtain better results while using Gabor bases in image processing [9], using this results in the Gabor Bases a minimum 3 dB bandwidth is observed. So, we use this value for implementation of the Modified Gabor transform in the NOFDM system. Also, it is observed that the BER-SNR performance improves as the value of sig decreases and an optimal performance is obtained at a sig value of $2/\pi$ as shown in **Figure 4**.

In the NOFDM system we modulate the symbols using the Modified Gabor basis signals which follow the equation,

$$s(n)=\sum_{k=0}^{N-1} x_k(n)w(n)e^{-j2\pi f_k t} \quad (5)$$

At the receiver, to demodulate the transmitted signal we use the Inverse Gabor transform whose bases are the conjugate of the Gabor basis [8]. The Gaussian window used in the inverse transform is given by,

$$W'(n)=(1/2\pi)^{1/2}\cdot e^{\frac{\sigma n^2}{2}} \quad (6)$$

The inverse transform is given by,

$$y(n)=\sum_{k=0}^{N-1} r(n)W'(n)e^{j2\pi f_k t} \quad (7)$$

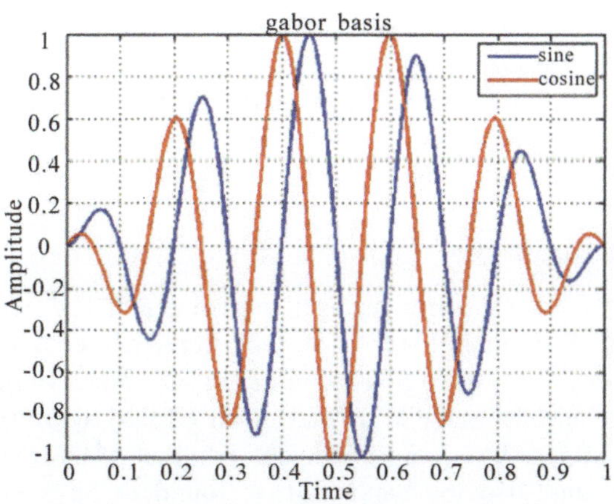

Figure 3. Gabor basis for a specific channel.

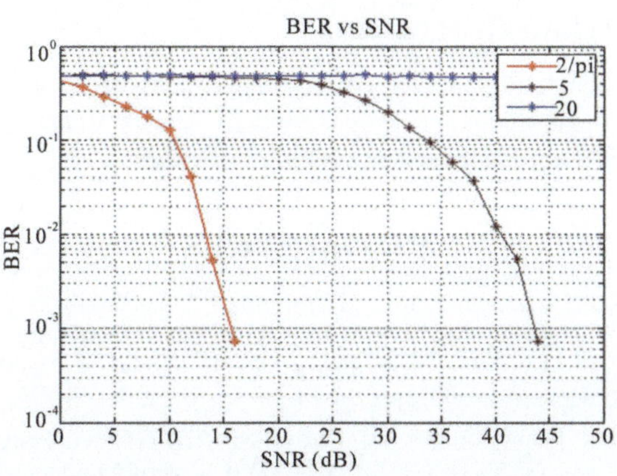

Figure 4. BER-SNR for $\sigma = 2/\pi$, 5, 20.

In the NOFDM system, we define a factor "p" that is a power of 2. This "p" is a ratio of the symbols to be transmitted to the window length. "N/p" subcarriers are used for the transmission of N symbols. Due to windowing operation (use of Gabor basis) the N/p subcarriers can be used to facilitate the transmission of the p sets of symbols. The Hanning window used ensures that the Gaussian window goes to zero outside the specified window length. This minimizes abrupt changes in the transmitted signal and hence leads to a much sharper frequency response. The transmitted signal in the NOFDM system is given by,

$$v(n)=\sum_{a=0}^{\frac{N}{p}-1}\sum_{k=0}^{p-1}\sum_{n=1}^{N/p}\{w(n-kN/p) \times X_a(n+kN/p)\cdot e^{-j2\pi(F_c+a/N)n}\} \quad (8)$$

where,

$$p=N/(\text{window length}). \quad (9)$$

4. Simulation Results and Conclusions

The following results are for an OFDM and NOFDM system designed for the Digital Video Broadcasting (DVB) standard of European terrestrial digital television (DTV) service based on [10]. Our simulations will focus in the 2 k mode of the DVB-T standard which uses 1705 carriers which carry symbols with a useful duration of 224 ms.

From the plots **(a)** and **(b)** of **Figure 5** we find that the bandwidth decreases from 8 MHz to 4 MHz as the factor p changes from 1 to 4. This implies that the bandwidth (B) used decreases with the window length *i.e.* B is proportional to N/p (9). From the plots **(b)** and **(c)** of **Figure 5** we observe that the bandwidth used is almost the same but there is a decrease in the power of the out of band emissions.

On transmitting the symbols on a set of subcarriers over a channel that imparts AWGN noise, one can observe that the recovery of the symbols is possible in both the OFDM and NOFDM systems and the data transmitted through the NOFDM system has a BER just slightly higher than that transmitted through an OFDM system for a given SNR (depicted in **Figure 6**).

It is also observed that the BER-SNR performance improves for the NOFDM system with a decrease in the window length used to develop the non-orthogonal basis set as shown in **Figure 7**. Since the main role of cyclic prefix is to maintain the orthogonality of the sub channels, it must be noted that no cyclic prefix is needed in the NOFDM system. Thus the NOFDM system designed does not have cyclic prefix. However, cyclic prefix can be incorporated if necessary in order to combat ISI.

Thus from the results obtained we can conclude that

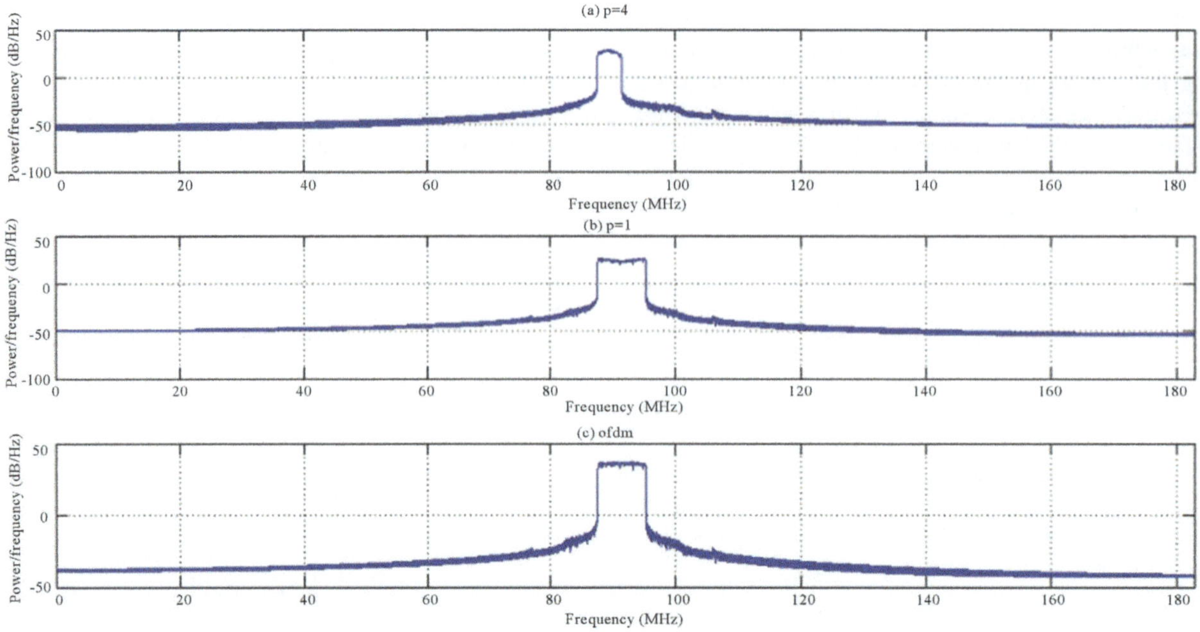

Figure 5. Power Spectral Density of the transmitted signal given $N = 4096$ for (a) $p = 4$ in an NOFDM system; (b) $p = 1$ in an NOFDM system; (c) OFDM system.

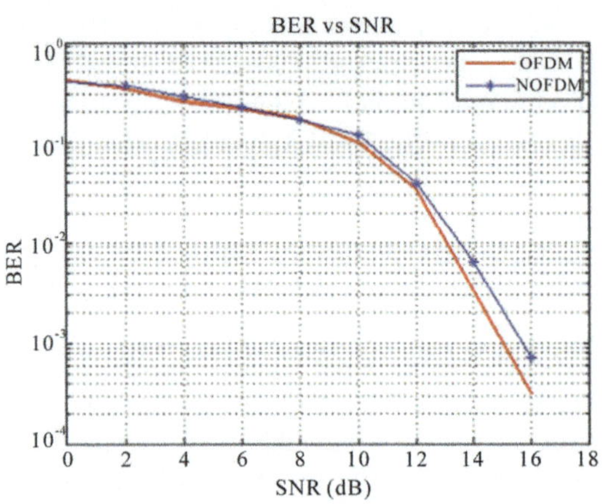

Figure 6. BER vs SNR for an OFDM and an NOFDM system.

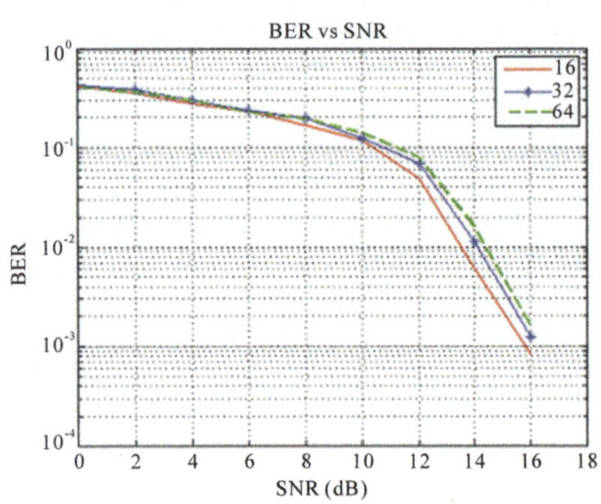

Figure 7. BER vs SNR for a NOFDM system for different window lengths.

the spectrum of the NOFDM signal transmitted is occupies lesser bandwidth when compared to that of the OFDM system when $p > 1$ (where, $p = 1, 2, 4, 8, \cdots$). This is a major advantage over the OFDM system. Since the BER-SNR performance of this system remains comparable with that of the baseline OFDM system, the developed system is an improvement over OFDM in DVB-T.

5. Summary

The NOFDM system has been implemented. The BER vs. SNR performance for a system implemented using a Gabor basis with sigma value of $2/\pi$ was found to be optimal and comparable to that of an OFDM system. The spectrum is found to be efficient than that of OFDM systems and also the bandwidth is found to be inversely proportional to the window length of the Gabor transform.

6. Acknowledgements

We thank the faculty of the department of Electronics and Communication Engineering of Amrita Vishwa Vidyapeetham for their support and motivation.

REFERENCES

[1] M. Rodrigues and I. Darwazeh, "A Spectrally Efficient Frequency Division Multiplexing Based Communication System," *8th International OFDM Workshop*, Hamburg, September 2003, pp. 70-74.

[2] S. I. A. Ahmed and I. Darwazeh, "IDFT Based Transmitters for Spectrally Efficient FDM System," London Communications Symposium, 2009.

http://www.ee.ucl.ac.uk/lcs/previous/LCS2009/LCS/lcs09_37.pdf

[3] S. Isam and I. Darwazeh, "Simple DSP-IDFT Techniques for Generating Spectrally Efficient FDM Signals," 2010 *7th International Symposium on Communication Systems Networks and Digital Signal Processing* (*CSNDSP*), London, 21-23 July 2010, pp. 20-24.

[4] Z. Zhang, W. Chen, Q. Liu and G. Xuan, "NOFDM System Based on Circular Conjugate Symmetry Properties of DFT," *4th International Wireless Communications*, *Networking and Mobile Computing*, Chongqing, 12-14 October 2008, pp. 1-4.

[5] W. Kozek and A. F. Molisch, "Nonorthogonal Pulseshapes for Multicarrier Communications in Doubly Dispersive Channels," *IEEE Journal on Selected Areas in Communications*, Vol. 16, No. 8, 1998, pp. 1579-1589.

[6] R. Prasad, "OFDM for Wireless Communication Systems," Artech House, Boston-London, 2004.

[7] H. Feichtinger and T. Strohmer, "Gabor Analysis and Algorithms, Theory and Applications," Birkhäuser, Berlin, 1998.

[8] A. Kliks, A. Zalonis, I. Dagres, A. Polydoros and H. Bogucka, "PHY Abstraction Methods for OFDM and NOFDM Systems," *Journal of Telecommunication and Information Technology*, No. 3, 2009, p. 116.

[9] S. Veni and K. A. Narayanankutty, "Vison-Based Hexaganol Image Processing Using Hex-Gabor," Springer, Berlin, 2012.

[10] G. Acosta, "OFDM Simulation Using Matlab," 2000. http://www.ece.gatech.edu/research/labs/sarl/tutorials/OFDM/Tutorial_web.pdf

2

Broadband Quasi-Yagi Antenna for WiFi and WiMax Applications

Jean-Marie Floc'h, Ahmad El Sayed Ahmad

Unité Mixte de Recherche 6164 (UMR), Institut d'Electronique et de Télécommunications de Rennes (IETR), Rennes, France.

ABSTRACT

In this paper, we present a broadband quasi-Yagi antenna. Good impedance matching is obtained by using parasitic elements. The antenna has been designed and successfully measured. Experimental results show that the 10 dB return loss bandwidth of this antenna is 50% operating from 2.3 GHz to 3.8 GHz. We obtain very flat gain (around 5 dB) over the entire bandwidth. For the design and optimization of antennas, we use HFSS CAD software from ANSOFT.

Keywords: Printed Dipole Antenna; Quasi-Yagi Dipole Antenna; Broadband; Director; Reflector

1. Introduction

Nowadays, quasi-Yagi antenna is widely used in wireless communication for its high directivity, high radiation efficiency, low cost, low profile, as well as easy to fabrication [1,2]. However, the main disadvantage of these antennas is narrow bandwidth and the bandwidth commonly achieves just 10%. To solve this problem, researchers have made great efforts to improve the bandwidth of Yagi antenna using many ways.

Several different techniques have been proposed in literature, concerning the enhancement of impedance bandwidth in microstrip antennas, mostly by increasing the substrate thickness, by decreasing the substrate dielectric constant [3], and by alternative feeding methods (aperture coupled or proximity instead of direct contacting feed). Moreover, by increasing the width of the printed dipole antennas, the bandwidth of antenna increases [4]. By using triangular instead of rectangular dipole arms (bow-tie-configuration), the bandwidth can be further expanded to 37% as presented by Bailey [5]. Coplanar parasitic elements have been proposed by Deal *et al.* [6] in a quasi-Yagi configuration, yielding an impedance bandwidth of 48%. A similar Yagi-like double sided antenna [7], achieved a bandwidth of 37%. Stacked parasitic elements (aperture attacked patches) are widely used for enhanced bandwidth, but their fabrication is relatively complicated and expensive. Additionally, a broad-band quasi-Yagi antenna achieving a measured 48% bandwidth is presented for radar systems and millimeter-wave imaging arrays in [8]. We have presented few year ago in [9] a multi-band frequency printed dipole antenna with operation at 2.45, 5.8 and 10 GHz.

The paper is organized as follows: first, we present the design concept of the proposed printed dipole antenna with reflector (with minimal complexity design). In Section 3, a parasitic element is embedded to increase the bandwidth of antenna. Finally, the printed dipole antenna with a three parasitic elements is studied to enable the operation of wideband. The antennas have been designed and successfully measured.

2. Basic Antenna

The quasi-Yagi printed dipole antenna is composed of the dipole and reflector as shown in **Figure 1**. The two arms of the dipole are printed on each side of a CuClad substrate of thickness 0.8 mm and permittivity of 2.17 (to overcome the complicated feeding technique). We use a microstrip line combined with a bifilar printed line. The length of these arms and the distance between them and reflector are around quarter-wavelength ($\lambda_0/4$). λ_0 is the wave-length in free space corresponding to resonant frequency around 2.45 GHz.

The reflector has to be longer than the length of the two arms of the dipole in order to reflect the radiated power in the front direction ($\theta = 0°$). In the case of a shorter reflector, the radiation is concentrated in back direction, it performs as a director ($\theta = 180°$).

The advantage of this structure is to increase the gain of the printed dipole and to decrease the backside radiation. The **Figure 2** shows the realized printed dipole antenna with reflector.

The simulated reflection coefficient evaluation is shown in **Figure 3**. The printed dipole antenna is matched to −29 dB in simulation around the operating frequency. The measurements were achieved in the laboratory anechoic chamber and have been compared to simulation. The return loss of the tested antenna is presented in **Figure 4**. The level of measured S_{11} is lower than −10 dB over the frequency bandwidth [2.3 GHz - 2.7 GHz].

We present the radiation pattern in 3D at 2.45 GHz obtained in simulation and measurement (**Figures 5** and **6**). We note a good comparison between simulation and measurement. The measured gain in 0° and 180° direction is respectively 5 dB and −8 dB at 2.45 GHz. The difference between the simulated realized gain and the measured realized gain is around 0.5 dB. The measurement accuracy of the anechoic chamber (±0.5 dB) can be responsible for this variation.

3. Antenna with Parasitic Elements

3.1. Antenna with One Parasitic Element

The addition of one parasitic close to the simple printed dipole creates a second resonance frequency (**Figure 7**). The resonant band is obtained through coupling between

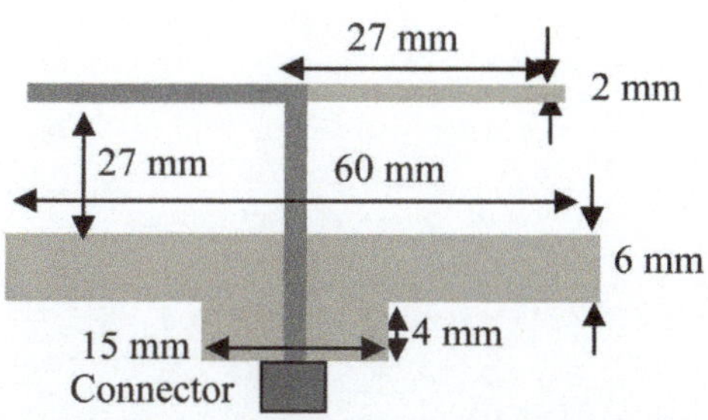

Figure 1. Geometry of the printed dipole antenna with reflector.

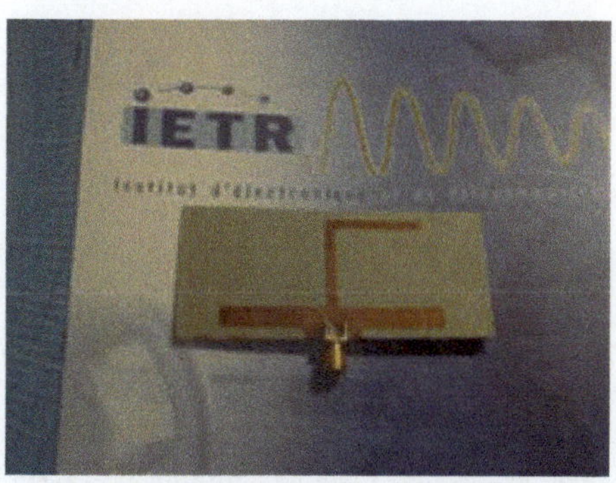

Figure 2. Photo of the printed dipole antenna with reflector.

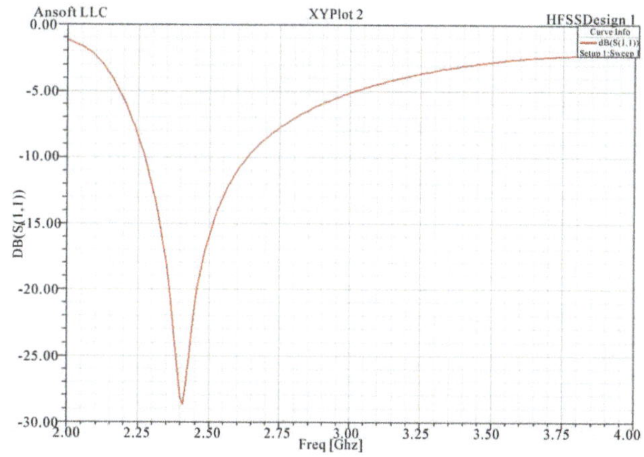

Figure 3. Simulated reflection coefficient of the simple printed dipole.

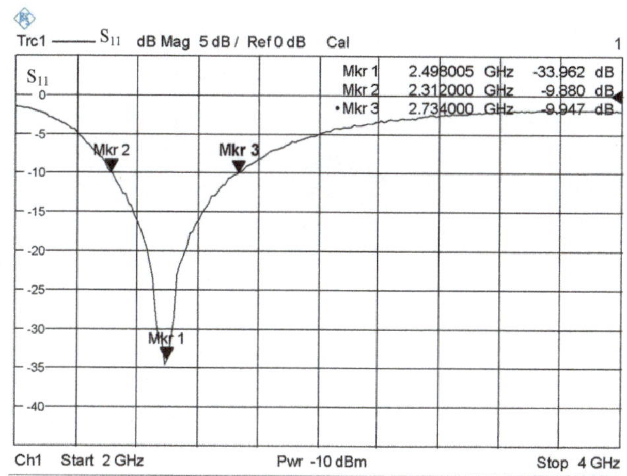

Figure 4. Measured reflection coefficient of the simple printed dipole.

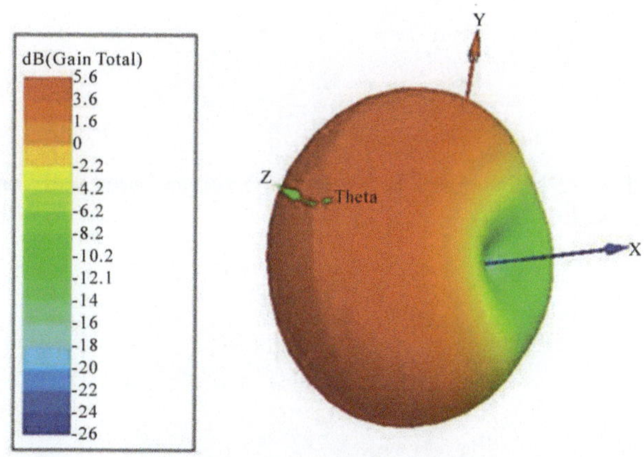

Figure 5. Simulated 3D radiation pattern of the dipole antenna with reflector.

the dipole arms and the parasitic strip. This frequency is determined by the length of the parasitic around $\lambda_0/2$. We have optimized the distance between dipole and reflector, the distance between dipole and parasitic strip (2 mm).

The distance between the dipole arms and the parasitic dipole is very short because the coupling between them is capacitive. The realized antenna is shown in **Figure 8**.

The simulated and measured S_{11} of this antenna is plotted in **Figure 9**. The addition of one parasitic close to the simple printed dipole permits to increase the bandwidth of the antenna. We note a good similarity between measurement and simulation. We obtain with this antenna around 1 GHz bandwidth for a $S_{11} < -10$ dB (2.3 GHz - 3.3 GHz).

We present the measured radiation pattern at 2.45 and 3 GHz. The maximum value of measured realized gain of the antenna with one parasitic is 5 dB in $\theta = 0°$ (**Figure 10**). In the **Figure 11**, we can see the measured maximum gain versus frequency. We observe a flat gain over the frequency band 2.3 GHz to 3.3 GHz at the value around 5 dB. The efficiency of this antenna is greater than 75% over the frequency band.

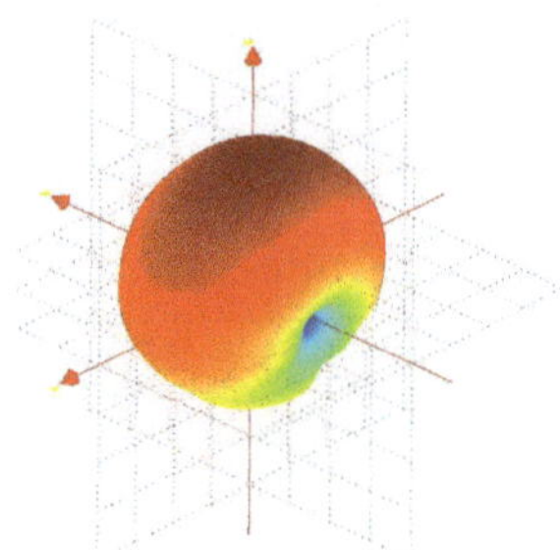

Figure 6. Measured 3D radiation pattern of the dipole antenna with reflector.

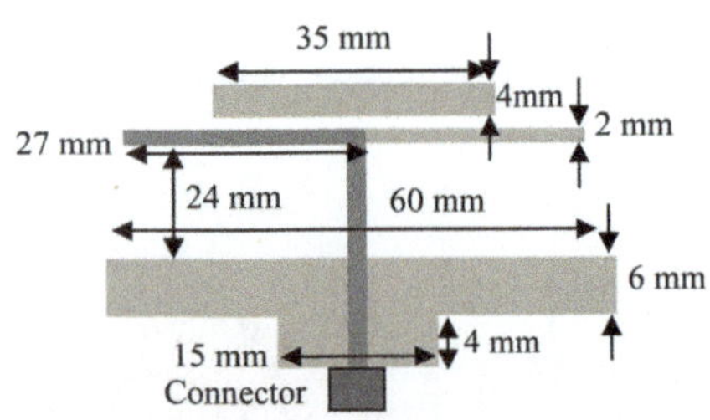

Figure 7. Geometry of the quasi-Yagi antenna with one parasitic.

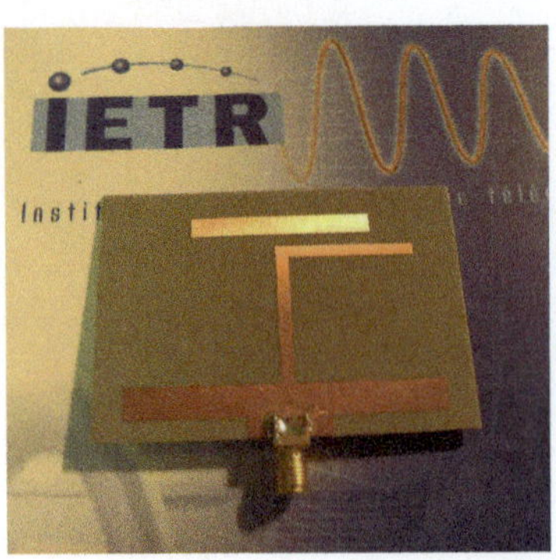

Figure 8. Photo of the quasi-Yagi antenna with one parasitic.

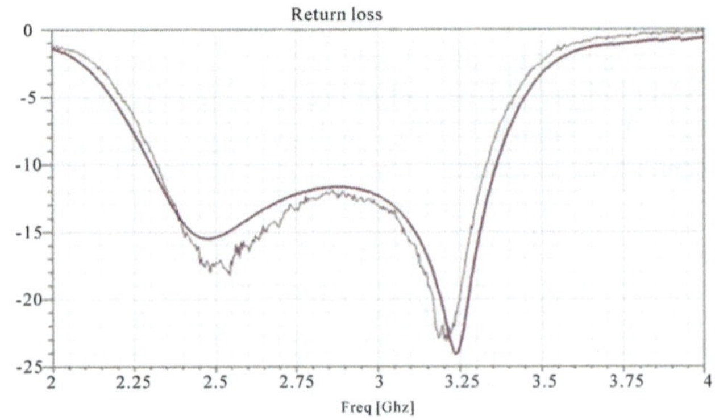

Figure 9. Simulated and measured return loss of the antenna.

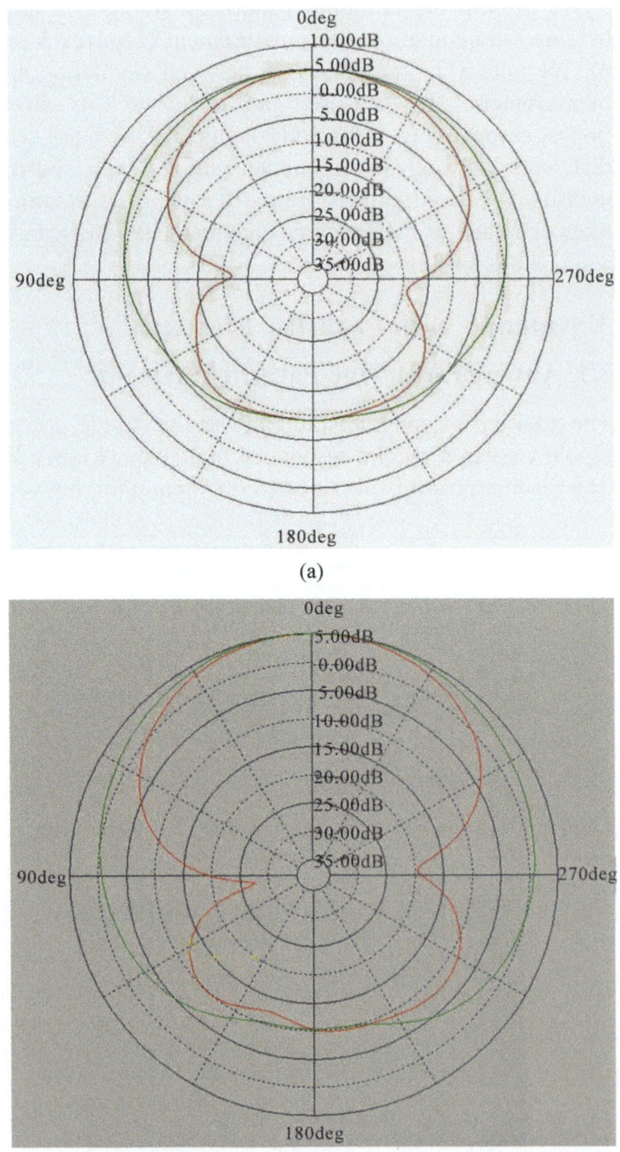

Figure 10. Measured radiation pattern (red curve: E plane; green curve: H plane) of the antenna at 2.45 GHz (a) and 3 GHz (b).

3.2. Antenna with Three Parasitic Elements

In order to increase more the antenna bandwidth (for WiFi and WiMax applications), we have added three parasitic elements on the bottom side of the substrate where one dipole arm; the reflector and the ground plane were printed (**Figure 12**). Each parasitic element should be shorter than the length of the two dipole-arms for create a resonances greater than the resonance of dipole antenna. Three parasitic elements are sufficient to cover the WiFi and WiMax bands. The antenna is designed using the same size of the prototype described in 3.1, with the new parameters of the three parasitic elements. The lengths of these parasitic elements are: 32 mm, 29 mm and 27 mm. The width is 4 mm. The distance between them is 2 mm. The tested antenna is shown in **Figure 13**.

The broadband printed quasi-Yagi antenna is matched in simulation over the operating frequency bandwidth. The return loss of the tested antenna is presented in **Figure 14**. The level of measured S_{11} is lower than −10 dB over the operating frequency bandwidth (2.3 GHz - 3.8 GHz). We note a good similarity between measurement and simulation. We obtain a bandwidth of 1.5 GHz (around 50%).

Figure 15 presents the measured radiation pattern at 2.2 GHz, 3 GHz and 3.8 GHz. We observe a very stable radiation pattern with less than −10 dB backside radiation versus the bore sight direction.

In terms of levels (**Figure 16**), the measured realized gain at $\theta = 0°$ is higher than 5 dB over the band [2.3 - 3.8] GHz. We can observe from the **Figure 16** that the gain is flat over the frequency band 2.3 GHz to 3.8 GHz. The efficiency of this antenna is greater than 70% over the frequency band.

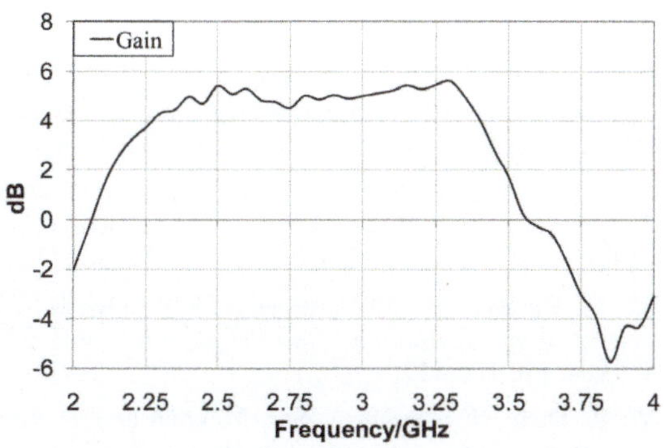

Figure 11. Measured gain versus frequency.

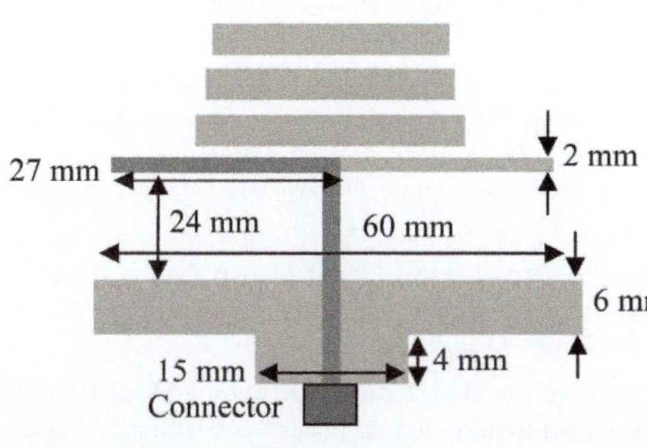

Figure 12. Geometry of the quasi-Yagi antenna with three parasitic elements.

Figure 13. Photo of the quasi-Yagi antenna with three parasitic elements.

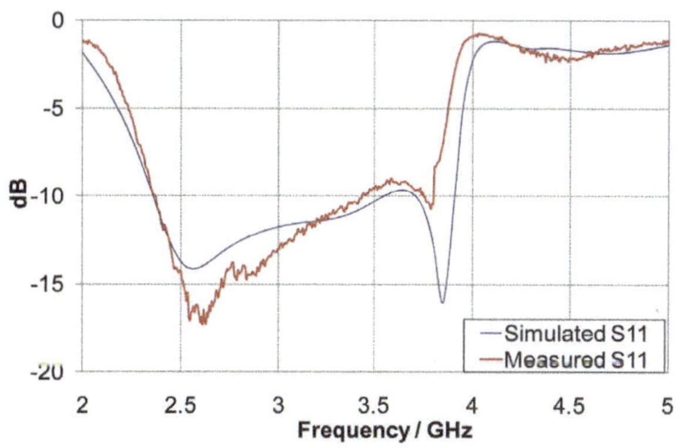

Figure 14. Simulated and measured return loss of the antenna.

4. Conclusion

In this paper, we have presented the design and the results obtained with broadband quasi-Yagi antenna. As a first step, a printed dipole was designed. As a second step, in order to increase the bandwidth of antenna, we have studied a printed dipole antenna with one parasitic element. Finally, we have developed a special broadband antenna for WiFi and WiMax applications. All the designed antennas were simulated, realized and characterized by measuring their return loss, radiation patterns and gain. We obtain good comparison between simulations and measurements. The proposed antenna operates from 2.3 GHz to 3.8 GHz. The average measured gain in this band is about 5 dB, and the antenna can achieve wide

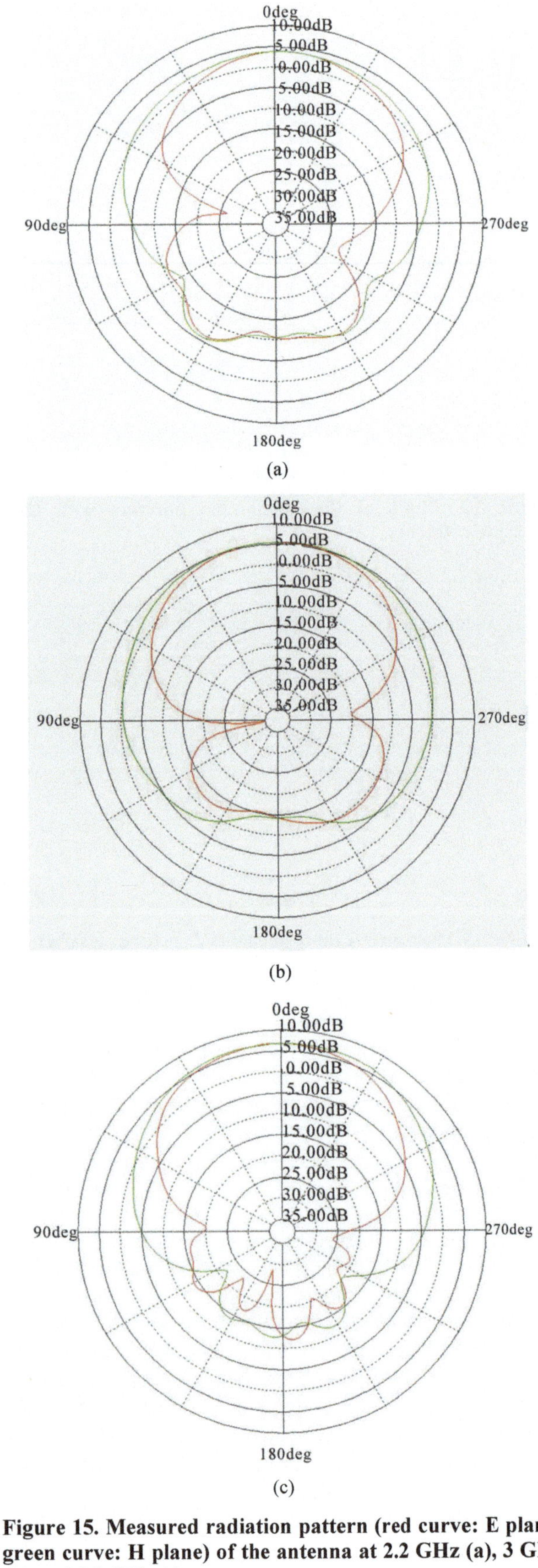

(a)

(b)

(c)

Figure 15. Measured radiation pattern (red curve: E plane; green curve: H plane) of the antenna at 2.2 GHz (a), 3 GHz (b) and 3.8 GHz (c).

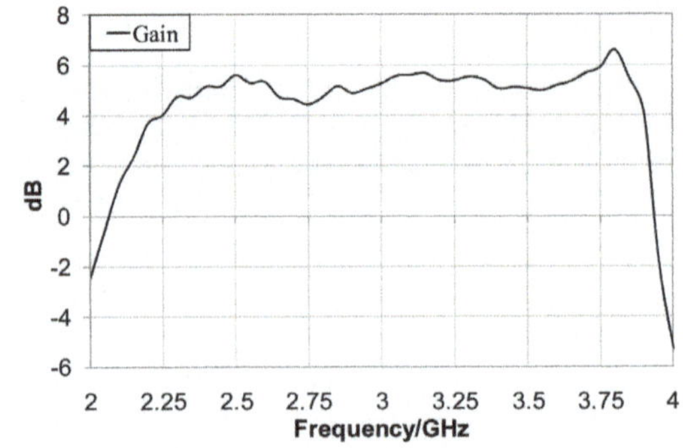

Figure 16. Measured radiation results (realized Gain) versus frequency of the broadband quasi-Yagi antenna.

impedance and gain bandwidths (50%). At the same time, this antenna possesses low profile and miniaturization characteristics. Furthermore, the wide beamwidth and high gain can promise the antenna to be widely applied in many communication systems.

REFERENCES

[1] G. R. DeJean, T. T. Thai, S. Nikolaou and M. M. Tentzeris, "Design and Analysis of Microstrip Bi-Yagi and Quad-Yagi Antenna Arrays for WLAN Applications," *IEEE Antennas and Wireless Propagation Letters*, Vol. 6, 2007, pp. 244-248.

[2] O. Kramer, T. Djerafi and K. Wu, "Vertically Multilayer-Stacked Yagi Antenna with Single Ande Dual Polarizations," *IEEE Transactions on Antenna and Propagation*, Vol. 99, 2010, pp. 1022-1030.

[3] E. Levine, S. Shtrikman and D. Treves, "Double-Sided Printed Arrays with Large Bandwidth," *IEEE Proceedings of Microwaves, Antennas and Propagation*, Vol. 135, No. 1, 1988, pp. 54-59.

[4] C. A. Balanis, "Antenna Theory-Analysis and Design," 2nd Edition, John Wiley & Sons, New York, 1997, pp. 454-456.

[5] M. C. Bailey, "Broad-Band Half-Wave Dipole," *IEEE Transactions on Antennas and Propagation*, Vol. 32, No. 4, 1984, pp. 410-412.

[6] W. R. Deal, N. Kaneda, J. Sor, Y. Qian and T. Itoh, "A New Quasi-Yagi Antenna for Planar Active Antenna Arays," *IEEE Transactions on Microwave Theory and Techniques*, Vol. 48, No. 6, 2000, pp. 910-918.

[7] G. Zheng, A. A. Kishk, A. B. Yakovlev and A. W. Glisson, "Simplified Feeding for a Modified Printed Yagi Antenna," *IEEE Antennas and Propagation Society, International Symposium Digest*, Vol. 3, 2003, pp. 934-937.

[8] N. Kaneda, W. R. Deal, Y. X. Qian, R. Waterhouse and T. Hoh, "A Broad-Band Planar Quasi-Yagi Antenna," *IEEE Transactions on Antennas and Propagation*, Vol. 50, No. 8, 2002, pp. 1158-1160.

[9] J. M. Floc'h and H. Rmili, "Design of Multi-Band Printed Dipole Antennas Using Parasitic Elements," *Microwave and Optical Technology Letters*, Vol. 48, No. 8, 2006, pp. 1639-1645.

3

A New Improved Symbol Mapper/8-Ary Constellation for BICM-ID

Slimane Benmahmoud[1,2], Ali Djebbari[1]

[1]Telecommunications and Digital Signal Processing Laboratory, Djillali Liabes University, Sidi Bel-Abbes, Algeria; [2]Department of Electronics Engineering, M'sila University, M'sila, Algeria.

ABSTRACT

This contribution proposes a new combination symbol mapper/8-ary constellation, which is a joint optimization of an 8-ary signal constellation and its symbol mapping operation, to improve the performance of Bit Interleaved Coded Modulation with Iterative Decoding (BICM-ID). The basic idea was to use the so called (1,7) constellation (which is a capacitive efficient constellation) instead of the conventional 8-PSK constellation and to choose the most suitable mapping for it. A comparative study between the combinations most suitable mapping/(1,7) constellation and SSP mapping/conventional 8-PSK constellation has been carried out. Simulation results showed that the 1st combination significantly outperforms the 2nd combination and with only 4 iterations, it gives better performance than the 2nd combination with 8 iterations. A gain of 4 dB is given by iteration 4 of the 1st combination compared to iteration 8 of the 2nd combination at a BER level equal to 10^{-5}, and it (iteration 4 of the 1st combination) can attain a BER equal to 10^{-7} for, only, a SNR = 5.6 dB.

Keywords: Bit Interleaved Coded Modulation; BICM; Iterative Decoding/Processing; Log-Likelihood Ratios (LLRs); 8PSK Constellation; Signal Mappings; Constellation Labeling Maps; Extrinsic Information; BCJR Algorithm

1. Introduction

A straightforward concatenation of a convolutional encoder and an M-ary modulator doesn't provide the best performance. A joint design of a convolutional encoder and a multi-level/phase modulator, called Trellis Coded Modulation (TCM) [1], maximizes the minimum squared Euclidian distance to provide a significant improvement over AWGN channels but it doesn't provide the same improvement over Rayleigh fading channels.

Symbol interleaving is used to make TCM system memoryless and its diversity order is then defined as the minimum number of distinct symbols between two codewords [2].

By replacing symbol interleaving by bit-level interleaving in TCM, Zehavi in [2], succeeded to increase the diversity order to the binary Hamming distance of the coded system.

The scheme proposed by Zehavi, called later in [3] Bit Interleaved Coded Modulation (BICM), achieves better performance than Ungerboeck's TCM over Rayleigh fading channels.

Bit interleaving in BICM causes random modulation which reduces the free Euclidian distance that makes BICM significantly worse than Ungerboeck's TCM over AWGN channels [2-4].

Turbo codes were introduced in 1993 by Berrou *et al.* [5]. The results of iterative decoding, applied for turbo codes, motivated authors of [6] to apply it (iterative decoding) for BICM which led to BICM with Iterative Decoding (BICM-ID). ID greatly improves the BICM's performance over both AWGN and Rayleigh fading channels.

With a suitable design of the mapping function for a given constellation, BICM-ID can attain an excellent performance [7]. In this later reference, Gray, SP and SSP mappings were considered and a comparison of these labeling schemes for 8PSK constellation has been done. It has been found that SSP mapping offers the best performance.

Authors of [8] presented various optimal non-orthogonal 16-QAM mappings and an improvement of 0.2 dB has been shown compared to the state of the art orthogonal QAM constellation with approx. the same error floor.

A new mapping scheme, called the Maximum Squared Euclidian Weight (MSEW), has been introduced in [9].

The objective of this paper was to jointly optimize an 8-ary constellation and its symbol mapping operation to improve the performance of BICM-ID. To do so, the BER performances of the seven proposed mappings for (1,7) constellation were examined by simulation in order to choose the most suitable mapping for this constellation.

In [10] we found that SSP is the suitable mapping for conventional 8-PSK constellation and for rectangular 8-QAM constellation also [11,12].

Based on these results, a comparative study between the combinations most suitable mapping/(1,7) constellation and Semi-Set Partitioning (SSP)/conventional 8-PSK constellation has been carried out .

In this paper we show that BICM-ID, using a new proposed symbol mapper/8-ary constellation's combination, can provide excellent performance compared to it using the well-known SSP mapping/conventional 8-PSK constellation's combination.

The rest of the paper is organized as follows. In Section 2, we present the BICM-ID system's model and its iterative processing process. Then in Section 3, we talk about (1,7) constellation and different mappings used in this study. We talk also about conventional 8-PSK constellation and SSP mapping. We present and discuss simulation results in Section 4. Section 5 concludes this paper. Finally, Appendix gives further simulation results on the performance of (1,7) constellation and its mappings.

2. BICM-ID System's Model

2.1. BICM-ID System's Overview

Figure 1 illustrates the bloc diagram of a BICM-ID's transceiver. Its transmitter (**Figure 1(a)**) consists of a serial concatenation of a conventional encoder, a bit-interleaver and an M-ary (8-ary in our case) constellation mapper.

Data bits u are first encoded by the convolutional encoder. Then, the bit-interleaver permutes the encoded bits c to form a sequence of interleaved bits v. Each $m = \log_2 M$ ($m = 3$ in our case) successive interleaved bits will be grouped in one symbol $v = (v_1, v_2, \cdots, v_m)$. The resulting m-bits symbols are then mapped into complex symbols s_i in the constellation set Ψ (with $M = 2^m$ symbols) according to the following rule

$$\mu : s_i = \mu(v), s_i \in \Psi, i = 1, 2, \cdots, M \qquad (1)$$

For conventional 8-PSK constellation: $M = 8$ symbols,

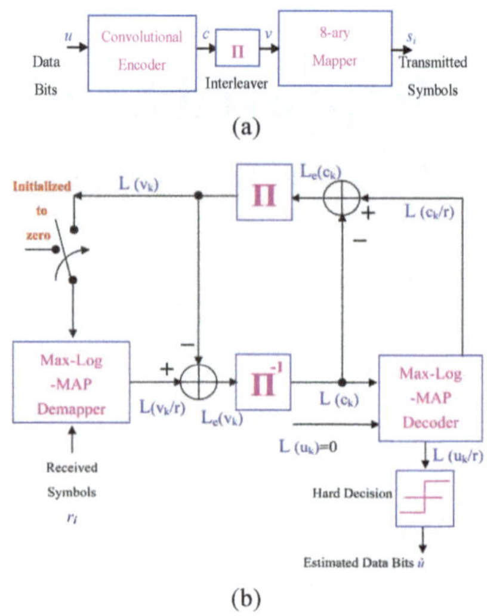

Figure 1. BICM-ID's, based 8-ary constellation, transceiver. (a) Transmitter; (b) Receiver.

$m = \log_2 M = 3$ bits, $v = (v_1, v_2, v_3)$, $\Psi = \{s_1, s_2, \cdots, s_8\}$, and

$$\Psi = \{\exp(j2\pi L/8)\}, L = 0, 1, \cdots, 7 \qquad (2)$$

The resulting complex symbols are then passed through a discrete-time channel (for a base band model). The received discrete-time signal, for a Rayleigh fading channel is given by

$$r_i = a_i \cdot s_i + n_i \qquad (3)$$

where a_i is the fading coefficient, which can be represented by a Rayleigh random variable, and n_i is a sample of an AWGN with a zero mean and a variance $\sigma_N^2 = N0/2$, where $N0$ is the unilateral power spectral density of a white Gaussian noise. For AWGN channel, we take $a_i = 1$.

2.2. Iterative Decoding with Soft Decision Feedback for BICM

Contrary to decoding algorithms that uses hard decision, like the Viterbi algorithm used for convolutional codes, a Soft-Input Soft-Output (SISO) module is used for BICM-ID. This SISO decoder uses a Maximum A Posteriori (MAP) algorithm to compute the Log-Likelihood Ratios (LLRs) of coded and data bits (**Figure 1(b)**). Practically, this MAP algorithm is implemented in the logarithmic domain to reduce numerical computations and complexity.

During the first iteration, the demapper is fed with the received signal r_i. The second input $L(v_k)$ (the a priori information of the interleaved bits) is set to zero during the first iteration, since no prior information is available on the input bits v (the interleaved bits) of the constellation mapper. The demapper computes the a posteriori LLRs $L(v_k/r)$ of the interleaved bits which will be used to calculate their extrinsic LLRs using the following equation

$$L_e(v_k) = L(v_k/r) - L(v_k) \quad (4)$$

This operation is shown by the lower subtraction path in **Figure 1(b)**.

The extrinsic LLRs $L_e(v_k)$ are then passed through the de-interleaver to create the a priori LLRs $L(c_k)$ of coded bits. The a priori LLRs $L(u_k)$ of data bits is always set to zero, which implies equally likely transmitted source's data bits. The a posteriori LLRs $L(u_k/r)$ of data bits will be used in final iteration to recover the data bits using hard decision according to the next rule

$$\hat{u} = \operatorname{sgn}\{L(u_k/r)\} \quad (5)$$

On the other hand, the a posteriori LLRs $L(c_k/r)$ of coded bits are used to calculate the extrinsic LLRs $L_e(c_k)$ of coded bits using the following equation

$$L_e(c_k) = L(c_k/r) - L(c_k) \quad (6)$$

The later operation is shown by the upper subtraction path in **Figure 1(b)**.

The extrinsic LLRs $L_e(c_k)$ are then fed back through the interleaver to create the a priori LLRs $L(v_k)$ of interleaved bits, which will be passed to the demapper to start a second iteration, and the cycle of iterations is repeated until a stopping condition.

Since the interleaver makes m bits in one symbol independent, the a priori information $p(s_i)$ of each transmitted signal $s_i \in \Psi$ is given as in [8] by

$$p(s_i) = p(v_1(s_i),\cdots,v_m(s_i)) = \prod_{k=1}^{m} p(v_k = v_k(s_i); I) \quad (7)$$

where, $v_k(s_i) \in \{0,1\}, 1 \le k \le m$ is the value of the kth bit in the label of s_i.

The demapper generates a posteriori LLRs $L(v_k/r)$ as follows

$$L(v_k/r) = \log\left\{ \sum_{s_i\in\Psi_k^1} p\left(\frac{r}{s_i}\right) \Big/ \sum_{s_i\in\Psi_k^0} p\left(\frac{r}{s_i}\right) \right\} + \sum_{k=1}^{m} L(v_k(s_i)) \quad (8)$$

By substituting Equation (8) in Equation (4), we find

$$\begin{aligned} L_e(v_k) &= \log\left\{ \sum_{s_i\in\Psi_k^1} p(r/s_i) \Big/ \sum_{s_i\in\Psi_k^0} p(r/s_i) \right\} + \sum_{k=1}^{m} L(v_k(s_i)) \\ -L(v_k) &= \left(-\min_{s_i\Psi_k^1} \|r - a_i \cdot s_i\|^2 \right) \\ &- \left(-\min_{s_i\Psi_k^0} \|r - a_i \cdot s_i\|^2 \right) + \sum_{\substack{j=1 \\ j\ne k}}^{m} L(v_k(s_i)) \end{aligned} \quad (9)$$

where, $\sum_{\substack{j=1\\ j\ne k}}^{m} L(v_k(s_i)) = \sum_{k=1}^{m} L(v_k(s_i)) - L(v_k)$. The amplitude's fading coefficient a_i is assumed perfectly estimated.

From (9), we can observe that $L_e(v_k)$ is computed from the a priori LLRs $L(v_k)$ of other interleaved bits ($j \ne k$) in the same symbol.

3. (1,7) Constellation and the Proposed Mappings

The bijection between the binary m-bits' symbols and the complex symbols of the constellation set is called symbol mapping.

For a given number of bits (the value of m) that should be transmitted per symbol, the signal constellation and the mapping method must be specified.

The (1,7) constellation's symbols set is given by

$$s_L = \begin{cases} \exp(j2\pi(L-1)/7) & \text{for } L = 1,2,\cdots,7 \\ 0 & \text{for } L = 8 \end{cases} \quad (10)$$

These symbols are shown in **Figure 2(a)** accompanied with the symbols of conventional 8-PSK constellation in **Figure 2(b)**.

In **Table 1**, we show the seven different mapping methods (called MAP1, MAP2, $\cdots$, MAP7) for (1,7) constellation and the SSP mapping used for conventional 8-PSK constellation. SSP mapping is widely used in the literature because it maximizes diversity by assigning binary labels which differ at least in two bits to adjacent symbols in the constellation diagram. For example, in the conventional 8-PSK constellation, the two binary labels 100 and 001 are assigned to adjacent points.

In practice, to compute the a posteriori LLRs $L(v_k/r)$ of the interleaved bits (demapping operation), we need to divide the constellation set Ψ into two subsets Ψ_k^b for each position k where $b = 0,1$ (the bit's binary value). For a MAP3 mapped (1,7) constellation (**Figure 2(c)**), we have the following constellation's sub-sets

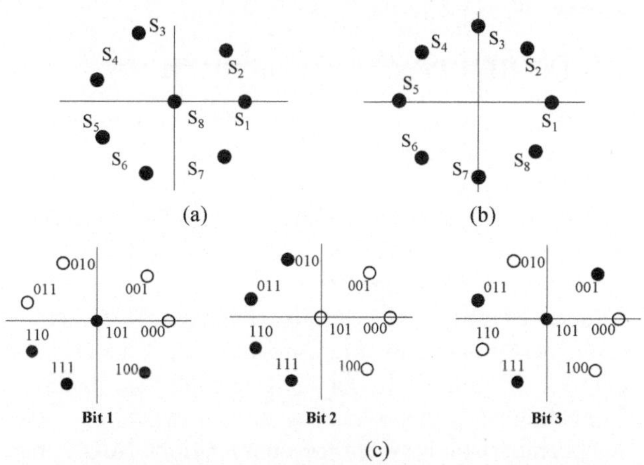

Figure 2. The two used constellations: (a) and (b) their symbols' positions; (c) Sub-sets partition for (1,7) constellation (MAP3 is considered).

Table 1. (1,7) constellation and the seven used mappings (MAP1, MAP2, ···, MAP7).

(1,7) constellation	S_1	S_2	S_3	S_4	S_5	S_6	S_7	S_8
MAP1	000	001	011	010	110	111	101	100
MAP2	000	001	010	011	100	101	110	111
MAP3	000	001	010	011	110	111	100	101
MAP4	000	101	010	011	100	001	110	011
MAP5	001	010	111	100	011	101	110	000
MAP6	001	111	100	101	011	110	010	000
MAP7	111	101	100	001	011	010	110	000
8-PSK constellation	S_1	S_2	S_3	S_4	S_5	S_6	S_7	S_8
SSP	000	101	010	111	100	001	110	011

$$\Psi_1^0 = \{s_1, s_2, s_3, s_4\},\ \Psi_1^1 = \{s_5, s_6, s_7, s_8\} \quad (11)$$

$$\Psi_2^0 = \{s_1, s_2, s_8, s_7\},\ \Psi_2^1 = \{s_3, s_4, s_5, s_6\} \quad (12)$$

$$\Psi_3^0 = \{s_1, s_4, s_5, s_8\},\ \Psi_3^1 = \{s_2, s_3, s_6, s_7\} \quad (13)$$

In **Figure 2(c)**, a filled (respectively an unfilled) circle represents a symbol for which the binary label has the value $b = 1$ (respectively $b = 0$) in the kth bit's position.

4. Simulation Results

4.1. Simulation Parameters

In this contribution, we use a 4-states, rate 1/2 non-recursive and non-systematic (NSC) convolutional encoder with the generator polynomials $G = [g_1, g_2] = [7,5]_8$. The random bit-interleaver is of length 6138 coded bits. For each BER point we simulate almost 10^{+8} data bits, and we count at least 100 erroneous bits. The Max-log-MAP version of the BCJR algorithm is used for the decoding of the NSC code. A Rayleigh fading channel model is considered.

4.2. The BER Performance of BICM-ID

Figure 3 explicitly compares the BER performances of these seven mappings after 8 iterations.

These results show that MAP3 is the most suitable (between all used mappings) mapping for (1,7) constellation. For example, with MAP3 we can achieve a BER equal to 5×10^{-8} for 6 dB (**Figure 3**).

From **Figure 3**, we can also remark that to achieve better performance we can use MAP4 for SNRs smaller than 3.2 dB and MAP3 for SNRs greater than this value (we call this technique *adaptive symbol mapping*).

A comparison between the combinations MAP3 mapping/(1,7) constellation and SSP mapping/Conventional 8-PSK constellation were carried out and its result is given in **Figure 4**. This figure shows that the first com-

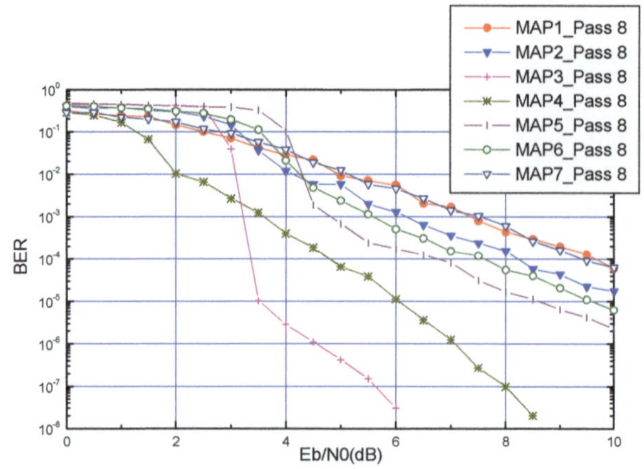

Figure 3. The BICM-ID's BER performance for all used mappings (for (1,7) constellation) after 8 iterations.

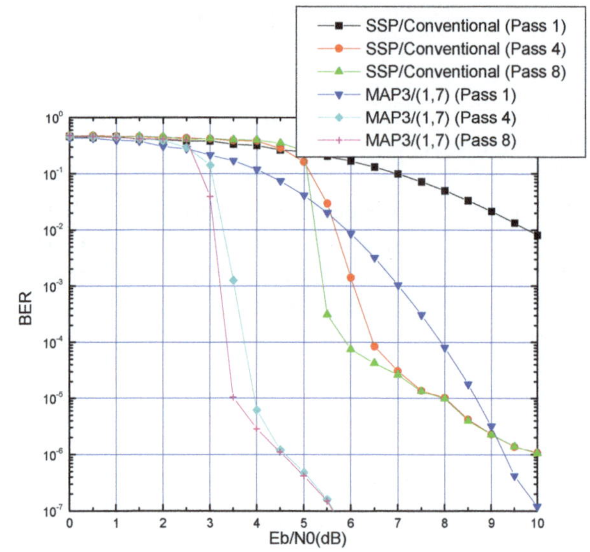

Figure 4. A comparison between the performances of the system using the two combinations.

bination gives a very significant gain compared to the second combination and this is from the 1st to the 8th iteration.

It is necessary to note that the 1st combination, with only 4 iterations, gives better performance than the 2nd combination with 8 iterations. For example, a gain of 4 dB (4.5 dB respectively) is given by iteration 4 (iteration 8 respectively) of the 1st combination compared to iterations 4 and 8 of the 2nd combination at a BER level equal to 10^{-5}, and they (iterations 4 and 8 of the first combination) can attain a BER equal to 10^{-7} for a SNR = 5.6 dB.

For further results on the performance of (1,7) constellation and these used mappings reader can refer to Appendix.

5. Conclusion

This paper proposes MAP3 as a very suitable mapping

for (1,7) constellation, and shows that the new combination MAP3/(1,7) constellation significantly outperforms the well-known combination SSP mapping/conventional 8-PSK constellation.

REFERENCES

[1] G. Ungerboeck, "Channel Coding with Multilevel/Phase Signals," *IEEE Transactions on Information Theory*, Vol. IT-28, 1992, pp. 55-67.

[2] E. Zehavi, "8-PSK Trellis Codes for a Rayleigh Fading Channel," *IEEE Transactions on Communications*, Vol. 40, No. 5, 1992, pp. 873-883.

[3] G. Caire, G. Taricco and E. Biglieri, "Bit Interleaved Coded Modulation," *IEEE Transactions on Information Theory*, Vol. 44, No. 3, 1998, pp. 927-946.

[4] C. E. Sundberg and N. Seshadri, "Coded Modulations for Fading Channels: An Overview," *European Transactions on Telecommunication*, Vol. 4, No. 3, 1993, pp. 309-324.

[5] C. Berrou and A. Glavieux, "Near Optimum Error-Correcting Coding and Decoding," *IEEE Transactions on Communications*, Vol. 44, No. 10, 1996, pp. 1261-1271.

[6] X. Li and J. A. Ritcey, "Bit Interleaved Coded Modulation with Iterative Decoding," *IEEE Communications Letters*, Vol. 1, No. 6, 1997, pp. 169-171.

[7] X. Li, A. Chindapol and J. A. Ritcey, "Bit Interleaved Coded Modulation with Iterative Decoding and 8PSK Signaling," *IEEE Transactions on Communications*, Vol. 50, No. 8, 2002, pp. 1250-1257.

[8] N. S. Muhammad and J. Speidel, "Joint Optimization of Signal Constellation and Bit Labeling for Bit Interleaved Coded Modulation with Iterative Decoding," *IEEE Communications Letters*, Vol. 9, No. 9, 2005, pp. 775-777.

[9] J. Tan and G. L. Stuber, "Analysis and Design of Symbol Mappers for Iteratively Decoded BICM," *IEEE Transactions on Wireless Communications*, Vol. 4, No. 2, 2005, pp. 662-672.

[10] S. Benmahmoud, A. Djebbari and A. Taleb-ahmed, "The Effects of Signal Mappings on the Performances of a BICM-ID System Based on 8PSK Constellation Operating over a Rayleigh Fading Channel," *Proceedings of the 2nd International Conference on Advanced Computer Theory and Engineering*, Cairo, 25-27 September 2009.

[11] S. Benmahmoud and A. Djebbari, "Signal Mappings of Rectangular 8-QAM Constellation for BICM-ID Systems over an AWGN Channel," *Proceedings of the International Conference on Image, Signal Processing and their Applications*, Mostaganem, 19-21 October 2009.

[12] S. Benmahmoud and A. Djebbari, "Seven Different 8-ary Constellations for BICM-ID," *Proceedings of the 9th International Conference on Electronics, Computer and Computation*, Ankara, 1-3 November 2012.

Appendix

Figures 5-11 give the BER versus SNR curves for each mapping for (1,7) constellation for iterations 1 until 8.

The following simulation curves can be used for comparison purposes.

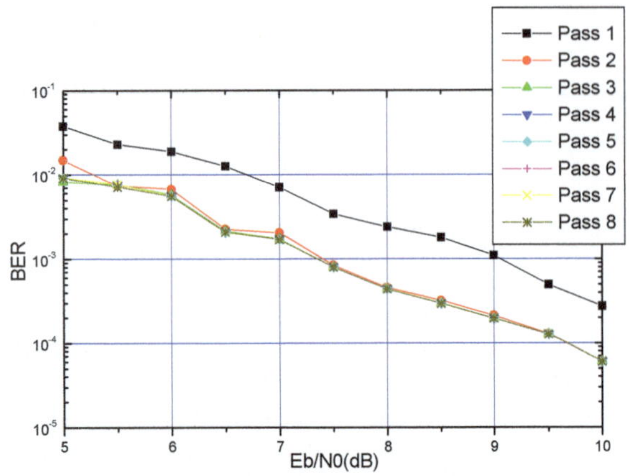

Figure 5. The BICM-ID's BER performance with MAP1/ (1,7) constellation for iterations 1, 4, ···, 8.

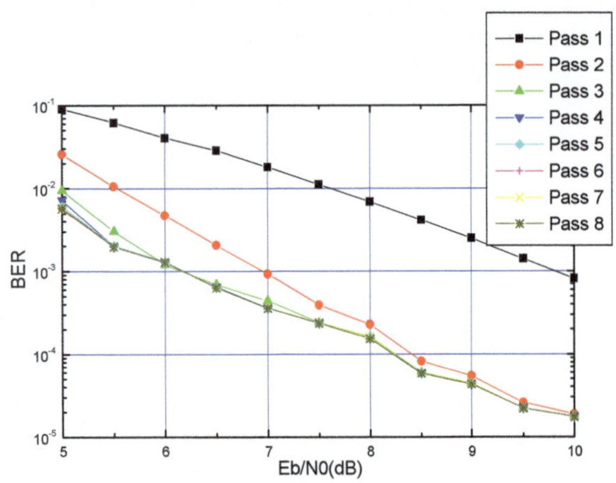

Figure 6. The BICM-ID's BER performance with MAP2/ (1,7) constellation for iterations 1, 4, ···, 8.

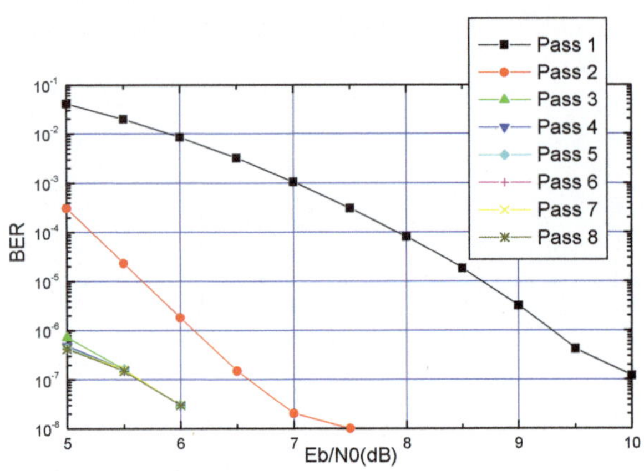

Figure 7. The BICM-ID's BER performance with MAP3/ (1,7) constellation for iterations 1, 4, ···, 8.

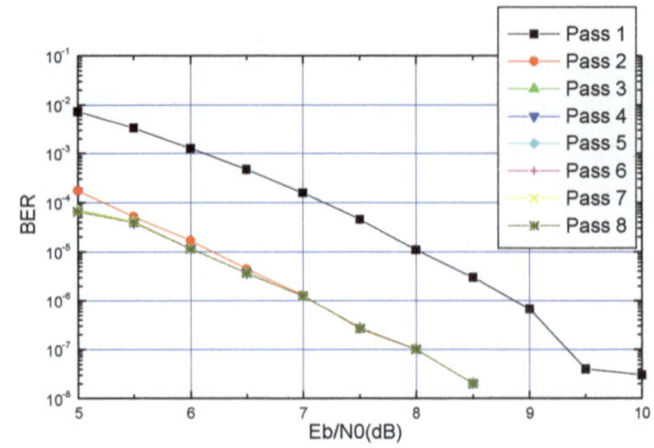

Figure 8. The BICM-ID's BER performance with MAP4/ (1,7) constellation for iterations 1, 4, ···, 8.

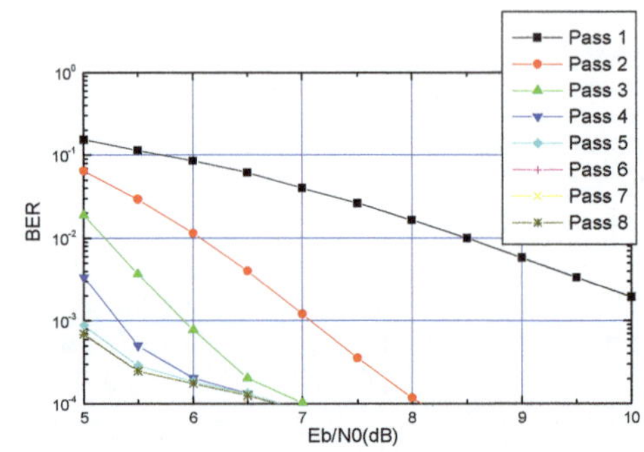

Figure 9. The BICM-ID's BER performance with MAP5/ (1,7) constellation for iterations 1, 4, ···, 8.

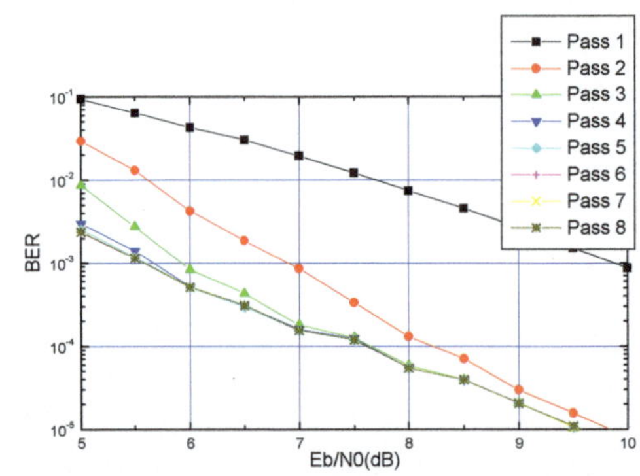

Figure 10. The BICM-ID's BER performance with MAP6/ (1,7) constellation for iterations 1, 4, ···, 8.

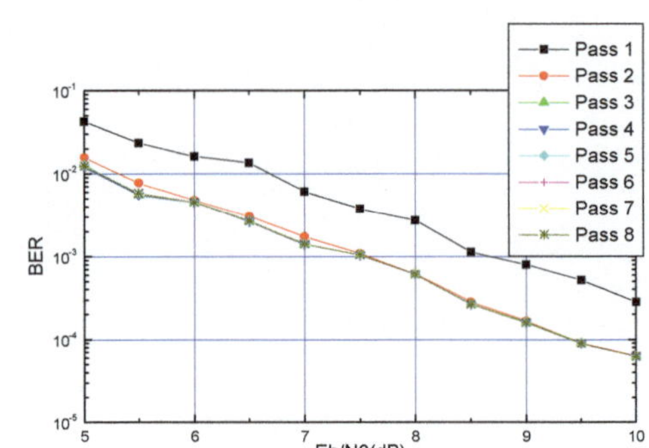

Figure 11. The BICM-ID's BER performance with MAP7/ (1,7) constellation for iterations 1, 4, ···, 8.

4

Performance Enhancement of SOVA Based Decoder in SCCC and PCCC Schemes

Ahmed A. Hamad

Department of Electrical Engineering, University of Babylon, Babel, Iraq.

ABSTRACT

This study proposes a simple scaling factor approach to improve the performance of parallel-concatenated convolutional code (PCCC) and serial concatenated convolutional code (SCCC) systems based on suboptimal soft-input soft-output (SISO) decoders. Fixed and adaptive scaling factors were estimated to mitigate both the optimistic nature of a posteriori information and the correlation between intrinsic and extrinsic information produced by soft-output Viterbi (SOVA) decoders. The scaling factors could be computed off-line to reduce processing time and implementation complexity. The simulation results show a significant improvement in terms of bit-error rate (BER) over additive white Gaussian noise and Rayleigh fading channel. The convergence properties of the suggested iterative scheme are assessed using the extrinsic information transfer (EXIT) chart analysis technique.

Keywords: Turbo Codes; SOVA; SCCC; Scaling Factor; EXIT

1. Introduction

Since the invention of turbo codes [1], a considerable interest had been devoted to reduce the complexity of the optimum MAP [2] decoding algorithm. This is to comply with the excessive need for low-power and low-cost decoder chips that are used in wireless mobile devices that pervade in recent years. The log-MAP, max-log-MAP, and SOVA algorithms [3-5] are some of the examples proposed for this target. The reduction in complexity that is achieved by near optimum algorithms like SOVA is accompanied by degradation in performance in terms of bit error rate.

Several papers have looked into the reasons behind the degradation in performance of these practical turbo-decoding algorithms, especially the one based on SOVA relative to the MAP algorithm [6-12]. A conviction has been created that the reason behind this degradation is the overestimation of reliability values generated by the SOVA decoder compared by those that would have been produced by the MAP decoder. It has been suggested in some of those papers that this optimistic extrinsic information may be due to the relatively high correlation between the intrinsic and extrinsic information [6]. As a result, various approaches were suggested to improve the performance of SOVA. Fixed and adaptive scaling factors are the more conventional approaches that are used to alleviate the distortion of the extrinsic information produced by SOVA [11].

This study proposes a simple approach for dealing with the exaggerated reliability values and the excessive correlation between the intrinsic and extrinsic information produced by the SOVA decoder. The suggested remedy is based on mathematical statistics, and it involves using two scaling factors, one is applied to the a-posteriori soft-output information of the SOVA and another is applied to the extrinsic information before it passes to the other decoder component in the iterative decoding process.

It is worth mentioning that the proposed approach could be applied to both SCCC and PCCC schemes based on MAP, log-MAP, max-log-MAP and SOVA decoders and it has almost the same complexity as that of the conventional schemes, which makes it quite attractive. The numerical results show that the turbo decoding algorithm based on SOVA that employs the proposed scaling factors can achieve a better performance compared to the one that does not employ scaling factors.

2. Improved SOVA

2.1. PCCC Scheme

In this work, it has been suggested that the turbo code consists of two recursive systematic convolutional codes (RSC) joined by an interleaver.

Let $u=[u_1,u_2,\cdots,u_k]$ where $u\in\{0,1\}$, is the binary information sequence. The modulated symbols corresponding to the coded bits are as follows

$$c=\left[c_{1,1},c_{1,2},c_{1,3},\cdots,c_{k,1},c_{k,2},c_{k,3}\right],$$

where $c\in\{-1,+1\}$, assuming binary phase shift keying modulation is applied. The noisy received sequence at the channel output is

$$y=\left[y_{1,1},y_{1,2},y_{1,3},\cdots,y_{k,1},y_{k,2},y_{k,3}\right]$$

and it is given by;

$$y_{k,i}=d_{k,i}c_{k,i}+\omega_{k,i}\quad i=1,2,3 \tag{1}$$

where $d_{k,i}$ is the fading magnitude, and $\omega_{k,i}$ is additive noise modeled as Gaussian with zero mean and variance $N_o/2$ (the two-sided power spectral density of the Gaussian channel). When the simulation is used over the additive white Gaussian channel (AWGN), $d_{k,i}$ is 1, while it is a Rayleigh random variable for the case of a Rayleigh flat-fading channel. It is assumed that coherent detection with a perfect estimation of $d_{k,i}$ is present at the receiver.

To avoid extra complexity when using the MAP algorithm, the SOVA decoder is proposed. Actually, SOVA has two essential modifications (over the conventional Viterbi Algorithm) which allows it to be used as a component decoder for turbo codes [4,13]. Firstly, the path metrics used are modified to take account of a-priori information when selecting the maximum likelihood path through the trellis. Secondly, the algorithm is modified so that it provides a soft output in the form of the a-posteriori LLR (L_{ap}) for each decoded bit. After each iteration, the LLR for first decoder (L_{ap}) can be represented as [14]

$$L_{ap}=L_c y_{k,1}+L_a+L_e \tag{2}$$

where $L_c y_{k,1}$ is the received systematic symbol scaled by the reliability of the channel L_c ($L_c=4dr\frac{E_b}{N_o}$, which can be set to 1 for SOVA [15]), where r is the rate of the code, E_b is the energy per information bit, and L_a is the a priori information achieved by interleaving (Π) or deinterleaving (Π^{-1}) of the extrinsic information L_e produced by the other decoder. For the first decoder (DEC1), $L_a^1=\Pi^{-1}\left[L_e^2\right]$, whereas for second decoder (DEC2), $L_a^2=\Pi\left[L_e^1\right]$.

To improve the reliability of L_{ap} for practical decoders like SOVA, two scaling factor α_1 and α_2 are proposed. The scaled L_{ap} is given by

$$\hat{L}_{ap}^j=\alpha_j L_{ap}^j\quad j=1,2. \tag{3}$$

The values of α_1 and α_2 are derived based on the minimum mean-square error (MMSE) criterion [16] as follows.

The threshold value of $\hat{L}_{ap}^1$, *i.e.*, $\operatorname{sgn}\left(\hat{L}_{ap}^1\right)$ produced by DEC1 is a growing estimate to the systematic coded bits $c_{k,1}$ with each iteration. Statistically, their mean-square difference (MSD)

$$\delta(\alpha_1)=E\left\{\left[\alpha_1\operatorname{sgn}\left(L_{ap}^1\right)-c_{k,1}\right]^2\right\}, \tag{4}$$

is a measure of how efficient the algorithm is with the proposed modification. $E[\cdot]$ denotes the expected value.

In a similar sense, the MSD related to DEC2 is

$$\delta(\alpha_2)=E\left\{\left[\alpha_2\operatorname{sgn}\left(L_{ap}^2\right)-\Pi\left[c_{k,1}\right]\right]^2\right\}. \tag{5}$$

It is obvious that to get better suboptimal decoding, the parameter α_j should be found to minimize the MSD $\delta(\alpha_j)$, which means that $\mathrm{d}\delta(\alpha_j)/\mathrm{d}\alpha_j=0$. From this equation, α_1 and α_2 are found to be

$$\alpha_1=E\left[\operatorname{sgn}\left(L_{ap}^1\right)\cdot c_{k,1}\right], \tag{6}$$

and

$$\alpha_2=E\left[\operatorname{sgn}\left(L_{ap}^2\right)\cdot\Pi\left[c_{k,1}\right]\right], \tag{7}$$

given that $E\left\{\left[\operatorname{sgn}\left(L_{ap}^j\right)\right]^2\right\}=1.$

We can describe the correlation between the received sequence and extrinsic information by their correlation coefficient [16]. Therefore, we have two correlation coefficients as follows

$$\rho_1\left(y_{k,1},L_e^1\right)=\frac{\boldsymbol{E}\left\{\left(y_{k,1}-\boldsymbol{E}\left[y_{k,1}\right]\right)\left(L_e^1-\boldsymbol{E}\left[L_e^1\right]\right)\right\}}{\sqrt{\boldsymbol{E}\left[\left(y_{k,1}-\boldsymbol{E}\left[y_{k,1}\right]\right)^2\right]\boldsymbol{E}\left[\left(L_e^1-\boldsymbol{E}\left[L_e^1\right]\right)^2\right]}}, \tag{8a}$$

$$\rho_2\left(\Pi\left(y_{k,1}\right),L_e^2\right)=\frac{E\left\{\left(\Pi\left(y_{k,1}\right)-E\left[\Pi\left(y_{k,1}\right)\right]\right)\left(L_e^2-E\left[L_e^2\right]\right)\right\}}{\sqrt{E\left[\left(\Pi\left(y_{k,1}\right)-E\left[\Pi\left(y_{k,1}\right)\right]\right)^2\right]E\left[\left(L_e^2-E\left[L_e^2\right]\right)^2\right]}}. \tag{8b}$$

To reduce the correlation between intrinsic and extrinsic information, it is proposed that we scale L_e^j by β_j

$$\hat{L}_e^j=\beta_j L_e^j, \tag{9}$$

where

$$\beta_j=\left(1-\rho_j\right). \tag{10}$$

The above method can be implemented as shown in **Figure 1**.

2.2. SCCC Scheme

A similar approach is considered for the case of the SCCC scheme with little modification. A simple block diagram for the conventional SCCC encoder and modified decoder is depicted in **Figure 2**. Simulation results show that better performance can be achieved by scaling the LLRs which are produced by outer decoder (ODEC) as shown in **Figure 2(b)**.

Following a similar procedure to PCCC, the LLR immediate produces by the ODEC scaled as:

$$\alpha_2 = E\left\{\operatorname{sgn}\left(L_{ap}^2\right)\cdot c_k^o\right\}. \tag{11}$$

Whereas, the extrinsic information for the same decoder is scaled by β_2 given by Equation (10). Here, the correlation coefficient ρ_2 may be represented as

$$\rho_2\left(L_a^2, L_e^2\right) = \frac{E\left\{\left(L_a^2 - E\left[L_a^2\right]\right)\left(L_e^2 - E\left[L_e^2\right]\right)\right\}}{\sqrt{E\left[\left(L_a^2 - E\left[L_a^2\right]\right)^2\right]E\left[\left(L_e^2 - E\left[L_e^2\right]\right)^2\right]}}. \tag{12}$$

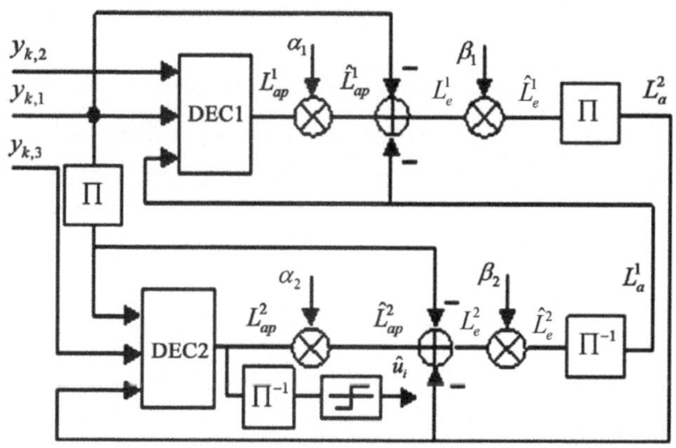

Figure 1. Implementation of modified SOVA.

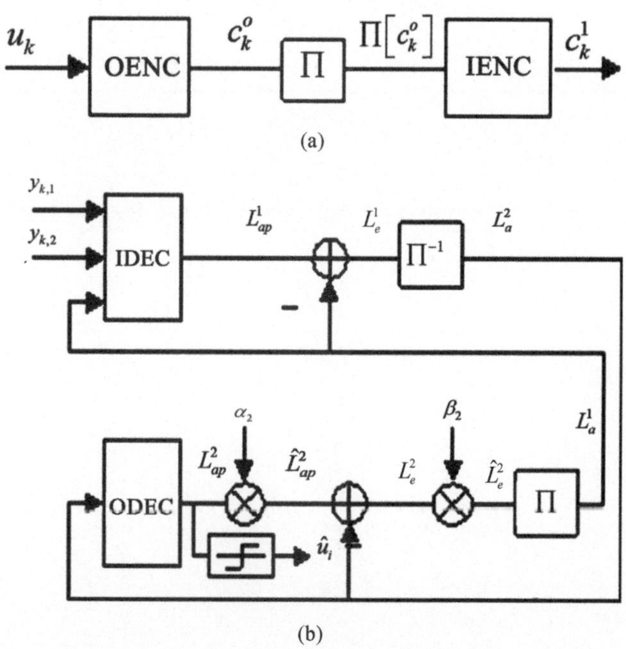

Figure 2. (a) SCCC encoder, (b) Modified SCCC decoder.

3. Numerical Results

This section presents the effectiveness of scaling factors on the performance of PCCC and SCCC schemes. The simulations were implemented over AWGN and flat fading Rayleigh channels. For the fading channel, the sampling time of the input signal is taken to be (1/50000) sec. and maximum Doppler shift of 100 Hz.

The 1/3 rate turbo code that is specified for high-speed downlink packet access (HSDPA) in UMTS [17] is considered here which consists of two 1/2 rate component RSC codes of memory 3, with polynomials (Gr, Gf) = (1 + D2 + D3, 1 + D + D3). The two RSCs are joined by interleaver of sizes 1024 bits, which is constructed for the same standard [17]. In SCCC, the same constituent codes are used, and to obtain a total coding rate of 1/3, the coded bits have been puncture using the pattern [10,11]. To examine the effectiveness of the four scaling factors (α_j and β_j, $j = 1,2$), different combinations of these factors are considered. **Figures 3** and **4** show the BER

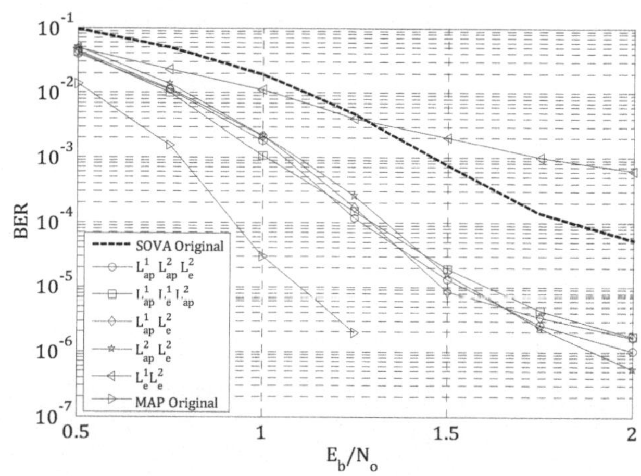

Figure 3. BER performance of modified PCCC systems over AWGN channel.

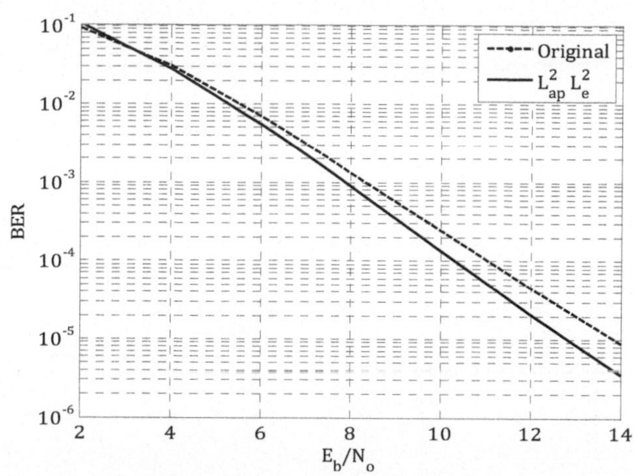

Figure 4. BER performance of modified system $L_{ap}^2 L_e^2$ over Rayleigh fading channel.

results for the proposed PCCC systems after five iterations over AWGN and Rayleigh fading channels, respectively. To avoid congestion, each figure is restricted to a number of curves, which demonstrate better improvement. The proposed schemes are also compared with the performance of turbo decoder based on log-MAP algorithm simulated over AWGN channel. Each curve is labeled with a group of symbols, which describe the simulated system. The letters L_{ap} and L_e refer to the scaled LLR and the numeral superscript indicates which decoder is belong (DEC1 or DEC2). The letters "O" and "I" refer to the outer (ODEC) and inner (IDEC) decoders respectively.

The simulation presents the gain achieved (about 0.9 dB over AWGN and 1 dB over faded channel at BER of 10^{-5}) by modified systems (with scaling) in comparison with the original system (without scaling). The performance of modified systems are become closer by about 0.35 dB (~0.75 dB for unmodified systems) to the performance of turbo code utilizes log-MAP decoder at BER of 10^{-4}. **Table 1** presents the average scaling factors (α_2 and β_2) derived offline for five iterations to system $L_{ap}^2 L_e^2$ at different signal to noise ratios (E_b/N_o) over flat fading channels. Referring to Equation (6) and (7), the values of α_j should be estimated offline, because the systematic coded bits ($c_{k,1}$) are not available at the decoder side, whereas the values of β_j can be obtained online. Utilizing the simulation results taken from the "free-running" iterative decoder, the average trajectory of the extrinsic information transfer chart (EXIT) [18] for system ($L_{ap}^1 L_{ap}^2 L_e^2$) is depicted in **Figure 5** at E_b/N_o = 0.5 dB. In comparison to the trajectory of original system, it is obvious from **Figure 5** that the modified systems present better convergence in mutual information.

Table 1. Average scaling factors, α_2 and β_2 applied on system $L_{ap}^2 L_e^2$ for different E_b/N_o and 5 iterations assuming Rayleigh fading channel.

	$L_{ap}^2 L_e^2$									
E_b/N_o (dB)	α_2					β_2				
	Iter1	Iter2	Iter3	Iter4	Iter5	Iter1	Iter2	Iter3	Iter4	Iter5
2.0	0.769	0.754	0.785	0.766	0.784	0.948	0.791	0.847	0.784	0.830
4.0	0.893	0.925	0.937	0.939	0.943	0.636	0.542	0.538	0.518	0.521
6.0	0.959	0.982	0.987	0.988	0.989	0.490	0.437	0.424	0.416	0.412
8.0	0.986	0.996	0.998	0.998	0.998	0.426	0.390	0.379	0.374	0.371
10.0	0.996	0.999	1.000	1.000	1.000	0.394	0.365	0.356	0.353	0.351
12.0	0.999	1.000	1.000	1.000	1.000	0.375	0.351	0.343	0.340	0.339
14.0	1.000	1.000	1.000	1.000	1.000	0.364	0.342	0.335	0.332	0.332

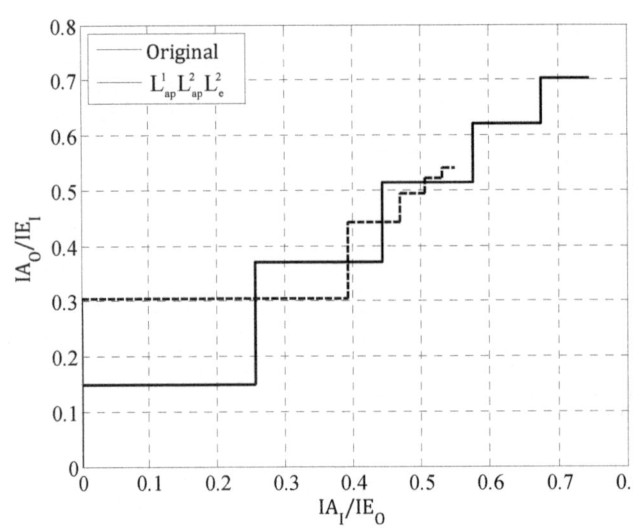

Figure 5. Average trajectory for system ($L_{ap}^1 L_{ap}^2 L_e^2$) at E_b/N_o = 0.5 dB in AWGN channel.

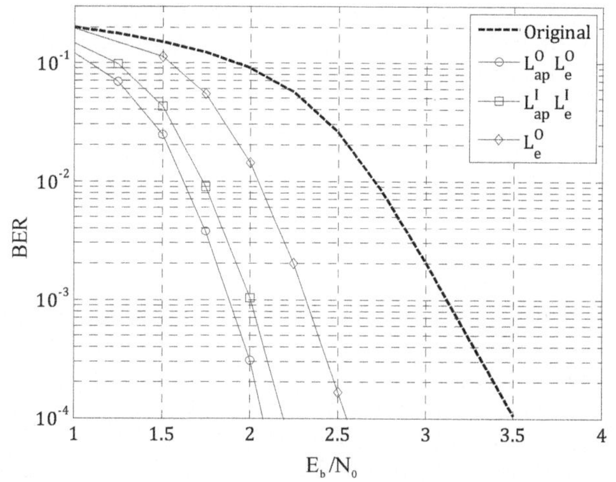

Figure 6. BER performance of modified SCCC systems over AWGN channel.

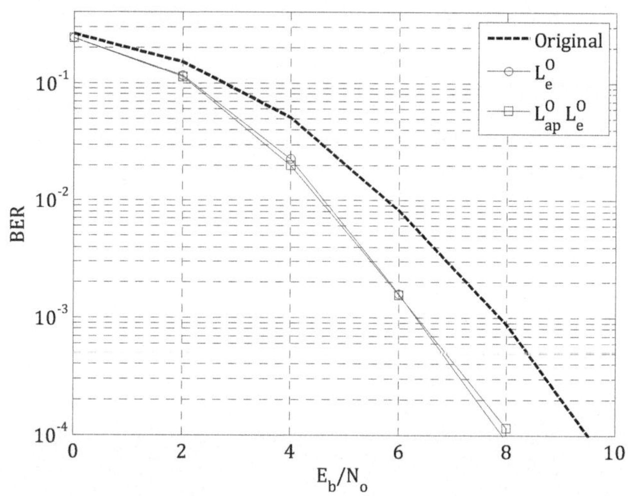

Figure 7. BER performance of modified SCCC systems over Rayleigh fading channel.

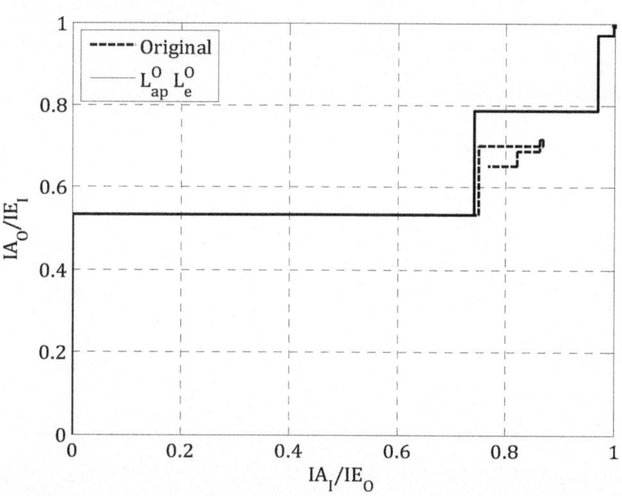

Figure 8. Average trajectory for system ($L^o_{ap}L^o_e$) at E_b/N_o = 2.25 dB in AWGN channel.

Figures 6 and **7** show the BER performance of modified SCCC schemes compared with the original SOVA over AWGN and flat fading channel respectively. The simulation of scheme $L^o_{ap}L^o_e$ reveals similar improvement of about 1.5 dB at BER 10^{-4} over the two channels. **Figure 8** shows a stall in the mutual information trajectory for conventional SOVA after two iterations, whereas $L^o_{ap}L^o_e$ continue in converge.

4. Conclusion

This paper introduces simple modifications to the conventional SOVA to alleviate the effect of optimistic a posteriori information and the strong correlation between the input and output of the SOVA. A method for offline and online computation of scaling factors has also been described. It has shown that the proposed scheme is significantly improved the performance of PCCC and SCCC schemes that are based on suboptimal decoders. The simulation shows that better improvement could be achieved by adding two or three simple multipliers to the traditional PCCC scheme, whereas two multipliers are sufficient to produce a modified SCCC scheme. The complexity resulting from incorporating scaling factors is almost the same as that of the traditional method without these factors. The convergence behavior of such decoders is investigated by using extrinsic information transfer (EXIT) charts.

REFERENCES

[1] C. Berrou, A. Glavieux and P. Thitimajshima, "Near Shannon Limit Error-Correcting Coding and Decoding: Turbo Codes," *Proceedings of the International Conference on Communications*, Geneva, 23-26 May 1993, pp. 1064-1070.

[2] J. Hagenauer and P. Robertson, "Iterative ("Turbo") Decoding of Systematic Convolutional Codes with the MAP And SOVA Algorithm," *Proceedings of the ITG Tagung, Codierung für Quelle, Kanal and Ubertragung*, Frankfurt, October 1994, pp. 21-29.

[3] P. Robertson, E. Villebrun and P. Höher, "A Comparison of Optimal and Sub-Optimal MAP Decoding Algorithms Operating in the Log Domain," *Proceedings of the International Conference on Communications*, Seattle, 18-22 June 1995, pp. 1009-1013.

[4] J. Hagenauer and P. Hoeher, "A Viterbi Algorithm with Soft-Decision Outputs and Its Applications," *Proceedings of IEEE Global Telecommunications Conference*, Dallas, 27-30 November 1989, pp. 1680-1686.

[5] C. Berrou, P. Adde, E. Angui and S. Faudeil, "A Low Complexity Soft-Output Viterbi Decoder Architecture," *Proceedings of IEEE International Conference on Communications, Geneva*, Vol. 2, 23-26 May 1993, pp. 737-740.

[6] L. Papke, P. Robertson and E. Villerbrun, "Improved De-

coding with the SOVA in a Parallel Concatenated (Turbo-Code) Scheme," *Proceedings of the IEEE International Conference Communications* (*ICC*), Dallas, 23-27 June 1996, pp. 102-106.

[7] C. H. Wang, W. T. Wang and C. C. Chao, "A Unified Structure of Trellis-Based Soft-Output Decoding Algorithms for Turbo Codes," *IEEE Transactions on Communications*, Vol. 52, No. 8, 2004, pp. 1355-1366.

[8] G. Colavolpe, G. Ferrari and R. Raheli, "Extrinsic Information in Iterative Decoding: A Unified View," *IEEE Transactions on Communications*, Vol. 49, No. 12, 2001, pp. 2088-2094.

[9] C. X. Huang and A. Ghrayeb, "A Simple Remedy for the Exaggerated Extrinsic Information Produced by SOVA Algorithm," *IEEE Transactions on Wireless Communications*, Vol. 5, No. 5, 2006, pp. 996-1002.

[10] S. Papaharalabos, P. Sweeney, B. G. Evans and P. T. Mathiopoulos, "Improved Performance SOVA Turbo Decoder," *IEE Proceedings—Communications*, Vol. 153, No. 5, 2006, pp. 586-590.

[11] C. X. Huang and A. Ghrayeb, "Improved SOVA and APP Decoding Algorithms for Serial Concatenated Codes," *IEEE Globecom*, Dallas, 29 November-3 December 2004, pp. 189-193.

[12] D.-W. Yue and H. H. Nguyen, "Unified Scaling Factor Approach for Turbo Decoding Algorithms," *IET Communications*, Vol. 4, No. 8, 2010, pp. 905-914.

[13] C. Berrou, P. Adde, E. Angui and S. Faudeil, "A Low Complexity Soft-Output Viterbi Decoder Architecture," *Proceedings of ICC* 1993, *Geneva*, May 1993, pp. 737-740.

[14] J. Hagenauer, E. Offer and L. Papke, "Iterative Decoding of Binary Block and Convolutional Codes," *IEEE Transactions on Information Theory*, Vol. 42, No. 2, 1996, pp. 429-445.

[15] L. Lin and R. S. Cheng, "Improvements in SOVA-Based Decoding for Turbo Codes," *Proceedings of the IEEE International Conference on Communications* (*ICC*), Montreal, Vol. 3, 8-12 June 1997, pp. 1473-1478.

[16] A. Papoulis, "Probability, Random Variables, and Stochastic Processes," McGraw-Hill Inc., Montreal, 1991.

[17] 3GPP, "Ts 25.212 v3.11.0 (2002-09)," Technical Specification Group Radio Access Network, Multiplexing and Channel Coding (FDD), 1999.

[18] S. T. Brink, "Convergence Behavior of Iteratively Decoded Parallel Concatenated Codes," *IEEE Transactions on Communications*, Vol. 49, No. 10, 2001, pp. 1727-1737.

5

Performance Evaluation of Adaptive Modulation Based MC-CDMA System

Iftekhar Alam[1], Vikas Srivastva[1], Arun Prakash[2], Rajeev Tripathi[2], A. K. Shankhwar[3]

[1]Wireless Communication and Computing, IIIT Allahabad, India; [2]ECED, MNNIT, Allahabad, India; [3]Electronics Engineering, HBTI, Kanpur, India.

ABSTRACT

In today's scenario mobile phones are not used only for messaging and talking, now we use mobiles also to access internet. For that we need higher data rates. Mobiles use wireless communication channel, which is very fast varying channel. To increase the data rate we have to utilize the full channel based on the channel condition. For that we have to estimate the channel condition and based on that we have to vary the data rate so that our communication system performance parameters (bit error rate, throughput) do not change beyond the accepted level. To increase the data rate we are also using the multi-carrier CDMA system in conjunction with the above described adaptive modulation. Adaptive modulation, in conjunction with MC-CDMA transmission system is very assuring way to increase data rate. In current scenario demand of very high data rates up to 100 Mbits/s may be achieved with this scheme. In this paper, adaptive modulation based M-ary QAM, M-PSK and M-ary CPM systems applied to a MC-CDMA system in Rayleigh fading channel environment have been investigated and BER performances of all these digital modulation schemes have been compared.

Keywords: Adaptive Modulation; CDMA; Digital Modulation; MC-CDMA; OFDM

1. Introduction

In wireless communication system spectrum bandwidth is the most precious commodity. The main problem to the wireless channel is Rayleigh fading channel that degrades the error performance of the wireless communication system. In wireless communication channel the signal to noise ratio (SNR) is time variant due to multipath fading and interference from other users, in direct contrast with the additive white Gaussian noise (AWGN) channel which is characterized by a constant signal to noise ratio.

We may utilize full channel capacity in two ways in presence of multipath fading and interference from other users: adaptation and diversity techniques. By adaptation we mean to change the modulation scheme and/or coding used by transmitter according to the channel condition.

Figure 1 shows that the channel estimation is needed for using different digital modulation techniques. As we increase the range of communication, we step down to lower modulations but as we reduce the communication range means we are closer to the base station we can utilize higher order modulations for increased throughput. Adaptive modulation also allows the system to overcome multipath fading and other interference. In essence, this is a way to optimize the modulation scheme for the better transmission of data at higher rate in accordance to the state of the channel for required fidelity. For example, when the channel condition is poor we may reduce the modulation index in order to improve fidelity. Conversely if the channel is in good condition we can increase the modulation index to increase the data rate.

The remainder of the paper is organized as follows. Author presents survey of some relevant adaptive modulation schemes in Section 2, followed by the system model discussed in Section 3 in which author presents a model and some formulas for calculating SNR. The paper continues with result and discussion in Section 4. Finally, the paper is concluded.

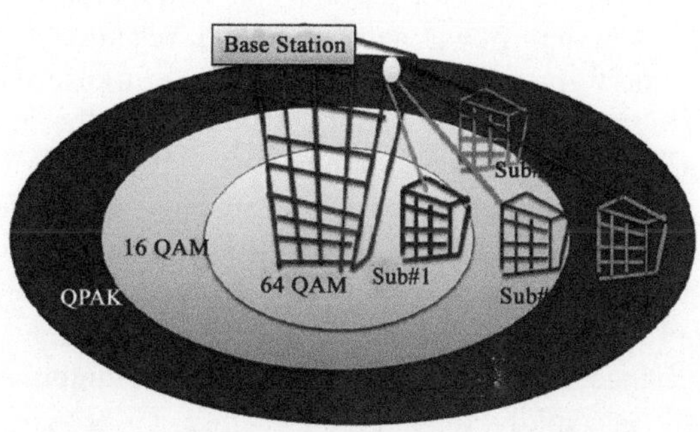

Figure 1. Illustration of adaptive modulation.

2. Related Work

Adaptive technology has been firstly investigated in late 60's and early 70's [1,2]. Interest in this modulation scheme was short lived, because of hardware constraints and lack of good channel estimation techniques. As technology improved these issues became less constraining, which resulted in a revived interest of researchers in adaptive modulation schemes for 3rd generation wireless systems [3].

In the adaptive techniques, transmission parameters, for example transmission power [4], symbol transmission rate [5], constellation size [6-8], coding rate/scheme [9], or any combination of these parameters [8,10], are changed in response to the estimated time-varying channel conditions. Many researchers have given their system model by using different combination of modulation schemes and evaluate its performance in terms of BER, throughput and data rate and showed improvement from the existing systems. Some of them we studies are:

2.1. Reliable Adaptive Modulation Aided by Observations of Another Fading Channel

Tung-Sheng Yang and Alexandra Duel-Hallen in [11] have proposed an approach to use adaptive transmission technique using past channel observations of 1 carrier to predict future channel state information and this technique performs adaptive modulation without feedback path for another correlated carrier. They derived a MMSE (Minimum mean-square error) long range channel prediction method which uses the time-domain and frequency domain correlation function of Rayleigh fading channel. They used a standard stationary fading channel model to test their algorithm and showed a significant gain over non adaptive techniques.

If the communication between 2 stations is bidirectional then each station can estimate the channel condition and adapt the parameters to this estimation. This is known as open-loop adaptation [12]. And if the communication between two stations is not bidirectional the receiver has to estimate channel condition and feedback it to the transmitter, resulting in closed loop adaptation. The feed back load of the channel estimation information consumes system power and bandwidth, resulting in degraded performance. For many adaptive transmission applications, it is not necessary to feed back the actual fading coefficient instead of this it is sufficient to feed back to the transmitter only the antenna selection or modulation index produced from the estimated or predicted values at the receiver [13]. The feedback delay, channel estimation and processing delay degrade the performance of this adaptive modulation technique, especially in rapidly time-variant fading. To realize the good performance improvement for adaptive transmission techniques, the channel variations should be predicted reliably at least several milliseconds ahead. They extended the LRP algorithm into frequency domain. This LRP method was proposed in [14,15]. This long range prediction (LRP) algorithm utilizes an autoregressive model for characterizing the fading channel and also computes the minimum mean square error to estimate the future fading coefficient on the basis of past observations. Particularly the concentrated on a scenario where they observed a received uplink signal at the carrier of frequency f^1 and attempted to predict the down link signal at carrier of frequency f^2 without any feedback from the mobile. For the accomplishment of this prediction, the predicted samples should be sufficiently correlated in both frequency and time.

2.2. Adaptive Modulation Systems for Predicted Wireless Channel

Sorour Falahati and Arne Svensson have studied an optimum design for an adaptive modulation technique based on encoded M-QAM, assisted with channel prediction for the Rayleigh fading channel in [15]. Signaling delays create problem with outdated Channel State Information (CSI) in adaptive modulation. This degrades the adaptive modulation system performance such as decrease in the throughput. The use of power prediction will improve the performance of adaptive modulation. Sorour Falahati and Arne Svensson have shown their interest to take the quality of these predictions into account in designing adaptive modulation scheme. The key issues they studied in [15] are how a known prediction error variance can affect the transmission properties, such as the signal-to-noise ratio (SNR) boundaries that determine when to change the modulation index, and to what extent it could affect the spectral efficiency. **Figure 2** shows the discrete model of the system they proposed. They considered fast link adaptation where the receiver estimates the received power and transfers this information to the transmitter site via feedback channel to modify the transmission parameters such as modulation index, estimates of the channel state information will be based on outdated information. Theoretically an optimized way of handling this problem could be to estimate the conditional probability density

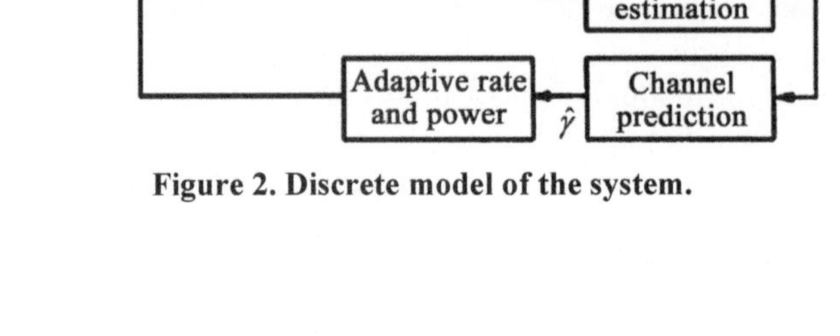

Figure 2. Discrete model of the system.

function of the BER optimally for a given modulation format form the all past received information signals and other relevant information. And then the modulation scheme is selected on the basis of this probability density function. But this proposed approach is hard to realize because the BER is a non-linear function of channel power gain. Therefore they considered a suboptimal approach that the channel power gain is predicted by a scheme and adaptive modulation system is adjusted on the basis of statistics of these predictions.

3. System Model

3.1. System Model and Block Diagram

Figure 3 shows the block diagram for the adaptive modulation based MC-CDMA system. Firstly the binary information is encoded with turbo codes, which is followed by a serial to parallel converter to produce a data stream of low bit rate in order to avoid the inter-symbol interference.

Then each stream is modulated with by a suitable digital method, in this thesis work we used BPSK, QPSK, 16-QAM, 64-QAM, M-ary MHPM etc., depending upon the current channel signal to noise ratio estimate information provided by the receiver. After this the modulated bit streams are passed through the Multi-carrier CDMA system transmitter then this is up-converted by an RF amplifier and finally transmitted through the channel. At the receiver site reverse operations are performed to demodulate and decode the original information. In this thesis we used pilot assisted channel estimation for estimating the quality of channel which estimates the carrier to noise ratio of the channel. Pilots symbols used are known QPSK symbols. The CDMA code used in this thesis is Walsh-Hadamard code.

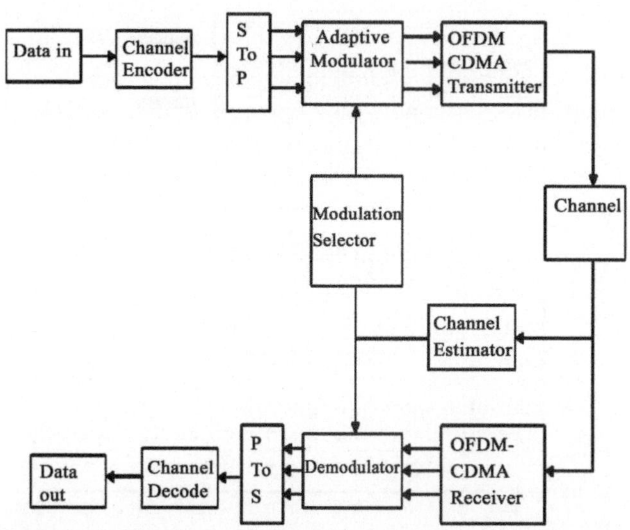

Figure 3. Block diagram for adaptive modulation base MC-CDMA system.

3.2. SNR Estimation

The basic idea for channel estimation in our thesis is that if we have the signal which is corrupted by a given channel we can estimate SNR of the channel through mean and variance calculation. Firstly, the average received power is calculated by taking the square of mean of magnitude of the received signal. This will be used as estimated signal power. And the noise power will be calculated by variance calculation on the received signal.

For PSK modulation scheme, we use the following scheme to find the mean and variance of a currently received block of data:

$$Z = \frac{1}{N}\sum_{i=1}^{N}|r_i|$$

$$T^2 = \frac{1}{N-1}\left(\sum_{i=1}^{N}\left(|r_i| - |c_i'|\right)^2\right)$$

where Z represents mean of the received signal and T^2 represents the variance of the received signal corrupted though the channel noise N represents the no of symbols over which SNR has to be estimated. c_i is the channel estimate coming from FFT estimator.

The SNR can be calculated as:

$$\Gamma = \left(\frac{N-3}{N-1}\right)\frac{|Z|^2}{T^2} - \frac{1}{N}$$

3.3. Adaptive Modulation Boundaries

In adaptive modulation system, we need a way to decide which modulation level is best suited for the present channel conditions. We decided to use the estimated SNR of the link as channel metric to decide switching levels. Pons and Dunlop in [16] proposed that bit error rate at the receiver would be really a good channel metric to decide modulation level switching. But our personnel experience says that BER estimation on small packet or for very short duration could not be reliably estimated.

Now the question arises how to decide that which modulation scheme is to be used for which range of SNR? Answer lies within the AWGN performance of every modulation scheme.

Received signal is given as $r(t) = c(t).s(t) + n(t)$. Where $c(t)$ is Rayleigh channel gain, $n(t)$ is additive Gaussian noise. In our system model we considered the signal power is to be the transmitted signal power multiplied with the Rayleigh channel gain. This signal power is the instantaneously received signal power, which could be directly compared to the Gaussian noise power. So, allowing us to consider the BER over an AWGN channel. Now consider the BER performance of the 3 modulation schemes used in our thesis. Form [17] equations for their BER are as follows:

$$P_{\text{QPSK}}(Y) = Q\left(\sqrt{Y}\right)$$

$$P_{16\text{ QAM}}(Y) = \frac{1}{4}\left[Q\left(\sqrt{\frac{Y}{5}}\right) + Q\left(3\sqrt{\frac{Y}{5}}\right) + \frac{1}{2}Q\left(\sqrt{\frac{Y}{5}}\right)\right]$$

$$\begin{aligned} P_{64\text{ QAM}}(Y) = \frac{1}{12}\Bigg[& Q\left(\sqrt{\frac{Y}{21}}\right) + Q\left(3\sqrt{\frac{Y}{21}}\right) + Q\left(5\sqrt{\frac{Y}{21}}\right) \\ & + Q\left(7\sqrt{\frac{Y}{21}}\right) + \frac{1}{6}Q\left(\sqrt{\frac{Y}{21}}\right) + \frac{1}{6}Q\left(3\sqrt{\frac{Y}{21}}\right) \\ & + \frac{1}{6}Q\left(5\sqrt{\frac{Y}{21}}\right) + \frac{1}{6}Q\left(7\sqrt{\frac{Y}{21}}\right) + \frac{1}{3}Q\left(\sqrt{\frac{Y}{21}}\right) \\ & + \frac{1}{4}Q\left(3\sqrt{\frac{Y}{21}}\right) - \frac{1}{4}Q\left(5\sqrt{\frac{Y}{21}}\right) - \frac{1}{6}Q\left(7\sqrt{\frac{Y}{21}}\right) \\ & + \frac{1}{6}Q\left(9\sqrt{\frac{Y}{21}}\right) + \frac{1}{12} + Q\left(11\sqrt{\frac{Y}{21}}\right) \\ & - \frac{1}{12}Q\left(13\sqrt{\frac{Y}{21}}\right)\Bigg] \end{aligned}$$

$$P_{\text{QPSX}}(Y) = Q0$$

In the above mentioned equations, γ is the SNR, & $Q(.)$ is the Q function defined as:

$$Q(x) = \frac{1}{\sqrt{2\pi}}\int_{\gamma}^{N} e^{-\frac{x^2}{2}}\,dx$$

On the basis of above formula we calculate the BER and run the MATLAB simulation. We observe a considerable change in throughput and data rate in case of adaptive modulation. This is discussed in next section.

4. Result and Discussion

The MATLAB simulations of MC-CDMA in Rayleigh fading channel have been carried out and for various M-ary QAM, M-ary CPM, M-ary MHPM. **Figure 4**, the curves from right to left represent the BER for 64 QAM, 16 QAM and QPSK in a Rayleigh fading channel, respectively. For deciding the proper switching levels we need to decide first what performance, in terms of BER we need from our system. For our system model we decided to choose BER performance of 10^-3 as operating point. This means that this system will try to keep the BER of the system lower than 10^-3 with most spectrally efficient modulation scheme whenever possible. Spectral efficiency is defined as the number of information bits encoded on a single modulated transmission symbol.

Therefore, with our operating point, and given BER plots in **Figure 4** we conclude to have the following SNR ranges for each modulation scheme:

QPSK	SNR < 16 dB
16 QAM	17 <= SNR <= 23 dB
64 QAM	SNR > 23 dB

We made this conclusion in the following way: from **Figure 4**, there is no modulation scheme that gives us performance below 10^-3 at an SNR below 10 dB. So, we choose QPSK as it was the most robust modulation scheme.

In the range of 10 to 17 dB, there was only one scheme that gives us our desired performance that was QPSK. In the range of 17 to 23 dB, 16 QAM gives us performance below 10^-3, at a better spectral efficiency. And for SNR higher than 23 dB, 64 QAM provides us our desired BER performance with the best spectral efficiency.

Figure 5 shows the BER performance under different channel SNR for a given number of users with M-ary MHPM as modulation technique. This figure also shows that the performance of the system degrades with the increase in the number of users due to multipath interference. The adaptive modulation with MC-CDMA technique gives a better BER performance as shown by green line.

5. Conclusion and Future Scope

In this paper we studied various adaptive modulated MC-CDMA systems in Rayleigh fading channel. The digital modulation schemes considered in our thesis are M-ary PSK, M-ary QAM, M-ary MHPM, M-ary CPM, and GMSK with varying bit-duration-bandwidth product. The results of our study of the adaptive modulated systems clearly show that the dynamic switching of the modulation orders can enhance the system performance and capacity per given bandwidth with the expected BER performance. With the adaptive modulation based systems,

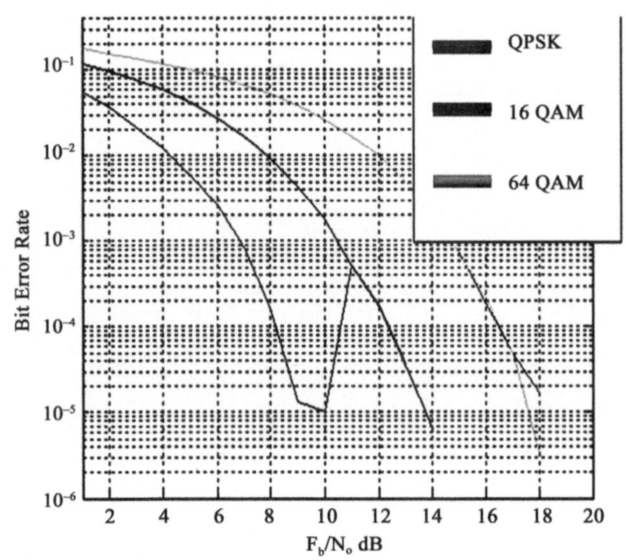

Figure 4. BER performances for different modulation scheme.

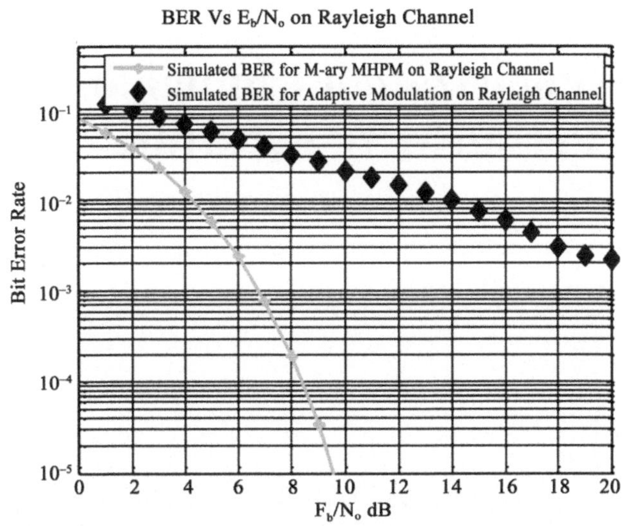

Figure 5. Comparison of overall performance of a system with adaptive modulated MC-CDMA system and M-ary MHPM in Rayleigh fading channel at CSNR = 15 dB.

the bit rate per given bandwidth does not remain constant as in fixed modulation based systems but it varies according to the channel condition.

We found that the system using M-ary MHPM modulation technique shows the best performance among the considered modulation schemes. So an adaptive M-ary MHPM modulated MC-CDMA system is able to provide highly effective system in supporting high data rate services with mobility in wireless networks, and takes substantial care for the inter-symbol interference generated due to multipath effect and provides adequate bandwidth for the intended services.

REFERENCES

[1] J. K. Cavers, "Variable-Rate Transmission for Rayleigh Fading Channels," *IEEE Transactions on Communication*, Vol. 20, No. 1, 1972, pp. 15-22.

[2] S. Otsuki, S. Sampei and N. Morinaga, "Square-QAM Adaptive Modulation/TDMA/TDD Systems Using Modulation Level Estimation with Walsh Function," *Electronics Letters*, Vol. 31, No. 3, 1995, pp. 169-171.

[3] A. Goldsmith and S. G. Chua, "Adaptive Coded Modulation for Fading Channels," *Transactions on Communication*, Vol. 46, No. 5, 1998, pp. 595-602.

[4] GSM 02.60, "GPRS Service Description—Stage 1," Version 7.5.0, 1998.

[5] P. Stuckmann and O. Paul, "Dimensioning GSM/GRPS Networks for Circuit-Switched and Packet-Switched Services," *Proceedings of WPMC*'01, Aalborg, Denmark, September 2001.

[6] H. Holma and A. Toskala, "WCDMA for UMTS—Radio Access for Third Generation Mobile Communications," John Wiley & Sons, New York, 2001.

[7] 3GPP TS 25.211, "Physical Channels and Mapping of Transport Channels onto Physical Channels (FDD)," 2002.

[8] T. Ue, *et al*., "Symbol Rate and Modulation Level Controlled Adaptive Modulation/TDMA/TDD for High-Bit-Rate Wireless Data Transmission," *IEEE Transactions on Vehicular Technology*, Vol. 47, No. 4, 1998, pp. 1134-1147.

[9] T.-S. Yang and A. Duel-Hallen, "Reliable Adaptive Modulation Aided by Observations of Another Fading Channel," *IEEE Transactions on Communications*, Vol. 52, No. 4, 2004.

[10] K. Miya, O. Kato, K. Homma, T. Kitade, M. Hayashi and T. Ue, "Wideband CDMA Systems in TDD-Mode Operation for IMT-2000," *IEICE Transactions on Communications*, Vol. E81-B, No. 8, 1998, pp. 1317-1326.

[11] A. Duel-Hallen, S. Hu and H. Hallen, "Long-Range Prediction of Fading Signals: Enabling Adaptive Transmission for Mobile Radio Channels," *IEEE Signal Processing Magazine*, Vol. 17, 2000, pp. 62-75.

[12] T. Eyceoz, A. Duel-Hallen and H. Hallen, "Deterministic Channel Modeling and Long-Range Prediction of Fast-Fading Mobile Radio Channels," *IEEE Communication Letters*, Vol. 2, No. 9, 1998, pp. 254-256.

[13] T. Eyceoz, S. Hu and A. Duel-Hallen, "Performance Analysis of Longrange Prediction for Fast-Fading Channels," *Proceedings of the* 33rd *Annual Conference on Information Sciences and Systems*, Vol. 2, 1999, pp. 656- 661.

[14] A. Duel-Hallen, S. Hu and H. Hallen, "Long-Range Prediction of Fading Signals: Enabling Adaptive Transmission for Mobile Radio Channels," *IEEE Signal Processing Magazine*, Vol. 17, 2000, pp. 62-75.

[15] S. Falahati and A. Svensson, "Adaptive Modulation Systems for Predicted Wireless Channels," *IEEE Transactions on Communication*, Vol. 52, No. 2, 2004, pp. 307-316.

[16] J. Pons and J. Dunlop, "Bit Error Rate Link Adaptation for GSM," *The 9th IEEE International Symposium on Personal, Indoor and Mobile Radio Communications*, Vol. 3, 8-11 September 1998, pp. 1530-1534.

[17] J. M. Torrence and L. Hanzo, "Upper Bound Performance of Adaptive Modulation in a Slow Rayleigh Fading Channel," *IEEE Electronics Letters*, Vol. 32, No. 8, 1996, pp. 718-719.

6

Broadband Feed for Low Cross-Polarization Uniplanar Tapered Slot Antennas on Low-Permittivity Substrate

Jan Schorer, Jens Bornemann

Department of Electrical and Computer Engineering, University of Victoria, Saanich, Canada.

ABSTRACT

A broadband feeding technique for the uniplanar tapered slot antenna (TSA) is presented. The TSA operates at a center frequency of 6.5 GHz with a 7 GHz bandwidth (107 percent). The antenna and feed are realized with a broadband microstrip-to-slotline transition on a low permittivity high frequency substrate, Rogers RT/Duroid 5880. The input impedance of the system is designed for 50 Ω compatibility with other system components, and the cross polarization is kept below –30 dB. The developed TSA system was simulated with commercially available electromagnetic software and manufactured. Measured results validate the design process and the antenna's performance.

Keywords: Tapered Slot Antenna; Ultra-Wideband; Microstrip; Slotline; Planar Transmission Lines

1. Introduction

Omnidirectional ultra-wideband (UWB) antennas and related applications attracted increasing attention during the past decade for use in personal area network (PAN) communication applications, e.g. [1].

Directional UWB antennas are employed in radio astronomy receivers where TSA antennas are used to collect extra terrestrial radiation over a large bandwidth. For this purpose, a TSA system is usually mounted within an array setup, e.g. [2-5]. Mounting TSA systems in an array puts certain limits on the dimensions of the single TSA. Together with the need for non-complicated and inexpensive manufacturing processes, due to the huge amount of single TSAs elements needed, printed-circuit board (PCB) technology is found to accommodate the encountered restrictions very well.

TSA systems are well-examined and known antenna types [6,7]. Therefore, several papers and books on the different components of a TSA system are available and discuss related planar transmission lines (PTLs) [8] and transitions involving microstrip and slotlines [9-14].

It has been demonstrated that the antipodal TSA, with fins on opposite sides of the substrate, is easily fed by a microstrip line over a wide bandwidth [15-17]. However, it is the uniplanar TSA, with both fins on the same substrate side, which is usually employed in large array applications. So far, though, its feed has limited the exploitation of an ultra-wide bandwidth [2,4].

The uniplanar TSA with a microstrip-to-slotline transition as feed is shown in **Figure 1**. Traditionally, the transition is realized on high-permittivity substrate [8,9] due to the availability of better slotline models. However, a TSA performs best on low-permittivity substrate. On high-permittivity material, its performance is known to be poor unless parts of the substrate in the antenna are removed. This puts an extra effort on fabrication which has to be avoided when TSA arrays are involved [2,4]. Another feed that can be considered is a transition from coplanar waveguide (CPW) to slotline. However, such a transition requires air bridges [18] which again add to additional fabrication effort.

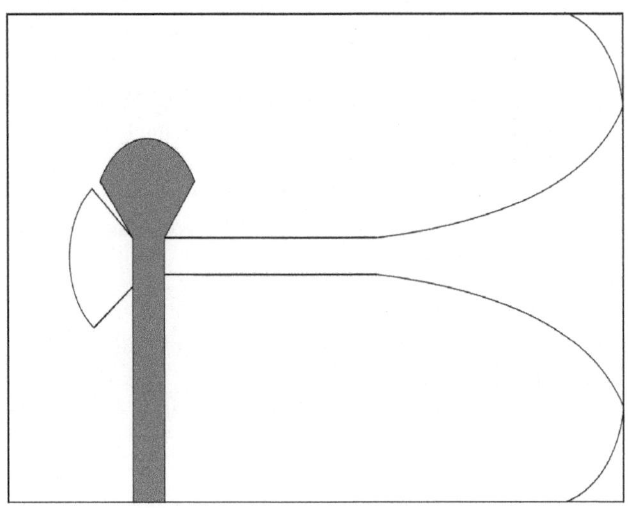

Figure 1. Circuit outline of broadband feed and uniplanar tapered slot antenna; microstrip line on top metallization, slotline with TSA on bottom metallization.

Therefore, several attempts have been made recently to integrate microstrip-to-slotline transitions on low-permittivity substrate [3,10,11,13]. While reasonable bandwidth has been obtained for the antenna system in [3], the application was limited to a one-dimensional array with no requirements for polarization purity. However, if uniplanar TSAs are used in two-dimensional arrays [2,4], then cross polarization presents a significant performance indicator. Other transitions either use a higher permittivity substrate [10,12,14] or are too narrowband [11,13] to accommodate a bandwidth in the order of 100 percent.

Therefore, this paper introduces an UWB feed for the uniplanar TSA on low-permittivity substrate that takes into consideration cross polarization as well as manufacturing limitations of a minimum slot width. The design is based on a modification of known structures but combines several aspects to obtain a UWB TSA system with a 7 GHz operational bandwidth at a center frequency of 6.5 GHz. Essential for this new design is the behavior of the electromagnetic field in the feed. It must be bound very tightly to the PCB and radiate well in the antenna part. This leads to critical design parameters such as characteristic impedance and substrate permittivity, which have to be chosen carefully.

The TSA system is examined theoretically, simulated with CST Microwave Studio, manufactured and measured. Good agreement between theory and experiment is obtained. This design is expected to significantly extend the bandwidth of TSA systems in polarization sensitive arrays for communications and radio astronomy applications.

2. Theoretical Considerations

To obtain a decision basis for the transmission-line components to be used in the TSA systems, several theoretical models for microstrip, slotline, coplanar waveguide (CPW) and coplanar stripline (CPS) have been calculated [8]. Only those satisfying constrains for UWB applications are considered here.

2.1. Planar Transmission Lines (PTLs)

Critical for our purpose to develop an UWB feed with low cross polarization is the distribution, with respect to frequency, of the electromagnetic field in the system. Such properties are dominantly influenced by the physical setup of the differently charged conductors, leading to a general simplification for the frequency-dependent relative permittivity:

$$\varepsilon_{re} = \frac{C_{sub}(f, d, \varepsilon_r)}{C_{\text{vacuum}}} \quad (1)$$

where d is connected to the dimension of the specific PTL type, ε_r the substrate's relative permittivity, and C_{sub} and C_{vacuum} the PTLs' capacitance with substrate or vacuum, respectively, between the conductors.

The field distribution is also significantly influenced by the orientation of the fields that determines the amount of energy traveling within the substrate or in air.

The PTLs have been categorized as follows: microstrip and CPW are considered as primary PTLs, whereas secondary PTLs are slotline and CPS. This classification is based on the fact that the primary group can easily be designed to match a characteristic impedance of $Z_{char,in}$ = 50 Ω on a wide variety of commonly used PCB substrates and that the secondary group is easily transformable to a TSA by simply widening the PTL.

Calculations of the primary PTLs [8] regarding the frequency-dependent relative permittivity show strong advantages for microstrip lines. Here, due to the physical conductor setup, the fields are propagating mainly within the substrate, which is advantageous in terms of dispersion and radiation loss. The low loss advantage is important to obtain the freedom to choose a substrate with an ε_r as low as possible in order to support radiation from the TSA without removing the substrate between the antenna fins.

Theoretical calculations [8] for the secondary PTLs lead to the result that the slotline outperforms the CPS when aiming for low characteristic impedance Z_{char}. However, the characteristic impedance of the slotline is usually limited to a lower boundary by manufacturing restrictions as will be considered in Section 2.3.

Based on these results, microstrip as primary PTL and slotline as secondary PTL are used for the TSA system.

2.2. Tapered Slot Antennas (TSAs)

An overview over the different TSA types is presented in [7]. To obtain small dimension, as mentioned above, the TSA profile is selected as an exponential taper, as first introduced in [6], which offers the shortest taper for broadband purposes. The taper in **Figure 1** is determined by

$$w_{TSA} = \pm A e^{px} \quad (2)$$

where w_{TSA} presents one half of the width between the fins, A adds an offset to the curve determined by the width of the secondary PTL (slotline with w_{sl} = 150 m), and p = 0.125 characterizes the slope of the taper influenced by the overall length (1.5 λ_{sl} at 6.5 GHz), the operational frequency range and the radiation characteristics of the TSA. Additionally, the sharp edges, which occur when the taper meets the radiating end of the substrate, are rounded to avoid disturbances and reflections in the electromagnetic fields (**Figure 1**). These features may later be replaced by corrugations [15] to limit the overall height of the antenna for array applications.

Moreover, it is found that the advantage of a low-permittivity substrate on the radiation characteristics compensates for the loss along the PTLs.

2.3. Transitions

The microstrip-to-slotline transition presented in [9] matches the basic requirements for this application. However, it is presented in a back-to-back setup on a high-permittivity substrate. Therefore, the single transition is redesigned on a low-permittivity substrate for which Rogers RT/Duorid 5880 material with ε_r = 2.2 and height h_S = 787 µm is selected. Ideally, the basic constraint for this redesign is:

$$Z_{char,ms} = Z_{char,sl} \quad (3)$$

The characteristic impedance of the microstrip line is designed to meet $Z_{char,ms}$ = 50 Ω. However, due to manufacturing limitations, in which the smallest slot width cannot fall below 150 µm, the lowest reachable value for the characteristic impedance of the slotline is $Z_{char,sl}$ = 81 Ω.

Figure 2(a) shows the top view of the TSA prototype which hosts the slot line components. The bottom view is depicted in **Figure 2(b)** and shows the microstrip circuitry. The two PTLs intersect at a 90° angle, and both lines are terminated with a 90° resonator whose length is $\lambda_m/4$ at midband frequency of 6.5 GHz. The fields are magnetically coupled, and the electric field rotates 90° in the transition.

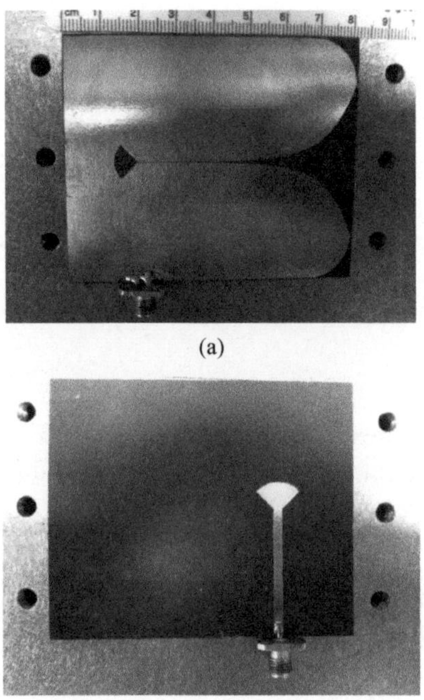

(a)

(b)

Figure 2. Photographs of the prototype TSA system: top view (a) With exponential taper and slotline with 90° stub resonator for termination; bottom view (b) With microstrip feed and 90° stub resonator.

3. Results

This section presents the performance of the UWB TSA system and compares simulated with measured results. All simulations are performed with CST Microwave Studio. Measurements use an Anritsu 37397C vector network analyzer (VNA).

Figure 3 presents a comparison of the reflection coefficient between the microstrip-to-slotline transition on high-permittivity substrate as proposed in [9] and the redesign on low-permittivity substrate used in this work. It is observed that our redesign is slightly better than the transition performance presented in [9] and that especially in the 4 GHz to 6.5 GHz range, the design used here is approximately 10 dB better than that in [9].

Figure 4 displays the effect of the rounded edges at the end of the taper by showing the input signal refection in the time domain. It is observed that the rounded edges are damping the reflected signal from 1.3 ns on and compressing the reflections between 0.7 ns and 0.5 ns.

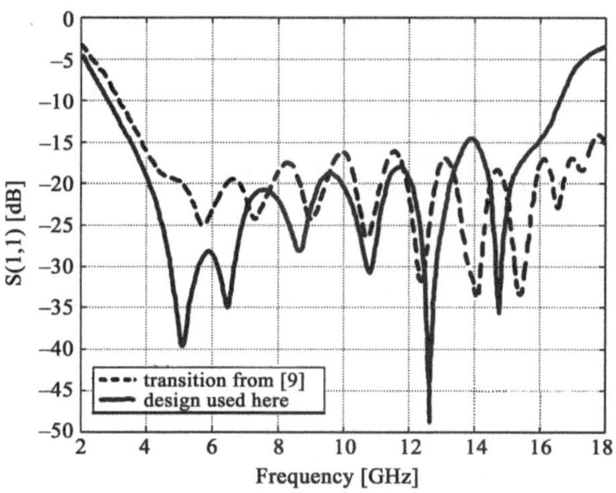

Figure 3. Single microstrip-to-slotline transition; redesign on low-permittivity substrate vs. transition introduced in [9] on high-permittivity substrate.

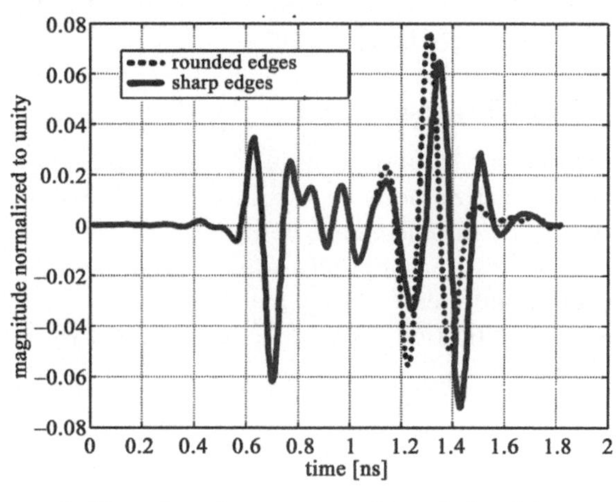

Figure 4. Time domain response of input signal taper with and without rounded edges.

Figure 5 compares measurements and simulations with respect to input impedance. Note that two prototype antennas have been fabricated which are indicated as ant #1 and ant #2 in **Figure 5**. For the input impedance in **Figure 5(a)**, the match between simulation and measurements is poor; even the two measured antennas do not agree. This is attributed to the influence of the SMA connectors and their difference in exact location and way of soldering.

Figure 5(b) shows good agreement between simulation and measurement for the reflection coefficient, especially the minimum at the center frequency of 6.5 GHz is clearly reproduced. The magnitude differs between simulation and measurement although the measured data is in the same range within ±5 dB for the two measured antennas. This deviation is also explained with the SMA connectors whose influences are not included in the simulation. Both measurements clearly show the broadband characteristic of the reflection coefficient which is below or only slightly above to –10 dB value between 2.5 GHz and 10.6 GHz.

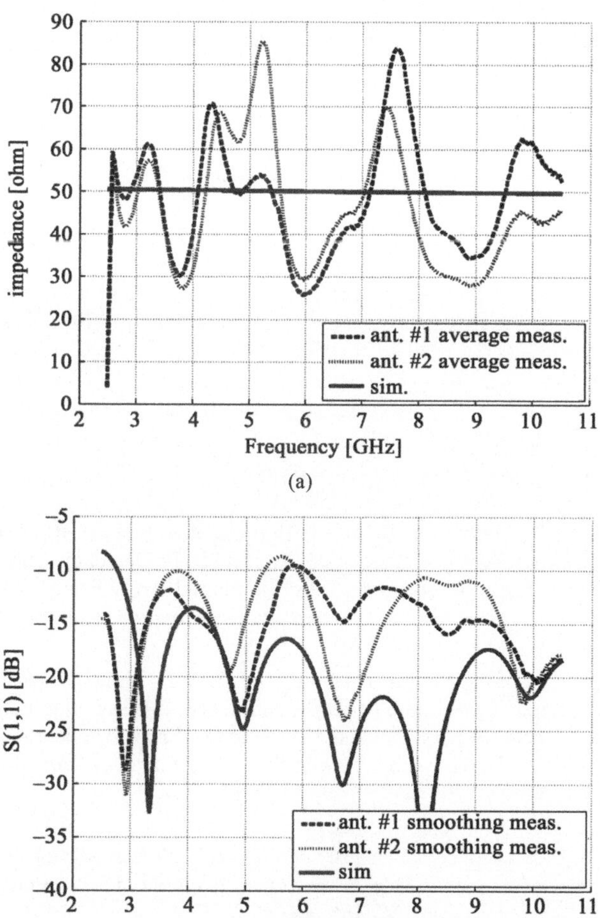

(a)

(b)

Figure 5. Comparison between measured and simulated results; (a) Input impedance; (b) Reflection coefficient in dB.

Radiation pattern measurements are performed in the anechoic chamber of the University of Victoria. The setup is presented in **Figure 6** and uses Orbit/FR positioners and controllers, a quad-ridge horn by ETS Lindgren, an HP amplifier and the Anritsu VNA.

Full pattern measurements are performed at 4.5 GHz, 6.5 GHz and 8.5 GHz in co- and cross-polarization with a 5° grid for 0° ≤ θ ≤ 180° and 0° ≤ ϕ ≤ 90°. Note that the range of –90° ≤ ϕ ≤ 0° was mirrored in the post processing process.

Figure 7 presents measurements for co- and cross-polarization for the three measured frequencies. A suppression of more than 30 db in the direction of the main lobe is observed. This is an excellent result that is due to the uniplanar arrangement of the TSA. Typical antipodal TSAs display cross-polarization levels closer to 20 dB [15]. It is expected that the cross-polarization value of 30 dB in this work can be further improved by using corrugations in the metallic fins such as in [15-17].

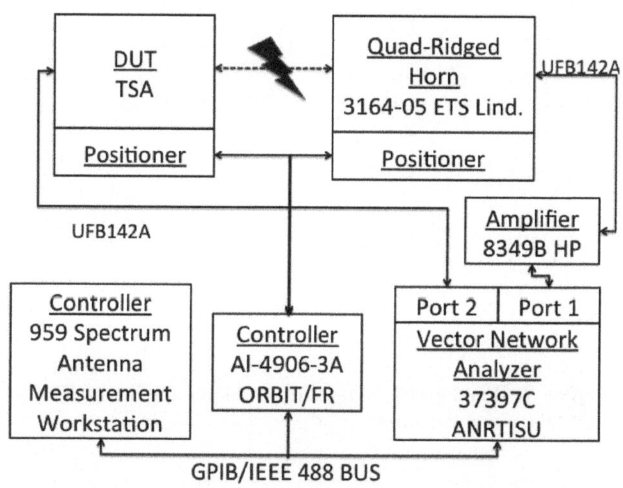

Figure 6. Setup for radiation pattern measurements.

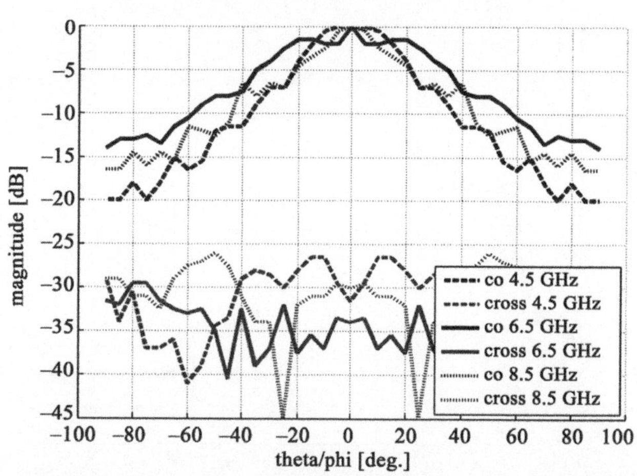

Figure 7. Measured co- and cross-polarization at 4.5 GHz, 6.5 GHz and 8.5 GHz with data for negative angles mirrored from that of positive values.

Figure 8 presents the simulated and measured radiation patterns for 6.5 GHz. A suitable agreement is observed, especially in the areas that matter for the cross polarization, namely the cardinal cuts, where the match is good. Moreover, the side lobes taper off even better in the measurements than in the simulations.

Unfortunately, calibrated gain measurements could not been carried out. Therefore, the simulated gain is shown in **Figure 9** and shows an increase from 4 dB at 3 GHz to 8.4 dB at 10 GHz. Such a gain variation is quite common for UWB antennas.

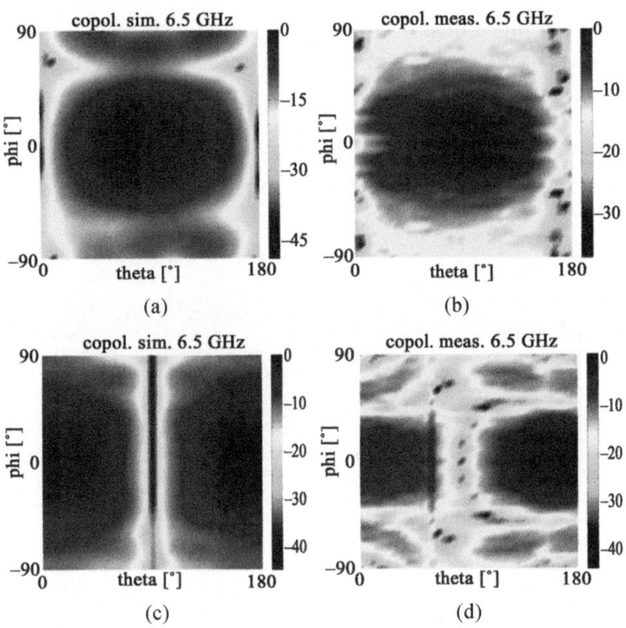

Figure 8. Beam pattern co- and cross-polarization: Simulation (a, c), measurement (b, d).

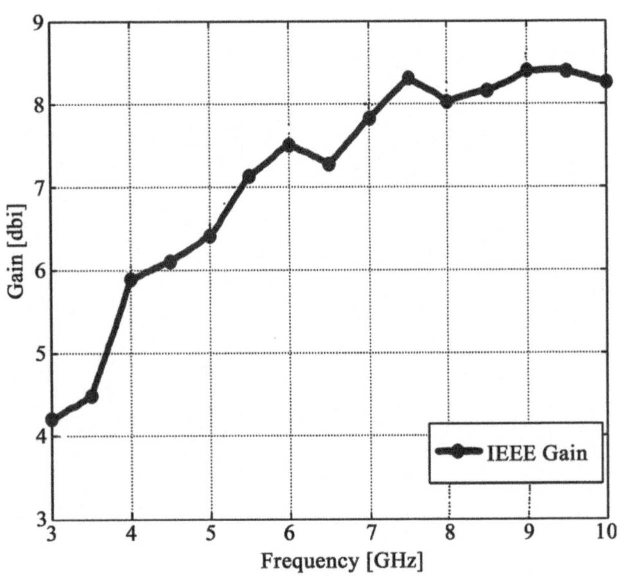

Figure 9. Simulated gain performance of unilateral TSA system.

4. Conclusion

A broadband feeding technique for the uniplanar TSA on low-permittivity substrate is presented. With a substrate permittivity of 2.2 it is possible to establish a good trade-off between radiation characteristics and losses in the feed system. It is demonstrated that low loss, low input reflection coefficient and sufficiently suppressed cross-polarization characteristics are attainable over an ultra-wide bandwidth in extent of 100 percent. Additionally, the TSA system shows sufficient gain values and directivity considering the wide 7 GHz bandwidth. Measurements on a prototype TSA system validate the design process, thus providing a solution for future UWB applications of the uniplanar tapered slot antenna in communications or radio astronomy applications.

5. Acknowledgements

The authors wish to acknowledge support for this work from the TELUS Research Grant in Wireless Communications.

REFERENCES

[1] Z. Irahhauten, J. Dacuña, G. J. M. Janssen and H. Nikookar, "UWB Channel Measurements and Results for Wireless Personal Area Networks Applications," *Proceedings European Conference on Wireless Technology*, Paris, 3-4 October 2005, pp. 189-192.

[2] H. Holter, T.-H. Chio and D. H. Schaubert, "Experimental Results of 144-Element Dual-Polarized Endfire Tapered-Slot Phased Arrays," *IEEE Transactions on Antennas and Propagation*, Vol. 48, No. 11, 2000, pp. 1707-1718.

[3] Y. Yao, M. Liu, W. Chen and Z. Feng, "Analysis and Design of Wideband Widescan Planar Tapered Slot Antenna Array," *IET Microwave, Antennas and Propagation*, Vol. 4, No. 10, 2010, pp. 1632-1638.

[4] B. Veidt, G. J. Hovey, T. Burgess, R. J. Smegal, R. Messing, A. G. Willis, A. D. Gray and P. E. Dewdney, "Demonstration of a Dual-Polarized Phased-Array Feed," *IEEE Transactions on Antennas and Propagation*, Vol. 59, No. 6, 2011, pp. 2047-2057.

[5] B. Y. El Khatib, T. Djerafi and K. Wu, "Three-Dimensional Architecture of Substrate Integrated Waveguide Feeder for Fermi Tapered Slot Antenna Array Applications," *IEEE Transactions on Antennas and Propagation*, Vol. 60, No. 10, 2012, pp. 4610-4618.

[6] P. J. Gibson, "The Vivaldi Aerial," *9th European Microwave Conference Proceedings*, Brighton, 17-20 September 1979, pp. 101-105.

[7] K. S. Yngvesson, D. H. Schaubert, T. L. Korzeniowski, E. L. Kollberg, T. Thungren and J. F. Johansson, "Endfire Tapered Slot Antennas on Dielectric Substrate," *IEEE Transactions on Antennas and Propagation*, Vol. 33, No. 12,

1985, pp. 1392-1400.

[8] K. C. Gupta, R. Garg, I. J. Bahl and P. Bhartia, "Microstrip Lines and Slotlines," Artech House, Boston, 1996.

[9] M. M. Zinieris, R. Sloan and L. E. Davis, "A Broadband Microstrip-Slot-Line Transition," *Microwave and Optical Technology Letters*, Vol. 18, No. 5, 1998, pp. 339-342.

[10] M. E. Bialkowski and Y. Wang, "Design of 180° Hybrid Employing Ground Slots and Microstrip-Slot Transitions," *Proceedings International Conference on Microwave Radar and Wireless Communications* (*MIKON*), Vilnius, 14-16 June 2010, pp. 1-4.

[11] W. Wen, W. Xuetian and F. Lili, "LTSA with Microstrip-Slotline Transition for MFPA Imaging Systems," *Proceedings of International Conference on Microwave Technology and Computational Electromagnetics*, Beijing, 22-25 May 2011, pp. 230-232.

[12] A. M. Abbosh, "Wideband Planar Crossover Using Two-Port and Four-Port Microstrip to Slotline Transitions," *IEEE Microwave and Wireless Components Letters*, Vol. 22, No. 9, 2012, pp. 465-467.

[13] D.-L. Zhai, C.-X. Zhang, Z.-M. Yang and S.-J. Hu, "Design of the Antenna Array with a Novel Feeding Network," *Proceedings of International Conference on Microwave and Millimeter Wave Technology* (*ICMMT*), Shenzhen, 5-8 May 2012, pp. 1-4.

[14] P. T. Nguyen, A. Abbosh and S. Crozier, "Ultra-Wide-Band Balun Using Microstrip to Slotline Transitions," *Proceedings IEEE Asia-Pacific Conference on Antennas and Propagation*, Singapore City, 27-29 August 2011, pp. 1-2.

[15] H. Sato, Y. Takagi and K. Sawaya, "High Gain Antipodal Fermi Antenna with Low Cross Polarization," *IEICE Transactions on Communications*, Vol. E94-B, No. 8, 2011, pp. 2292-2297.

[16] J. Bai, S. Shi and D. W. Prather, "Modified Compact Antipodal Vivaldi Antenna for 4-50-GHz UWB Application," *IEEE Transactions on Microwave Theory and Techniques*, Vol. 59, No. 4, 2011, pp. 1051-1057.

[17] M. Sun, X. Qing and Z. N. Chen, "60-GHz Antipodal Fermi Antenna on PCB," *European Conference on Antennas and Propagation Proceedings*, Rome, 11-15 April 2011, pp. 3264-3267.

[18] R.-Y. Fang, J.-K. Chuang and C.-L. Wang, "Coplanar Waveguide-to-Rectangular Waveguide Transition Using Meander Slotline," *Proceedings of Asia-Pacific Microwave Conference*, Melbourne, 5-8 December 2011, pp. 399-402.

7

Design and Parametric Simulation of a Miniaturized PIFA Antenna for the PCS Band

Abdelhakim Elouadih[1], Ahmed Oulad-Said[1], Moha Mrabet Hassani[2]

[1]Department of Electrical and Telecommunications Engineering, Royal Air Academy (ERA), Marrakesh, Morocco; [2]Department of Physics, Semlalia University of Sciences (FSSM), Marrakesh, Morocco.

ABSTRACT

This paper describes the design and simulation by HFSS simulator of a probe-fed Planar Inverted-F Antenna (PIFA) for the use in PCS band [1850 MHz - 1990 MHz]. A methodology based on parametric simulations (parameters are ground plan dimensions, height of radiating plate, feeding point position, shorting plate dimensions and positioning) was used to design optimized antenna. The simulation allowed the characterization of the designed antenna and the computing of different antenna parameters like S_{11} parameter, resonant frequency, SWR, bandwidth impedance in feeding point, gain, 2D and 3D diagram pattern, Fields distribution. The simulation results are interesting and respect the mean PCS requirements.

Keywords: PCS; HFSS; PIFA; Patch Antenna; Parametric; Simulation

1. Introduction

Wireless communications have progressed very rapidly in recent years, and many mobile units are becoming smaller and smaller. To meet the miniaturization requirement, the antennas employed in mobile terminals must have their dimensions reduced accordingly. Planar antennas, such as microstrip and printed antennas have the attractive features of low profile, small size, and conformability to mounting hosts and are very promising candidates for satisfying this design consideration. Planar antennas are also very attractive for applications in communication devices for wide mobile telecommunications like GSM 1900 called PCS (Personal Communications System), wireless local area network, aeronautics and embedded systems [1].

For optimum system performance, the antennas must have high radiation efficiency, small volume, isotropic radiation characteristics, small backward radiation and SAR, simple and low-loss impedance matching to patches. The major types of configurations of low-profile antennas with enhanced bandwidth performance include planar inverted F Antennas.

The PIFA consists in general of a ground plane, a top plate element, a feed wire attached between the ground plane and the top plate, and a shorting wire or strip that is connected between the ground plane and the top plate.

The antenna is fed at the base of the feed wire at the point where the wire connects to the ground plane. The PIFA is an attractive antenna for wireless systems where the space volume of the antenna is quite limited. It requires simple manufacturing, since the radiator must only be printed. The addition of a shorting strip allows good impedance match to be achieved with a top plate that is typically less than $\lambda/4$ long. The resulting PIFA is more compact than a conventional half-wavelength probe-fed patch antenna [2].

The miniaturization can affect radiation characteristics, bandwidth, gain, radiation efficiency and polarization purity. The miniaturization approaches are based on either geometric manipulation (the use of bend forms, meandered lines, PIFA shape, varying distance between feeder and short plate [3]) or material manipulation (loading with a high-dielectric material, lumped elements, conductors, capacitors, short plate [4]) or the environment characteristics (ground plane dimensions, coupling, measurement and fabrication errors [3]). In this case, the designed antenna is shorted to the ground plane by a plate, uses regular shapes and a high dielectric thin substrate under the radiating plate not above the ground plane.

If all precedent works are concentrated on studying the

effects of these elements (material, geometry, environment), the choice of a PIFA element is so improvised in the design. In this paper, a methodology based on parametric simulation is used to choose simultaneously or independently different PIFA elements.

In the next section, the author presents the methodology and detailed parametric simulations to optimize the antenna design. After, the dimensions and parameters are then chosen. The author exposes the characteristics of the designed antenna in the third section.

2. Design Methodology

2.1. The Description of the Studied Antenna

As shown in **Figure 1**, the designed antenna has a rectangular radiating patch witch length is L_p = 11 mm and width W_p = 30 mm. The patch in placed at a height h from the ground plan. The last has a length L_g and a width W_g. The patch is matched to the ground plan via a rectangular shorting plate. The shorting plate has a width W_s and a length L_s. The shorting post of usual PIFA types is a good method for reducing the antenna size, but results in narrow impedance bandwidth. It is placed in the (yz) plan at a distance D from the edge centre and a distance p from the feeding point. The patch is fed by a 50 Ω wire; a semi-rigid coax with a centre conductor that extends beyond the end of the outer conductor is used to form the PIFA feed wire. The outer conductor of the coax is soldered to the edge of a small hole drilled in the ground plane at the feed point. The volume between the radiating plate and the ground plan is filled by air except a thin region 0.8 mm under the radiating patch who is composed of FR4_epoxy (ε_r = 4.4).

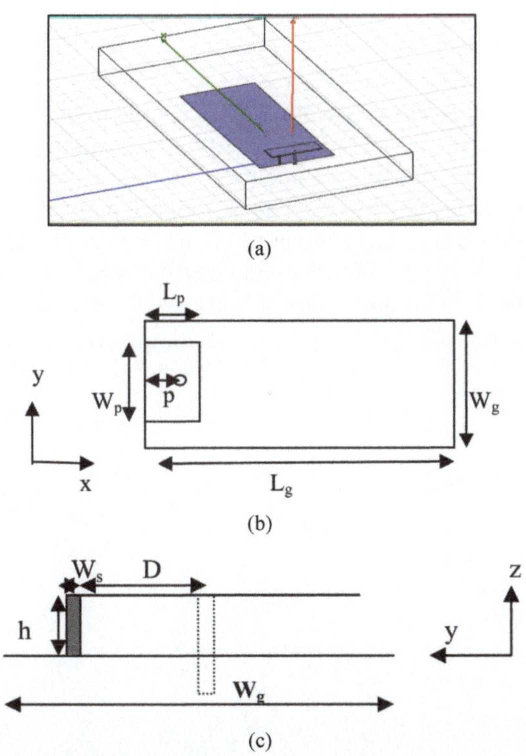

Figure 1. The designed antenna by HFSS. (a) Perspective view; (b) Top view; (c) Side view.

2.2. The Choice of the Patch Dimensions

It is very important for simulation by HFSS to estimate the resonant frequency that help the simulator to make a refinement mesh in a band around the resonant frequency and then give more precise values. The resonant frequency of a PIFA is approximated by the Equation (1) where F_r is the resonant frequency, C is the light velocity.

$$F_r = \frac{C}{4\left(L_p + W_p - W_s\right)} \quad (1)$$

If there is another substrate different from Air, C will be $C_0/\sqrt{\varepsilon_r}$, where $C_0 = 3\times10^8$ m/s. For our case, the space between the patch and the ground plan is essentially air minus a 0.8 mm FR4 epoxy layer.

To compute the resonant frequency, we have the following values: L_p = 11 mm, W_p = 30 mm, $W_{s_{\min}}$ = 1 mm. We obtain: F_r = 1875 MHz.

The obtained frequency is then not far from 1920 MHz the central frequency of PCS band. In fact, there is no equation to determine the resonant frequency for a PIFA that contains not only the patch dimensions but also the other parameters that can affect the antenna characteristics. For this, the author will make constant the patch dimensions that are the mean parameters can furnish the resonant frequency and he will vary undependably the other parameters (ground plan dimensions L_g and W_g, height of radiating plate h, feeding point position p, shorting plate width W_s and position D).

2.3. The Choice of the Simulator

HFSS (High Frequency Simulator Structure) is a high performance full wave electromagnetic (EM) field simulator for arbitrary 3D volumetric passive device modeling that takes advantage of the familiar Microsoft Windows graphical user interface. It integrates simulation, visualization, solid modeling, and automation are easy to learn environment where solutions to 3D EM problems are quickly and accurate obtained. Ansoft HFSS employs the Finite Element Method (FEM), adaptive meshing, and brilliant graphics to give unparalleled performance and insight to all of 3D EM problems. Ansoft HFSS can be used to calculate parameters such as S-Parameters, Resonant Frequency and Fields. Typical uses include Package Modeling, PCB Board Modeling, Mobile Communications (Patches, Dipoles, Horns, Conformal Cell Phone Antennas), Specific Absorption Rate (SAR), Infinite Arrays, Radar Section (RCS), Frequency Selective Surface (FSS) and filters such Cavity Filters, Microstrip, Dielectric. HFSS is an interactive simulation system whose basic mesh element is a tetrahedron T. In industry, Ansoft HFSS is the tool of choice for High productivity research, development, and virtual prototyping [5,6]. The HFSS is then used in our simulation; the author proposes after to expose the results of the HFSS simulations.

2.4. The Choice of the Height *h*

The height h is the distance between the top plate and the ground plane. In order to eliminate the effects of the ground plane effects, we place the patch on the edge of an infinite ground plane at a height varying from 6 mm to 13 mm. From the simulation result shown by **Figure 2**, the optimal height h is for h = 10 mm because it gives values S_{11} most important and closer to central frequency.

2.5. The Choice of the Ground Plan Width W_g

For the height equal to 10 mm, we will vary the groundplane width by having a very important length L_g to eliminate its effect. In general, the width W_g is not so far to W_p that is the minimum. For this, W_g will vary from 30 mm to 60 mm. the simulation result is shown by **Figure 3**. We can consider W_g = 45 mm an optimal width because it presents the minimum S_{11} for all the band and

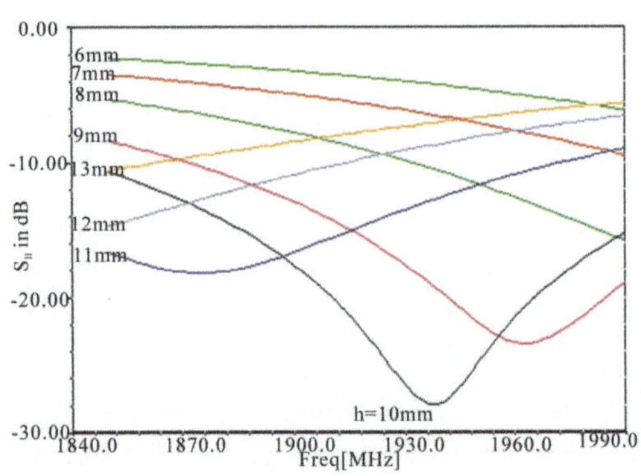

Figure 2. Parametric simulation by varying *h*.

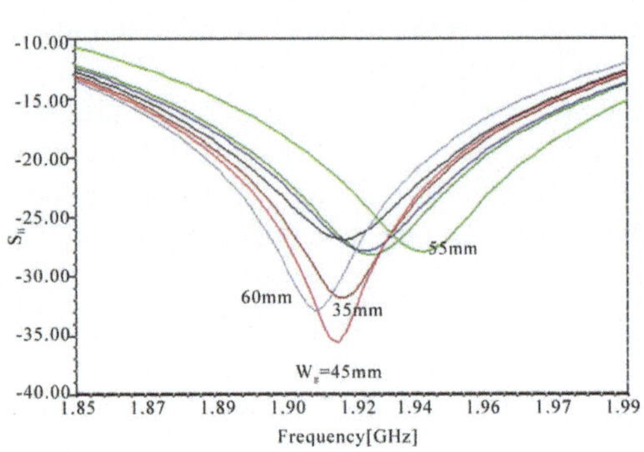

Figure 3. Parametric simulation by varying W_g.

in the same time the minimum is the closest to the central frequency in the PCS band.

2.6. The Choice of the Ground Plan Length L_g

The height h in now 10 mm and the W_g is 45 mm. We will vary the ground plane length L_g from 60 mm to 140 mm. This sweep is taken because the designed antenna will be mounted on a GSM handset. The **Figure 4** shows the result of the parametric simulation. We can see from the curve that the length 100 mm is the most adapted. In reality, the simulation shows close results for the lengths 100 mm, 110 mm and 120 mm. A sweep refinement is introduced and the optimum was 108 mm. Also, 120 mm and over can be considered not interesting values for a ground plane in a handset. We have then h = 10 mm, W_g = 45 mm and L_g = 100 mm. We will after simulate the effect of the shorting element width W_s and distance D from the patch edge centre.

2.7. The Choice of the Shorting Plate Position *D*

We will vary the shorting plate position between all possible values from 0 to 14 mm, we width W_s is in general close to 1 mm. we obtain the result shown by **Figure 5**. Both of positions 11 and 12 mm gibe good result. We can choose D = 11 mm, it's closer to the central frequency.

2.8. The Choice of the Shorting Plate Width W_s

We will vary the width W_s from 0.5 to 3 mm with step of 0.5 mm. We will obtain the result shown by **Figure 6**. We can consider the width W_s = 1 mm the most convenient for the bandwidth and the resonant frequency. From the curve, the width W_s is close to 1 mm. If we make a sweep refinement by choosing 0.1 mm step instead of 0.5 mm in order to hope having a S_{11} peak near the central frequency if possible. We obtained the result shown by **Figure 7**. We can remark a net peak for the particular

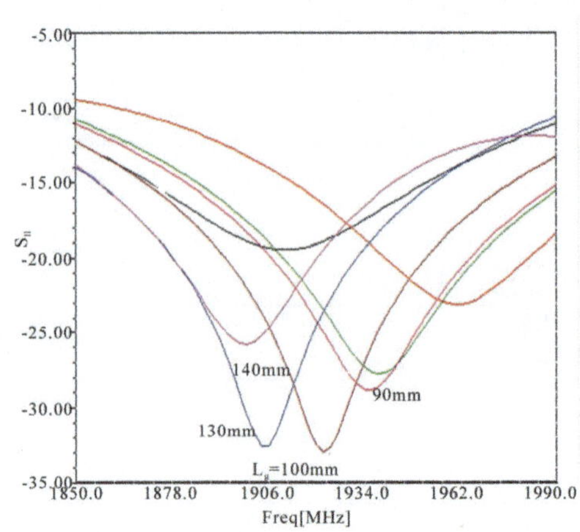

Figure 4. Parametric simulation by varying L_g.

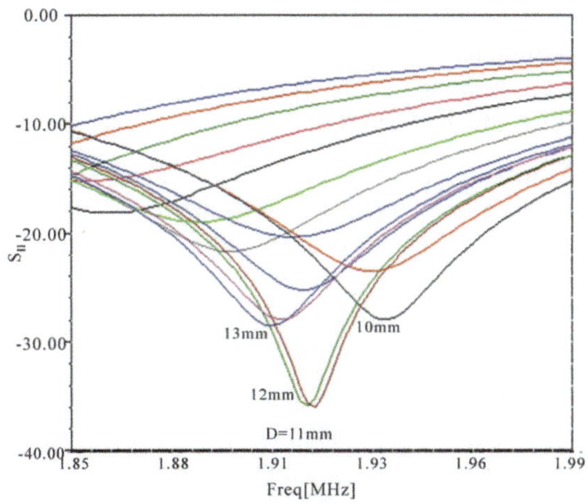

Figure 5. Parametric simulation by varying *D*.

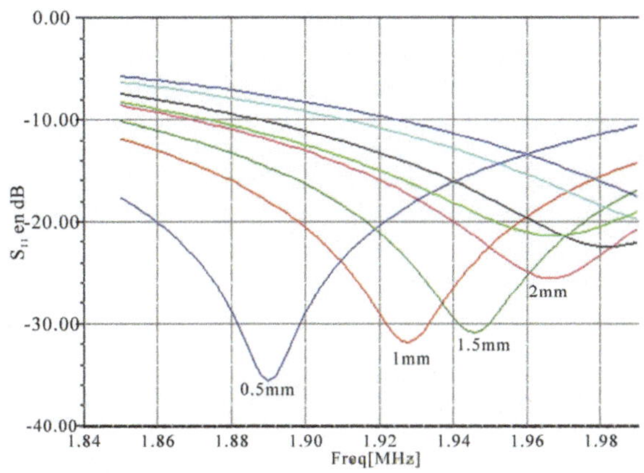

Figure 6. Parametric simulation by varying W_s (step = 0.5 mm).

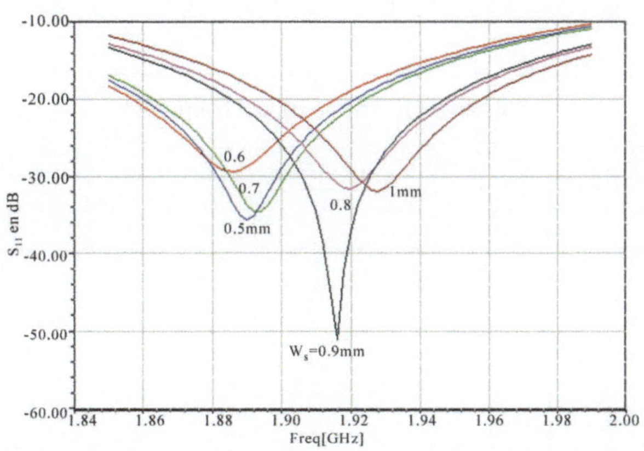

Figure 7. Parametric simulation by varying W_s (step = 0.1 mm).

value of 0.9 mm. For all studied parameters, any linear relation can't be established between bandwidth enhancement and the parameter variation. We have then chosen the parameters *h*, L_g, W_g, *D* and W_s. We can now look for the effect of the feeding point position *p* to enhance the bandwidth or the input impedance.

2.9. The Choice of the Feeding Point Position *p*

The feeding point position *p* is calculated from the rear edge of the patch. We will vary the *p* parameter from 2 mm to 10 mm we mean all possible cases. The values 1 and 11 mm can't be taken because the feeding point is theoretically a circle that has a radius. The result is shown on **Figure 8**. We can consider the peak for $p = 4$ mm an interesting position for feeding.

3. The Designed Antenna Characterization

The dimensions of the designed antenna for the PCS band are summarized in the **Table 1**. We will now present the antenna characteristics.

3.1. The Reflection Loss

The **Figure 9** shows the S_{11} parameter function of frequency in [0 - 3 GHz]. We can see a peak of S_{11} parameter around 1924 MHz very close to the PCS central frequency 1920 MHz. Also, the S_{11} values out of the PCS band are near 0 that means the designed antenna can't interfere with other radiations. We can also run the si-

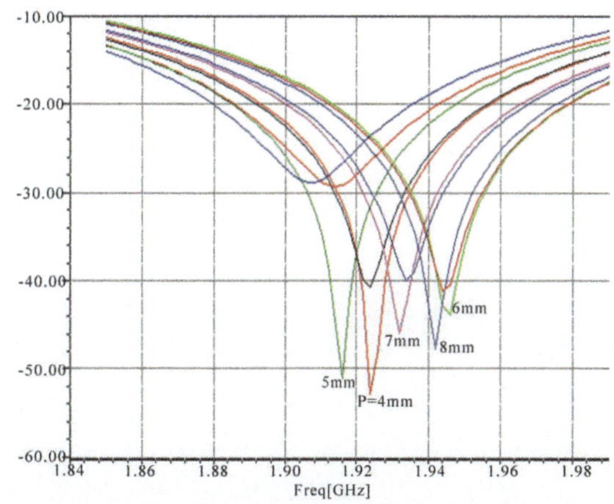

Figure 8. Parametric simulation by varying *p*.

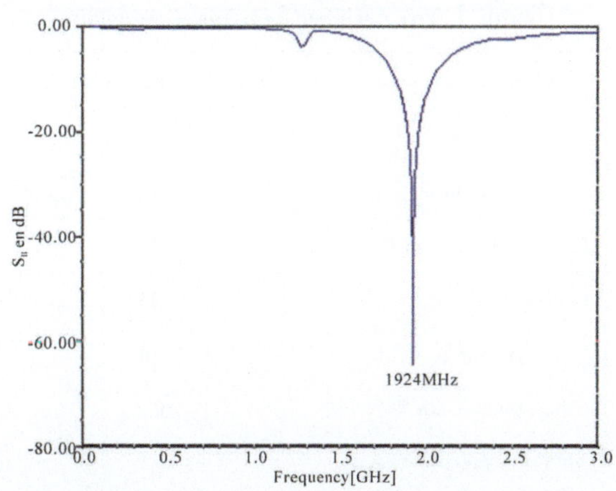

Figure 9. S_{11} depending on frequency for [0 - 3 GHz].

mulation by refining the sweep interval for more precision and as shown in **Figure 10**. The **Figure 11** furnished by the simulation gives exactly S_{11} = −12.43 dB for 1850 MHz (the low frequency of the PCS band), S_{11} = −60 dB for 1925 MHz (the peak frequency), the simulation tables give exactly a peak of −64.46 dB for the frequency 1924.7 MHz and S_{11} = −12.49 dB for 1990 MHz (the high frequency of the PCS band). For the whole PCS band, the designed antenna has a S_{11} litter than −12.4 dB. It is very interesting result.

3.2. The VSWR and Bandwidth

We obtain as shown in **Figure 12(a)** VSWR = 1.62 for 1850 MHz (the lowest frequency of the PCS band), $VSWR_{min}$ = 1.001 for 1924.7 MHz (the peak frequency), VSWR = 1.49 for 1990 MHz (the highest frequency of the PCS band). The VSWR is at its minimum, it's a very interesting result. Also, the PCS bandwidth (140 MHz) is for the designed antenna a 1:1.62 VSWR bandwidth. It is considered a very interesting result.

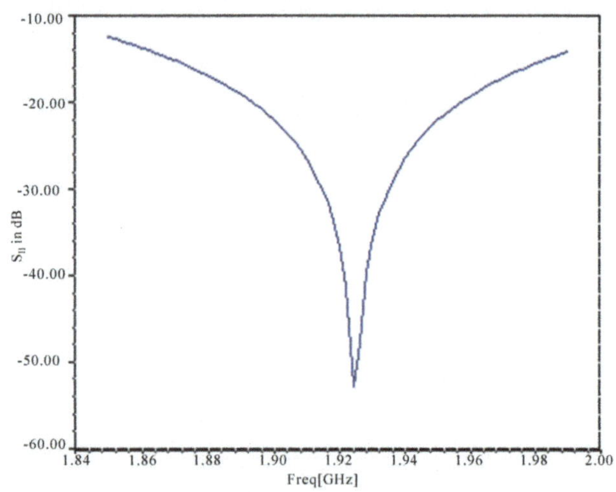

Figure 10. S_{11} **depending on frequency for PCS band.**

Table 1. The designed antenna dimensions.

Antenna parameter	Value of the parameter (mm)
Patch length L_p	11
Patch width W_p	30
Ground plan length L_g	108
Ground plan width W_g	45
Height h	10
Short plate width W_s	0.9
Short plate position D	11
Feeding point position	4

The designed antenna presents a 1:1.5 VSWR bandwidth equal to 130 MHz.

3.3. The Smith Chart

The **Figure 13** shows a regular smith chart with interesting parameters of reflection, impedance, VSWR, and Q. The feeding point impedance for extremes and central frequencies are given in **Table 2**.

3.4. Antenna Parameters

The simulations results shown in **Figure 14** give the antenna parameters. The obtained gain G is 1.16 dBi and the radiation efficiency is 1.0085.

3.5. The Diagram Pattern

We can confirm by the **Figure 15** that (*xz*) is the E-plane and its maximum is for (phi = 0 deg et theta = 56 deg). Also, we can verify that via the 3D polar diagram shown in **Figure 16**.

31	1.924500	-61.004117
32	1.924550	-62.153483
33	1.924600	-63.233862
34	1.924650	-64.076436
35	1.924700	-64.463742
36	1.924750	-64.262188
37	1.924800	-63.544375
38	1.924850	-62.519234
39	1.924900	-61.379179
40	1.924950	-60.239626

Figure 11. The resonant frequency.

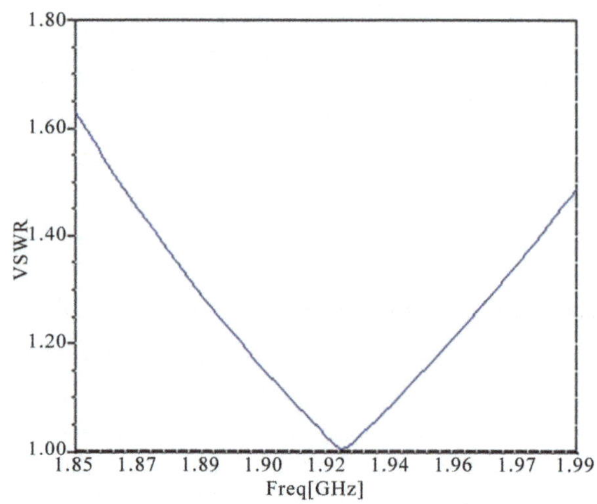

Figure 12. The VSWR depending on the frequency for the PCS band.

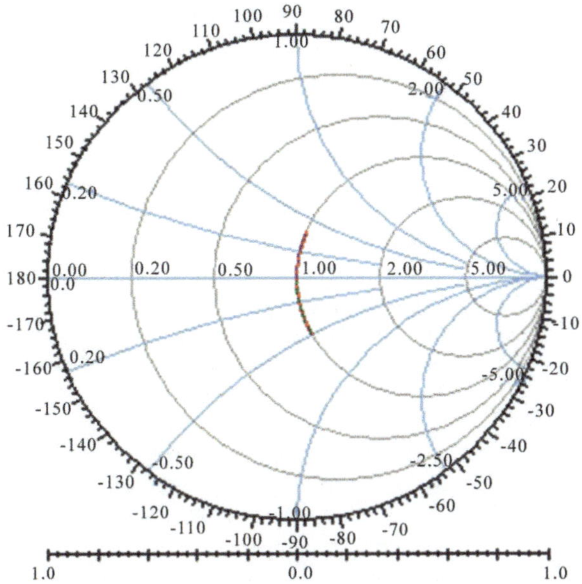

Figure 13. The smith chart for PCS band.

Antenna parameters:

Quantity	Value	Units
Max U	0.92134	W/sr
Peak Directivity	1.1483	
Peak Gain	1.1581	
Peak Realized Gain	1.1578	
Radiated Power	1.0083	W
Accepted Power	0.99978	W
Incident Power	1	W
Radiation Efficiency	1.0085	
Decay Factor	0	

Figure 14. Antenna parameters as given by HFSS.

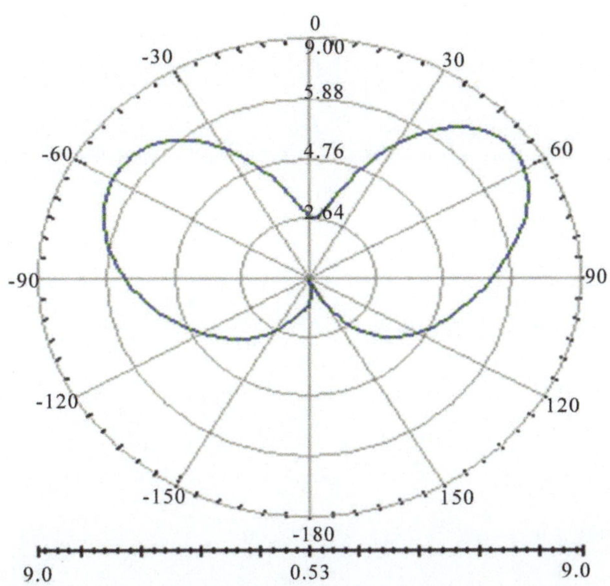

Figure 15. The E-field diagram pattern (for phi = 0 deg).

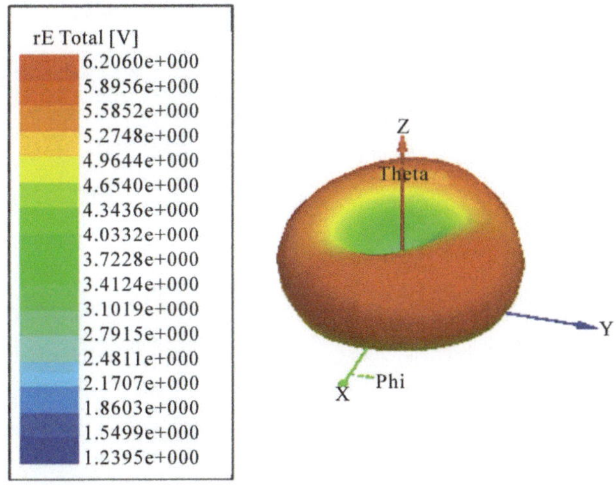

Figure 16. The E-field 3D polar diagram pattern.

Table 2. The impedances table in PCS band.

Frequency (MHz)	Impedance Z (Magnitude en Ω, Phase)	Port Z_0 (Magnitude en Ω, Phase)
1850	(39.592, −26.1)	(35.239, 0)
1924.7	(35.281, 0)	(35.239, 0)
1990	(38.438, 21.5)	(35.239, 0)

4. Conclusions

To design a PIFA for PCS band, we followed a methodology based on parametric simulations. During the design process, we can make some conclusions:

1) The PIFA characteristics are affected by each of the changed parameter (the height of the radiating plate, the ground plane dimensions, the position and width of the short plate, the position of the feeding point).

2) When we work in a medium bandwidth (like our case 140 MHz), any linear relation can't be established between the way the characteristics of the antenna change and the way the parameter change. The parametric simulations are very interesting to make effective choice.

3) It's interesting in the design of the PIFA to search an effective solution not the optimal one. In fact, difference will be very little, and also the values of the PIFA dimensions must be practical not theoretical. The important fact is that the solution respects the requirements.

4) In comparison with precedent designed PIFA [4,5] for different frequency bands, the antenna presents a very high radiation efficiency and larger bandwidth.

5) The designed PIFA for the PCS use respects the requirements especially for the resonant frequency, the VSWR, the bandwidth, the reflection loss, the anisotropy and the miniaturization.

The **Table 3** summarizes the characteristics and dimensions of the designed antenna. The analysis of those

Table 3. The design antenna parameters and dimensions.

Antenna parameter	Value of the parameter
Patch Length L_p	11 mm
Patch width W_p	30 mm
Ground plan length L_g	108 mm
Ground plan width W_g	45 mm
Height h	10 mm
Short plate width W_s	0.9 mm
Short plate position D	11 mm
Feeding point position	4 mm
Resonnat frequency	1924.7 MHz
S_{11} in 1850 MHz	−12.43 dB
S_{11} in 1924.7 MHz	−64.5 dB
S_{11} in 1990 MHz	−12.49 dB
$VSWR_{min}$	1.001
1:1.5 VSWR bandwith	130 MHz
1:1.62 VSWR bandwith	140 MHz (PCS band)
Gain	1.16 dB
Radiation efficiency	1.009
E total max	phi = 0 deg, theta = 56 deg

results makes from our designed antenna a succeeded trade-off that respects PCS requirements.

REFERENCES

[1] K. L. Wong, "Introduction and Overview," In: K. L. Wong, Ed., *Planar Antennas for Wireless Communications*, John Willy and Sons, Hoboken, 2003, p. 1.

[2] K. L. Virga and Y. Rahmat-samii, "Low-Profile Enhanced Bandwith PIFA for Wireless Communications Packaging," *IEEE Transactions on Microwave Theory and Techniques*, Vol. 45, No. 10, 1997, pp. 1879-1888.

[3] W.-J. Liao, T.-M. Liu and S.-Y. Ho, "Miniaturized PIFA Antenna for 2.4 GHz ISM Band Applications," *IEEE Proceedings of the 6th European Conference on Antennas and Propagation (EUCAP)*, Prague, 26-30 March 2012, pp. 3034-3037.

[4] A. K. Skrivervik, J.-F. Zürcher, O. Staub and J. R. Mosig, "PCS Antenna Design: The Challenge of Miniaturization," *IEEE Antennas and Propagation Magazine*, Vol. 43, No. 4, 2001, pp. 12-27.

[5] Ansoft Corporation, "HFSS 10.0 User's Guide," Ansoft Corporation, Pittsburg, 2005.

[6] J. McClure, "Statistics for Microarrays: Design, Analysis, and Inference," Rev.1.06, Ansoft Corporation, Pittsburg, 2005.

8

A U-Shaped UWB Antenna with Band-Notched Performance

Aliakbar Hosseinzadeh, Mirhamed Mirmozafari, Mohammad Vatankhah Varnoosfaderani, Changiz Ghobadi, Javad Nourinia

Department of Electronics and Communication Engineering, Urmia University, Urmia, Iran.

ABSTRACT

An ultra-wideband antenna with controllable band-notched is presented. Two semi-ellipses with different radiuses are subtracted to result in the main patch. By varying inner and outer radiuses, much more enhancement in bandwidth occurred. A U-shaped slot is used to make band-stop performance. Measured S11 is ≤ -10 dB over 2.3 - 5 GHz and 6.1 - 15.1 GHz.

Keywords: Ultra-Wideband Antenna (UWB); Slot; Slit; Band-Notched

1. Introduction

As stated in Federal Communications Commission (FCC), the ultra-wideband (UWB) radio systems use the designated 3.1 - 10.6 GHz frequency band. Due to this large frequency range, there would be interference with WLAN technologies such as IEEE 802.11 a (5.15 - 5.35 GHz, 5.725 - 5.825 GHz) [1]. Therefore, it is desirable to attenuate this interference by inserting some frequency selectivefilters. Various UWB antennas with band-notched characteristic have been reported to avoid interferences, such as TwoL-shaped quarter-waveguide resonators coupled to the ground plane with two shorting tracks at the sides of the antenna [2], split ring resonator [3], U-shaped slots and E-shaped slot [4], rectangular slots and 4-shaped slots [5], inverted U-shaped slots and H-shaped slot [6], C-shaped slots [7], and multi-U-shaped slots [8].

In this structure, Coplanar Waveguide Grounded (CPWG) transmission line is used due to its characteristics, such as low radiation loss, low dispersion and ability to integration with active solid-state devices.

In this paper, a U-shaped patch with inner radius of R_1 and outer radius of R_2 is presented. By adjusting different values of R_1 and R_2, bandwidth enhancement is achieved. A U-shaped slot is used to obtain band-stop performance. By properly adjusting the parameters, it is possible to find desirable bandwidth and to center frequency of notched band.

This paper includes following sections. In the next section, the geometry of the proposed antenna is described. Discussions and antenna performance are indicated in Section 3 and the measured results are provided in Section 4.

2. Antenna Configuration

Geometry of the proposed antenna is illustrated in **Figure 1**. A smaller semi-ellipse with radius of R_1 is subtracted from a larger one with radius of R_2 that results in a U-shaped patch. The inner radius (R_1) and the outer one (R_2) are equal to 8 mm and 18 mm, respectively. CPWG structure is used for antenna feedline that makes it suitable for integrating with the other components. The parameters of this feedline are chosen so that the impedance of feedline is equal to 50 Ω and is matched with SMA connector.

This structure is symmetrical to the longitudinal axis that exhibits appropriate radiation patterns.

The antenna is printed on an inexpensive FR4 with dielectric constant of 4.4 and thickness of 1 mm. A U-shaped slot is inserted in the patch that provides band-notch performance. The width of slot has been optimized and is equal to 0.5 mm. The overall size of the antenna is 40×40 mm^2. A conducting ground plane with size of 20×40 mm^2 is also printed on the other side of the substrate.

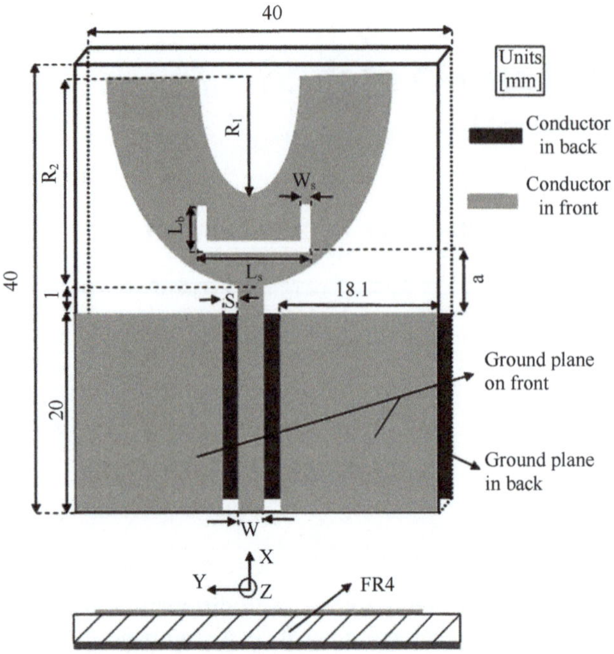

Figure 1. Geometry of the proposed CPWG-fed antenna with U-shaped slot.

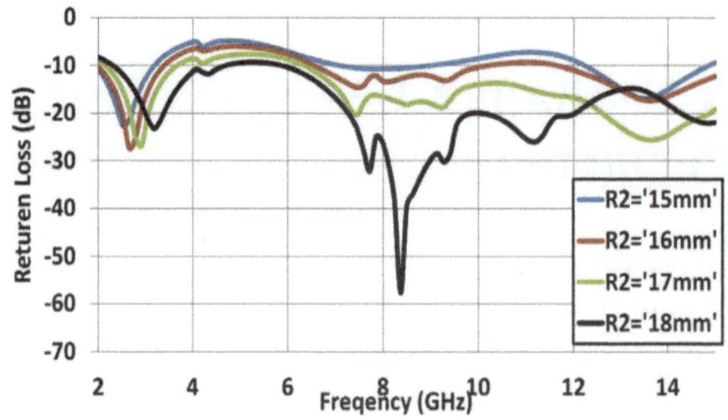

Figure 2. Simulated return loss versus frequency characteristics for various sizes of R_2.

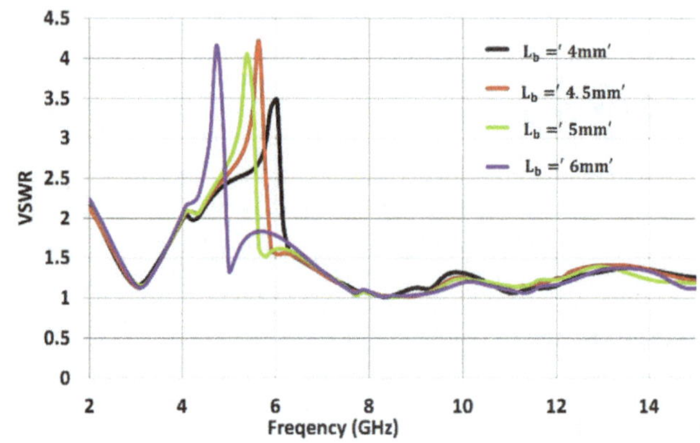

Figure 3. Simulated VSWR vs. frequency for different L_b.

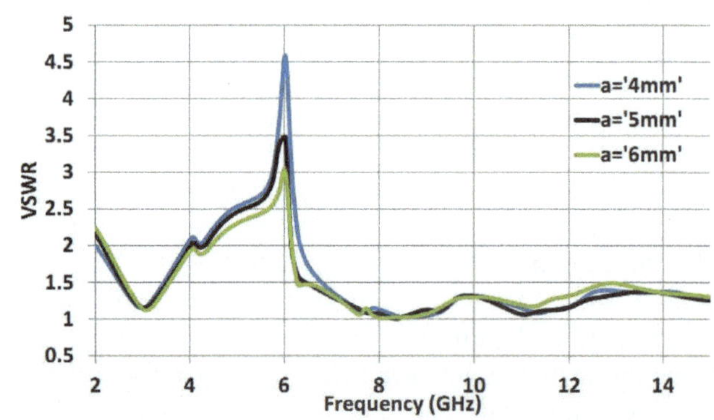

Figure 4. Simulated VSWR vs. frequency for differenta.

3. Antenna Discussion

The parameters of the proposed antenna are simulated and optimized using HFSS v13. By applying U-shaped patch, two important parameters (R_1, R_2) are obtained. Different values of these parameters result in different lengths of current paths that can affect upper and lower frequencies of bandwidth. It is clear that variation of outer radius (R_2) can alter the area of patch more than the inner one. In other case R_2 variations alter the gap between lower edge of radiating patch and top edge ofground plane and affect impedance matching. **Figure 2** shows R_2 effect on antenna bandwidth. It can be observed; by increasing R_2 larger bandwidth can be obtained. In this study the outer radius is set equal to 18 mm.

In order to achieve notched frequency at 5.5 GHz, a U-shaped slot is inserted in the patch. The width has been optimized, and then is set equal to 0.5 mm. Its length is about half of the wavelength of the notched frequency [9]. As an important parameter to control the center frequency of the notched band, the length of the U-shaped slot is investigated in **Figure 3**. It can be seen that by increasing the total length of slot, the notch frequency decreases monotonically. Position of slot plays an important role on the antenna performance. **Figure 4** illustrates the effect of slot position on the bandwidth of notch. This figure shows that if slot position moves downward, it causes wider notch bandwidth. **Figure 5** shows current distribution on the feedline, side grounds and main patch at 5.9 GHz and 8.4 GHz frequency. In comparison with 8.4 GHz, it is obvious that at 5.9 GHz (notch frequency) the current paths around the U-shaped slot are in opposite directions that results in a notch performance at this frequency.

The antenna has been fabricated according to its final optimized parameters. These parameters are provided in **Table 1**. The proposed antenna structure was fabricated, as shown in **Figure 6**. The fabricated prototype is measured by Agilent Technologies, E8361Cnetwork analyzer.

The measured and simulated return loss results are indicated in **Figure 7**. It is evident from this figure that antenna can cover 2.3 - 15.1 GHz with stop band from 5

Table 1. Geometrical parameters of the proposed antenna.

Parameter	value (mm)	Parameter	value (mm)
R1	18 mm	Lb	4 mm
R2	4 mm	S	1 mm
Ws	0.5 mm	W	1.8 mm
Ls	9.8 mm	a	5 mm

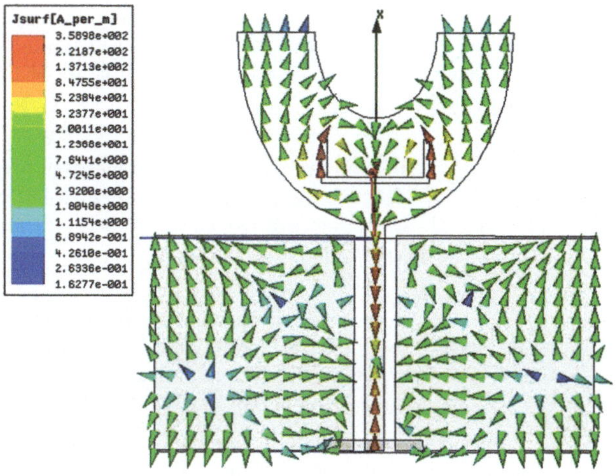

(a)

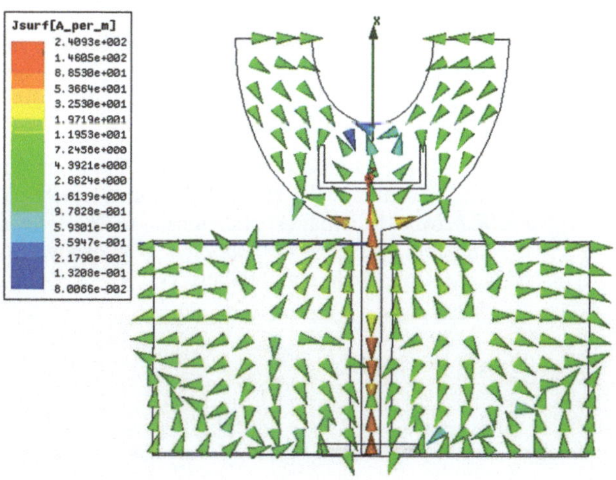

(b)

Figure 5. Simulated current surface distribution at a) 5.9 GHz, b) 8.4 GHz.

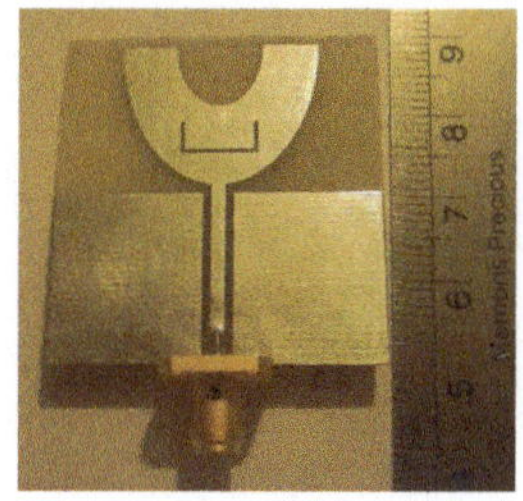

(a)

(b)

Figure 6. Photograph of proposed U-shaped antenna (Fabricated prototype). (a) Front View (b) Back View.

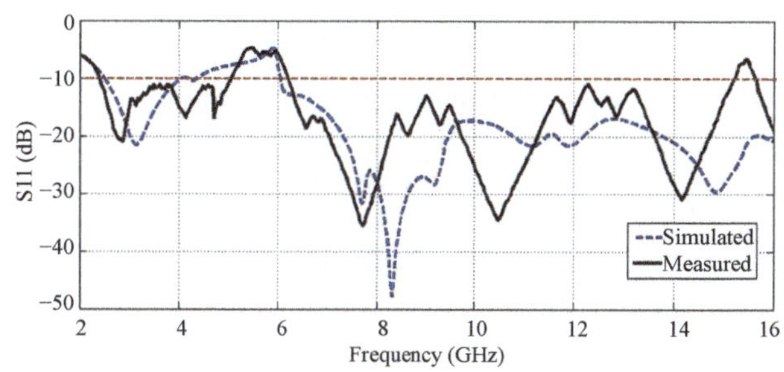

Figure 7. Simulated and measured S_{11} of the proposed U-shaped antenna.

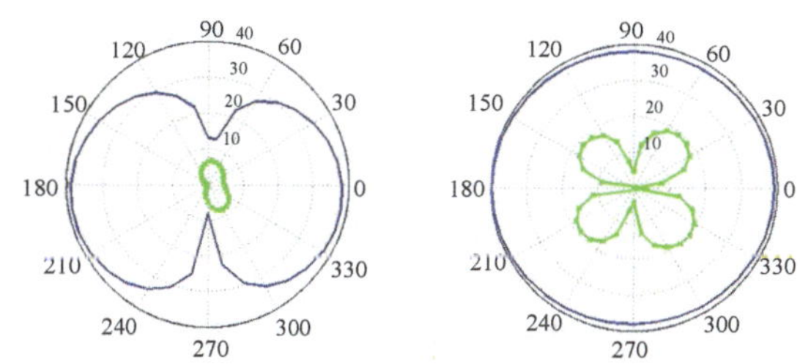

(a)

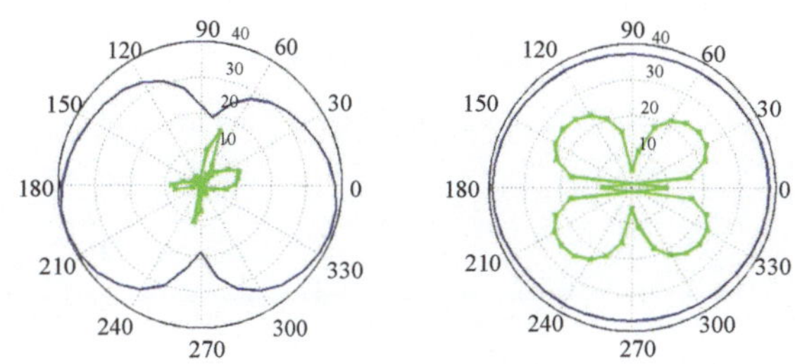

(b)

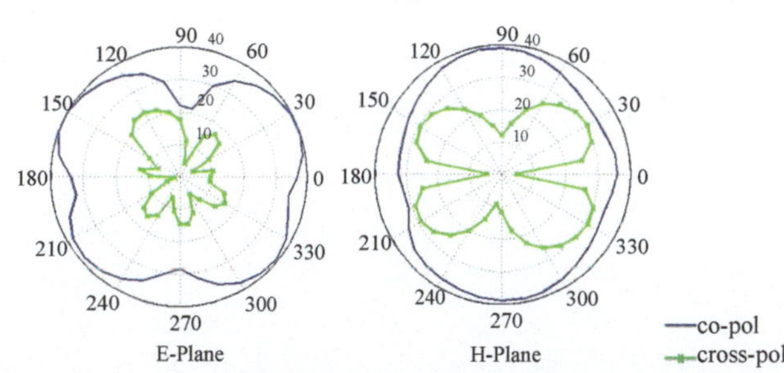

(c)

Figure 8. Radiation pattern of the proposed antenna (a) f = 3 GHz, (b) f = 5 GHz, (c) f = 8 GHz.

GHz to 6.1 GHz.

Radiation patterns of the proposed antenna in H-plane and E-plane at sampling frequencies of 3, 5, and 8 GHz are plotted in **Figure 8**. These patterns are measured in the anechoic chamber of Prof. Morshed Antenna laboratory. At lower frequencies, it is seen that our proposed design exhibits an omni-directional profile for the H-plane and a bi-directional one for the E-plane. With the increase of frequency, the proposed antenna becomes more directive.

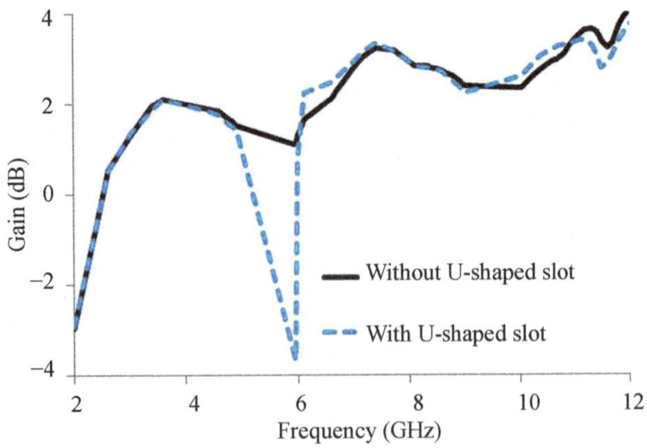

Figure 9. Gain of proposed antenna.

The gains of the proposed antenna with and without U-shaped slot areplotted in **Figure 9**, showing an acceptable flat gain over 3 - 10 GHz while low radiation level in the notched band.

4. Conclusion

An UWB U-shaped antenna with band notch characteristic has been presented and discussed. U-shaped patch has been used to improve antenna bandwidth. By embedding a U-shaped slot, band notch performance is achieved. The stop band is over 5 - 6.1 GHz. The center frequency of stop-band and its bandwidth can be controlled by length and width of slot, respectively. The antenna was fabricated and its performance has been tested. Good agreement is achieved between the simulated and measured results.

REFERENCES

[1] FCC First Report and Order on Ultra-wideband Technology, FCC, 2002.

[2] H. J. Lak, C. Ghobadi and J. Nourinia, "A Novel Ultra-Wideband Monopole Antenna with Band-Stop Characteristic," *Wireless Engineering and Technology*, Vol. 2, NO. 4, 2011, pp. 235-239.

[3] Q. X. Chu and Y. Y. Yang, "A Compact Ultra Wideband Antenna with 3.4/5.5 GHz Dual Band-Notched Characteristics," *IEEE Transactions on Antennas and Propagation*, Vol. 56, No. 12, 2008, pp. 3637-3644.

[4] Y. S. Li, X. D. Yang, C. Y. Liu and T. Jiang, "Compact CPW-Fed Ultra-Wideband Antenna with Dual Band-Notched Characteristics," *Electronic Letters*, Vol. 46, No. 14, 2010, pp. 967-968.

[5] J. B. Jiang, Z. H. Yan and J. Y. Zhang, "Dual Band-Notched Ultra-Wideband Printed Antenna with Two Different Typed Slots," *Microwave Optical Technology Letters*, Vol. 52, No. 9, 2010, pp. 1930-1933.

[6] W. S. Lee, D. Z. Kim, K. J. Kim and J. W. Yu, "Wideband Planar Monopole Antennas with Dual Band-Notched Characteristics," *IEEE Transactions on Microwave Theory Technology*, Vol. 54, No. 3, 2008, pp. 2800-2806.

[7] J. M. Beygi, J. Norinia and C. Ghobadi, "Compact Ultra-Wideband with Dual Band-Stop Characteristic," *IEICE Electronic Express*, Vol. 7, No. 9, 2010, pp. 596-600.

[8] L. Luo, Z. Cui, J. P. Xiong, X. M. Zhang and Y. C. Jiao, "Compact Printed Ultra-Wideband Monopole Antenna with Dual Band-Notch Characteristic," *Electronic Letters*, Vol. 44, No. 44, 2008, pp. 1387-1388.

[9] Y. Y. Lu, H.-J. Lam and J. Bornemann, "Coplanar Printed-Circuit Antenna with Band-Rejection Elements for Ultra-Wideband Filtenna Applications," *IEEE Antennas and Propagation Society International Symposium*, *AP-S* 2008, San Diego, 5-11 July 2008, pp. 1-4.

9

Design of Wide Band CMOS VCO with Common Source Transformer Feedback Topology

Meng-Ting Hsu, Ying-Hsiang Huang, Cheng-Chuan Chung

Microwave Communication and Radio Frequency Integrated Circuit Lab, Department and Institute of Electronic Engineering, National Yunlin University of Science and Technology, Douliou, Chinese Taipei.

ABSTRACT

This paper demonstrates wide-band CMOS VCO based on the transformer feedback from traditional circuit to our proposed work. The start up condition of the traditional cross-coupled pair is expressed by the high frequency model. The wide band technique of this structure is derived with the help of the high frequency model of the transistor. Therefore, the wide band CMOS VCO based on the common source transformer feedback topology can achieves the high performance in the low voltage and low phase noise. The measurement result of the VCO exhibits the figure of merit, core power consumption and output power at supply voltage 0.8 V are –193.1 dBc/Hz, 4.4 mW and –2.3 dBm, respectively. The phase noise is –124.3 dBc/Hz at 1 MHz offset under the operation frequency 5.8 GHz. And the tuning range of the circuit can obtain 28%, this VCO is fabricated in TSMC 0.18 μm 1P6M CMOS process.

Keywords: Voltage Controlled Oscillator (VCO); Transformer Feedback; Low Phase Noise; Wide Tuning Range; Low Voltage

1. Introduction

As RFIC integration requirements become ever more compact, each RF block will be integrated into a single chip. Therefore, the noise must be reduced or the mutual influences between the circuit blocks become stronger, reducing each element's performance. If the elements do not function well, then the circuit will also not function properly. Among the building blocks of a RF transceiver, the voltage controlled oscillator (VCO) is an important block. If the frequency setting and noise control are precisely work, then the system performances of the transceiver will work well marvelously. A fully integrated voltage controlled oscillator (VCO) is one of the most challenging parts of a radio transceiver in standard CMOS technologies. VCO has many demanding parameters, and such as large frequency tuning range, low power consumption, low cost and low phase noise. A wide tuning range is often important because it can accommodate more process and temperature variations.

Traditionally, transformer-based LC tank were reported recently [1-3]. In the meantime, the current driven transformer load [4,5] or switched coupled-inductors [6] have been used to implement wide tuning ranges, but the performance of phase noise is still to be improved with advanced investigation. And high performance of low voltage and low power operation circuit is recently demand for saving energy which reported in some researchers [7-9]. In this paper, the on chip transformers in positive feedbacks are used to increase the output voltage swings and also increase the quality factor of the resonator tank, therefore, reducing phase noise. A symmetric inductor introduced by a transformer on the feedback loop from the active components is employed to generate the oscillation frequency. And the terminal gate and drain of the MOS is paralleled with transformer and varactor to build up the bandwidth extension by small signal analysis. Therefore the tuning range of frequencies can be expanded when the ratio C_{max}/C_{min} is increased. A transformer-based LC VCO can provide a smaller area than the traditional LC resonator VCO. The size and the width of this transformer were optimized to improve parasitic capacitance and increase the self-resonant frequency. Furthermore, the proposed transformer, which is based on the layout optimization to minimize the series resistance of the inductor coil, taking into account both ohmic losses, and magnetically induce losses. The method optimizes the width of the metal strip-width layout.

In this paper, we propose a VCO topology using a transformer feedback technique with a gate to drain coupling transformer. A 5-GHz CMOS VCO based on this topology in 0.18-μm CMOS technology was implemented. The design and analysis of the proposed VCO topology are presented in Section 2. The oscillation frequency and start-up erudition is expressed by low frequency model, the other wide band technique derived by high frequency model is also discussed in this section. Measurement results are shown in Section 3. Finally, the conclusion is summarized in Section 4.

2. Circuit Design

The high performance of low voltage and low power VCO using transformer feedback is shown in **Figure 1(a)** [7]. The wide band VCO with switch capacitor module based on transformer feedback is shown in **Figure 1(b)** [10]. For the compact and simple circuit of our proposed wide band VCO with a symmetry transformer-feedback topology is also shown in **Figure 1(c)** [11].

The M_1 and M_2 transistors are active circuit components and provide the negative resistances of the VCO core to cancel the resistive loss in the LC tank circuit. Transistors M_3 and M_4 are an output buffer circuit. We adopted all PMOS transistors in our proposed VCO, because PMOS transistors have lower flicker noise than that do NMOS transistors. The LC tank includes varactors C_{var} and transformers L_1~L_4. The symmetric transformer is placed between the gate and drain of the M_1 and M_2 transistors. This topology forms an oscillating feedback loop that reduces current achieving low power operation and expand the tuning range.

2.1. Startup Condition of Cross Coupled Pair

The simplified parallel LC oscillator model is shown in **Figure 2(a)**, where R_p represents the tank and output buffers loss. The active circuit must generate a negative resistance in order to compensate the loss from the tank.

The simple schematic of cross-couple pair with **Figures 1(a)** and **(b)** is shown in **Figure 2(b)**. M_1 and M_2 indicate NMOS transistors and the simple small signal model of MOSFET is shown in **Figure 3(a)**. The high frequency equivalent circuit with capacitive parasitic is shown in **Figure 3(b)**.

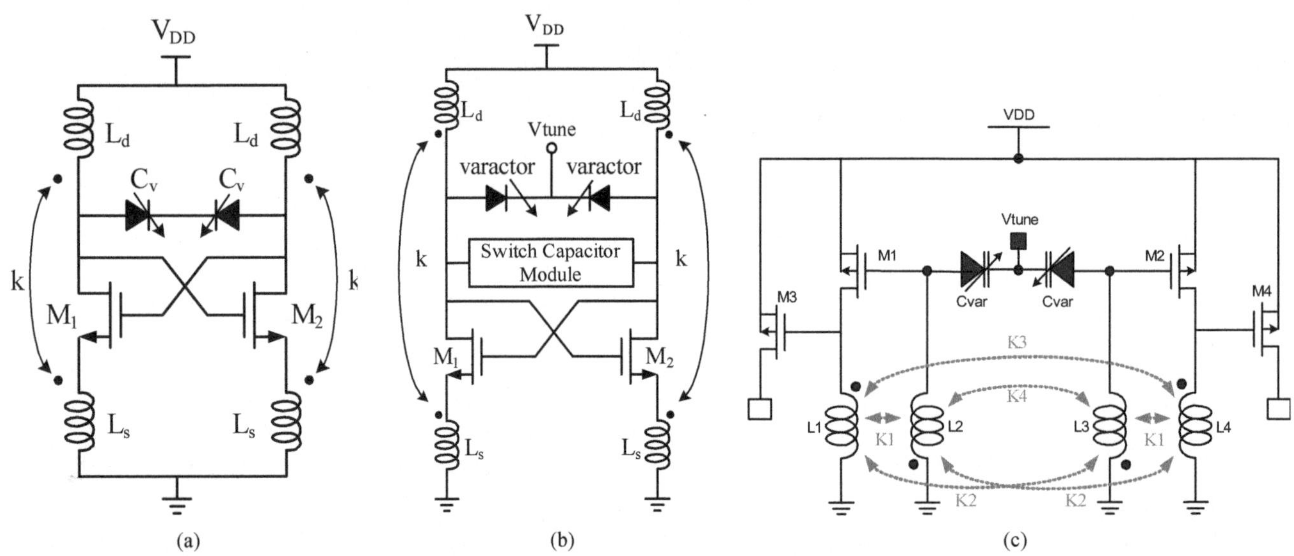

Figure 1. Schematics of (a) Transformer-feedback VCO; (b) Transformer-feedback wide band VCO; (c) Proposed common source transformer-feedback wide band VCO.

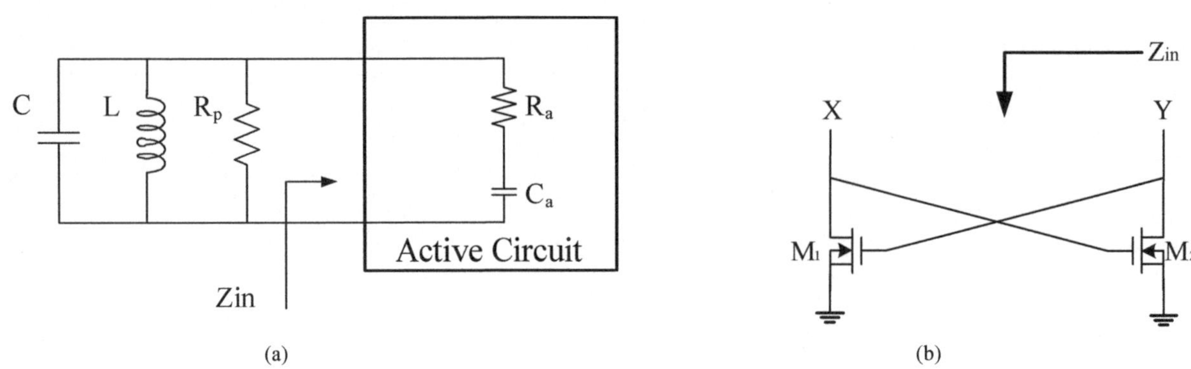

Figure 2. (a) Parallel LC oscillator model; (b) Cross-couple pair.

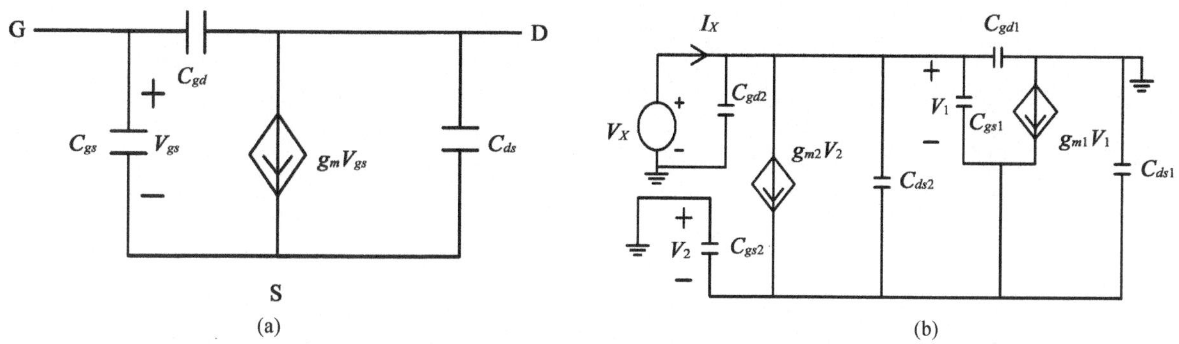

Figure 3. Small signal model of (a) MOSFET for (b) High frequency equivalent circuit.

We can obtain the input impedance Z_{in} as following:

$$Z_{in} = \frac{V_X}{I_X} = \frac{-2g_m}{g_m^2 + \omega^2\left(2C_{gd} + C_{ds}\right)^2} - j\frac{2\omega\left(2C_{gd} + C_{ds}\right)}{g_m^2 + \omega^2\left(2C_{gd} + C_{ds}\right)^2} \quad (1)$$

$$= R_a a + C_a$$

where, R_a is the real part and C_a is the imaginary part, respectively. The detailed calculation can be referred the paper [12].

To satisfy the stable oscillation condition, the circuit exhibiting the absolute value of negative resistance $\left|\mathrm{Re}\left(Z_{in}\right)\right|$ should be larger than R_p.

2.2. Oscillation Frequency

The determination of frequency is controlled by the LC tank. Therefore, we needed to derive oscillation frequencies; the equivalent half-circuit of the proposed VCO is shown in **Figure 4**, where R_1 and R_2 represent the series resistance of the inductor L_1 and inductor L_2, respectively. K is the coupling coefficient between the inductor L_1 and the L_2, and C is the effective tank capacitance.

In order to understand how the proposed circuit can achieve the wide tuning range, the equivalent circuit with high frequency model is shown in **Figure 5**. The equivalent circuit of the linear transformer can be expressed as the equation:

$$\begin{bmatrix} V_1 \\ V_2 \end{bmatrix} = \begin{bmatrix} sL_1 & sM \\ sM & sL_2 \end{bmatrix} \begin{bmatrix} I_1 \\ I_2 \end{bmatrix} \quad (2)$$

where the matrix is denoted by phasor analysis, M is the mutual inductance $M = K\sqrt{L_1L_2}$. In the circuit, R_1, R_2 and Rg (gate resistance) are small enough to be ignored for simplified manipulation. From the kirchhoff's current law with vg = vgs in the circuit, we can obtain the equation of I_1 and I_2 as the following:

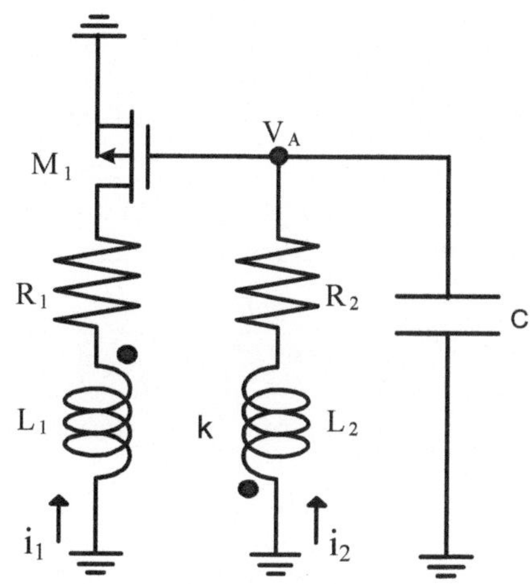

Figure 4. Equivalent half-circuit of the proposed VCO.

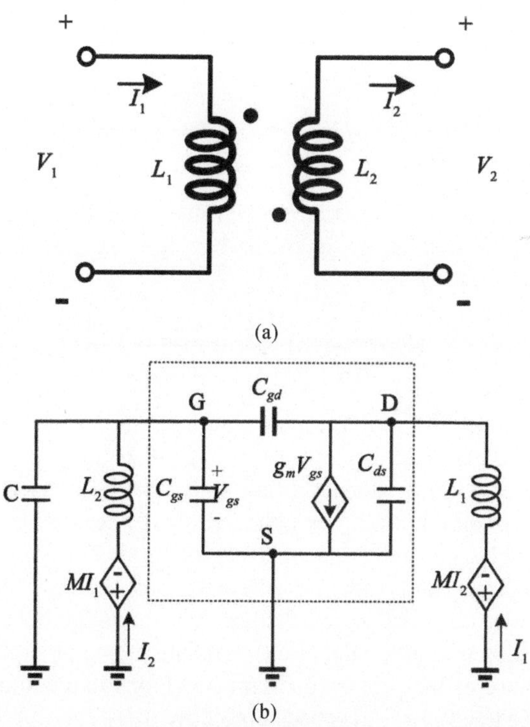

Figure 5. (a) Equivalent Circuit of a transformer; (b) Equivalent Circuit of a transformer with high frequency model in Figure 2.

$$I_1 = sC_{ds}V_d + gmV_g + sC_{gd}\left(V_d - V_g\right) \quad (3)$$

$$I_2 = sCV_g + sC_{gs}V_g + sC_{gd}\left(V_g - V_d\right) \quad (4)$$

The voltage of terminal gate and drain can be expressed as the following:

$$v_g = -\left(sL_2 I_2 + sMI_1\right) \quad (5)$$

$$v_d = -\left(sL_2 I_2 + sMI_2\right) \quad (6)$$

If the (e.g., see Equations (3) and (4)) is taken into the (e.g., see Equation (5)), then we can obtain voltage gain Av1 of the following equation:

$$Av_1 = \frac{v_d}{v_g} = \frac{s^2 MC_{gd} + sM\ gm - 1 - s^2 L_2\left(C + C_{gs} + C_{gd}\right)}{s^2 M\left(C_{ds} + C_{gd}\right) - s^2 L_2 C_{gd}} \quad (7)$$

If the (e.g., see Equations (3) and (4)) is taken into the (e.g., see Equation (6)), then we can obtain the other voltage gain Av2 of the following equation:

$$Av_2 = \frac{v_d}{v_g} = \frac{gm - sC_{gd}}{s^2 L_1\left(C_{ds} + C_{gd}\right) - 1} \quad (8)$$

The oscillation condition of the oscillator is given by [13]

$$v_d / v_g = -1 \text{ at } s = j\omega_o \quad (9)$$

Where ω_o is the oscillation frequency of the VCO. The (e.g., see Equation (9)) can be applied for both (e.g., see Equations (7) and (8)). From the simple manipulation and set the real part to be zero. The oscillation frequency is given by

$$\omega_o = \frac{1}{\sqrt{L_2\left(C + C_{gs} + 2C_{gd}\right) - M\left(C_{ds} + C_{gd}\right)}} \quad (10)$$

If the mutual inductance between L_1 and L_2 is gradually decreased to be ignored, then the oscillation frequency is given by

$$\omega_o = \frac{1}{\sqrt{L_2\left(C + C_{gs} + 2C_{gd}\right)}} \quad (11)$$

Therefore, the wide tuning range can be achieved by the total effective capacitance ($C + C_{gs} + 2C_{gd}$).

For the low frequency small signal model is used for consideration in (e.g., see Equation (11)). Meanwhile, the terminal gate and drain of MOS is paralleled with the transformer and varactor. The terminal source is also grounded for small signal analysis. This topology not only builds up oscillation but also expand the tuning range. The structure of the proposed circuit just looks like common source transformer feedback topology.

The traditional cross-coupled topology has more parasitic capacitances. **Figure 6** shows the comparison of the general LC VCO schematic and proposed LC VCO schematic.

Figure 7 shows the equivalent capacitances versus V_{tune} which is tuned from –2 V to 2 V. For the simulation results of the general LC VCO, the capacitances can be tuned from 290.9 fF to 543.7 fF over the entire tuning voltage, which corresponds to a tuning range of C_{max}/C_{min} ratio = 1.87. For the simulation results of the proposed LC VCO, the capacitance can be tuned from 194.1 fF to 446.9 fF over the entire tuning voltage, which corresponds to a tuning range of C_{max}/C_{min} ratio = 2.3. Therefore, the proposed VCO has a larger frequency tuning range than the general LC VCO.

2.3. Transformer

Transformer feedback currents are used to increase the voltage swing in the resonator and also increase the quality factor of the resonator tank. **Figure 8** shows the comparison of the amplitudes between with and without transformer feedback, when it is placed at the gate node of the PMOS (M_1). It is clear that the circuit transformer feedback has larger voltage swings than without the transformer feedback.

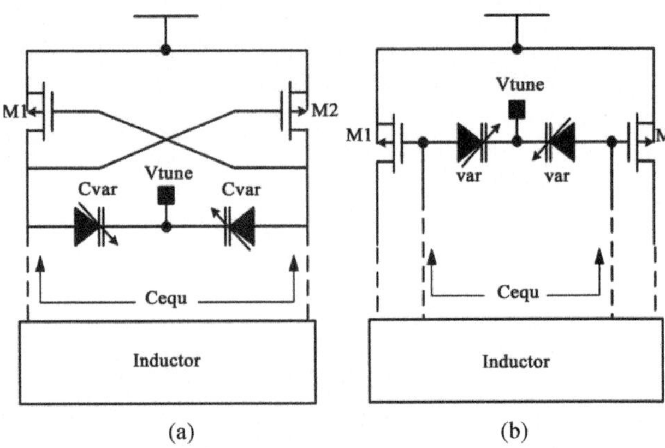

Figure 6. (a) General LC VCO; (b) Proposed LC VCO.

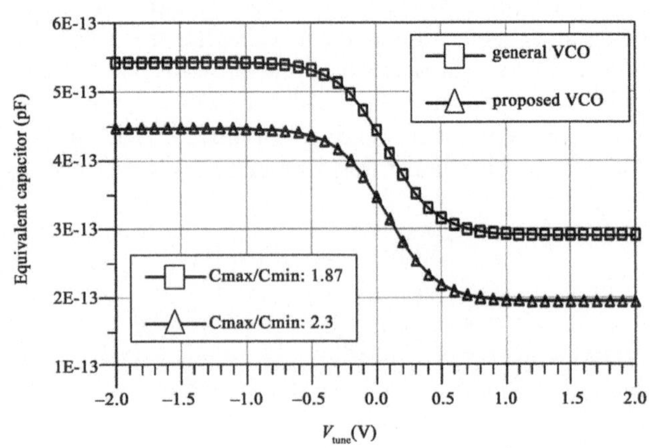

Figure 7. Equivalent capacitance of general LC VCO and proposed LC VCO versus V_{tune} which is tuned from −2 V to 2 V.

The transformer is designed from the inner to outer turn for different width, 11-μm, 13-μm, 15-μm, respectively. The line spacing is 2-μm, and 276-μm is the outer dimension. The layout of the transformer is shown in **Figure 9**. The top layer with AlCu metal is used for the windings to increase the quality factor. The metal width of the inductor increases gradually from the inner to outer turn, which can improve the quality factors of the inductor. The turns-ratio and size of the transformer, which is designed for optimization by EM simulation tool, has been investigated. The transformer turns-ratios were determined by varying the number of turns for each topology. Approximately, a 3:2 turns-ratio was chosen to supply sufficient feedback energy in the VCO. Since there are two symmetrical inductors instead of the single transformer, the chip area can be reduced. **Figure 10** depicts the lumped model of transformer, R_s is the series resistance of inductor, C_{ox} is the gate oxide capacitance, C_k is the capacitance of Number_ of_ Turns, R_{sub} and C_{sub} are Resister and Capacitance from inductance to substrate, respectively.

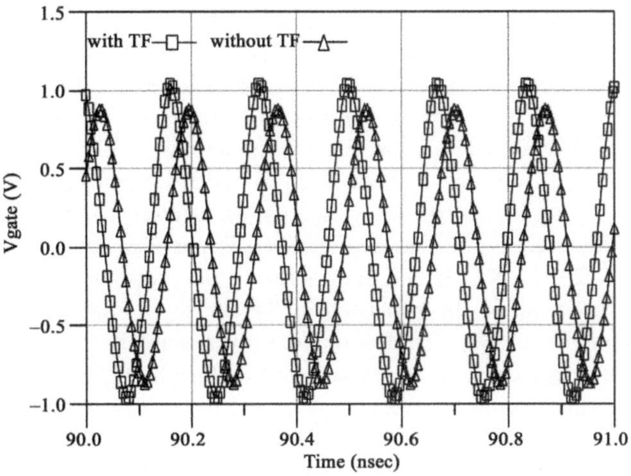

Figure 8. Simulated resonator tank amplitudes with and without transformer feedback.

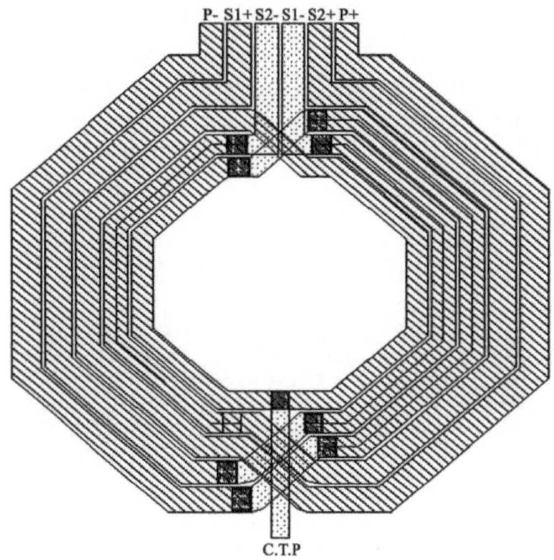

Figure 9. Layout schematic of transformer.

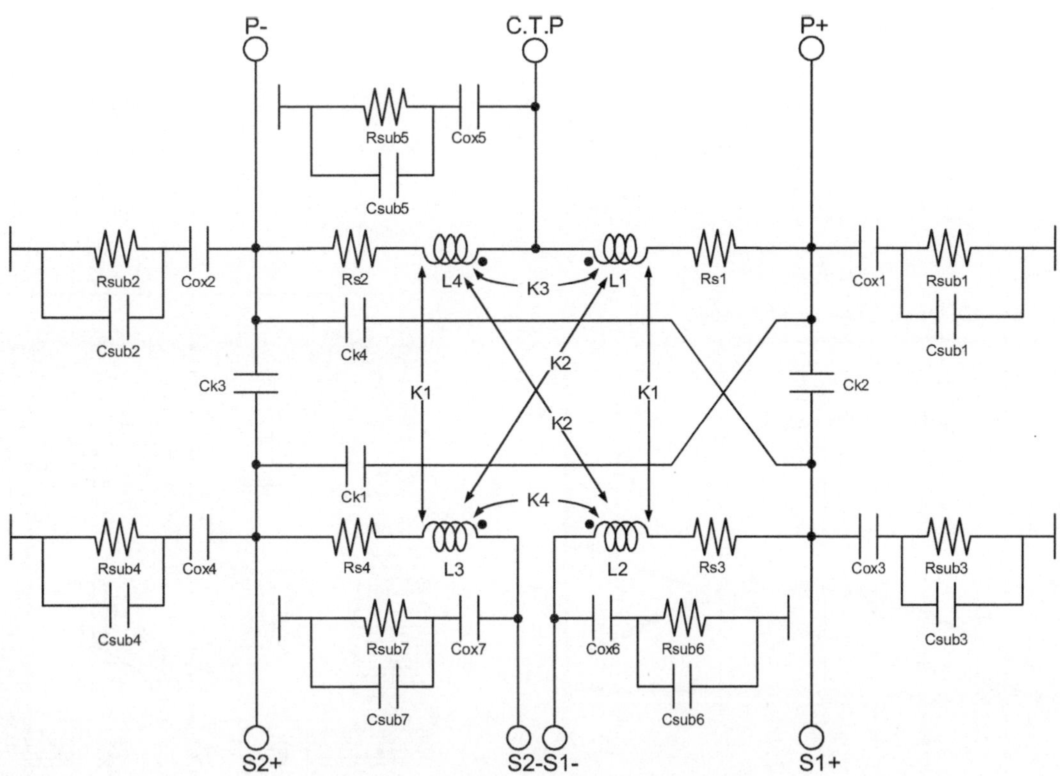

Figure 10. Lumped model of transformer.

The simulation-tool ADS Momentum was used to simulate the parameters of the inductors. The simulated primary inductor (L_2) and secondary inductor (L_1) values are 2.356 nH and 1.37 nH as shown in **Figure 11**. The quality factors of L_1 and L_2 are 7.3 and 12.2 as shown in **Figure 12**. The transformer coupling coefficients K_1, K_2, K_3 and K_4 are approximately 0.66, 0.187, 0.012 and 0.072 at 5 GHz as shown in **Figure 13**.

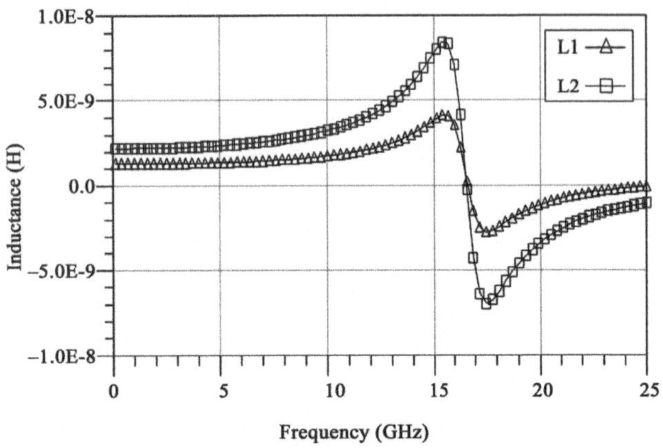

Figure 11. Inductance of L_1 and L_2.

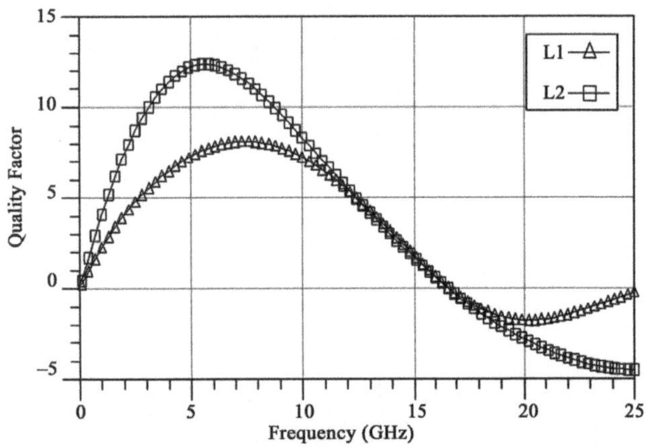

Figure 12. Quality factors of L_1 and L_2.

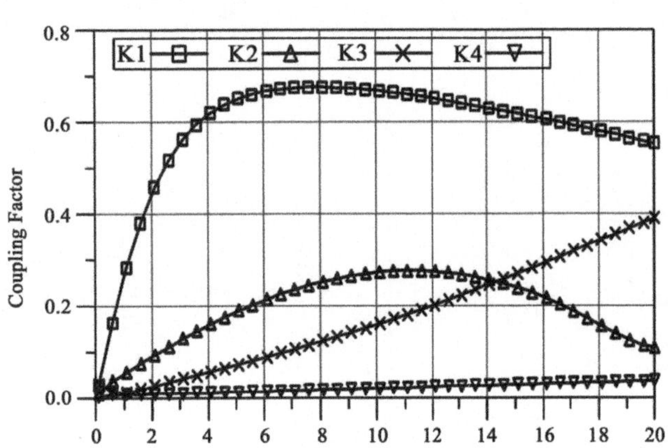

Figure 13. Coupling factors of transformer.

2.4. Circuit Devices Sizes

The size of the devices used in our proposed VCO circuit is shown in **Table 1**. Our work has optimized the size of the devices to maximize the quality factor and to generate enough resistance to the oscillation, and also improve the performance of this proposed VCO.

3. Measurement Results

The proposed transformer feedback VCO is fabricated using a 0.18-μm CMOS process with six metal layers. **Figure 14** shows the die micrograph, which occupies a chip area of 766 × 608 μm^2. The layout was made as symmetrical and compact as possible to ensure differential operation and reduce parasitic inductances or capacitances. Wafer measurements of output spectra were obtained using the Agilent E5052A spectrum analyzer. By varying bias across the varactors, we can tune the oscillation frequency. The tuning range is operated from 4.39 GHz to 5.8 GHz. **Figure 15** shows the measured output spectrum of the proposed VCO whose oscillation frequency is 5.8 GHz. The output power is about –2.3 dbm. In **Figure 15**, there are two spurious signals around the center oscillation frequency, this phenomena is the extra noise coming from power supplies that coupled to the main oscillation signal on both sides.

Table 1. Devices sizes.

0.18 μm process	MOSFET Size (L × W × M)
M_1, M_2	0.18 μm × 2 μm × 40
M_3, M_4	0.18 μm × 3 μm × 20
Varactor C_{var}	146 ~ 418 fF (G = 2, B = 30)
Transformer	L_1 = 1.37 nH @5 GHz Q_1 = 7.3 @5 GHz L_2 = 2.356 nH @5 GHz Q_1 = 12.2 @5 GHz

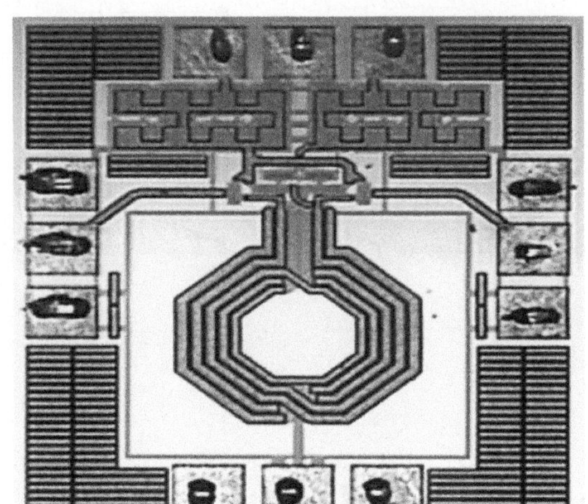

Figure 14. Chip photograph.

The phase noise spectrum measured, by a Agilent E5052A signal source analyzer, is shown in **Figure 16**. The phase noise is –124.3 dBc/Hz at 1-MHz frequency offset from 5.8 GHz carrier. The total current dissipation is in the vicinity of 11 mA. With a supply voltage of 0.8 V, the core current and power consumption are 5.5 mA and 4.4 mW, respectively.

The general performance when comparing the *FOM* and FOM_T of the VCO is defined as follows [14,15].

$$FOM = 10\log\left[\left(\frac{\omega_o}{\Delta\omega}\right)^2 \cdot \frac{1}{L\{\Delta\omega\}\cdot P_{dc}}\right] \quad (12)$$

$$FOM_T = FOM - 20\log\left[\frac{FTR}{10}\right] \quad (13)$$

where ω_o is the center frequency, $\Delta\omega$ is the frequency offset, $L\{\Delta\omega\}$ is the phase noise at $\Delta\omega$, P_{dc} is the dc power consumption in mW, and FTR is the frequency tuning range in percent. Based on this calculation, the

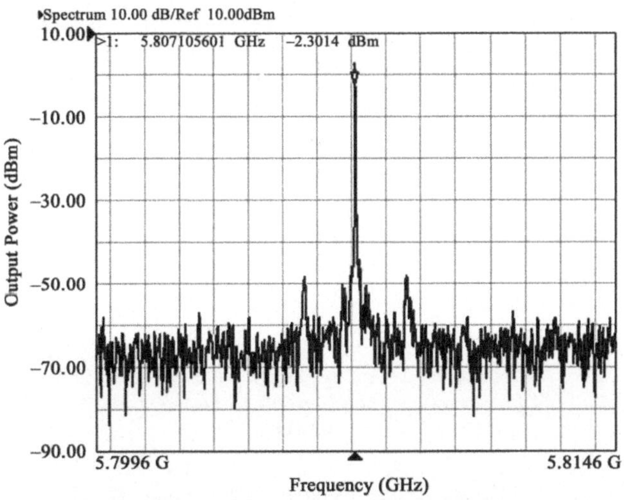

Figure 15. Output spectrum of the proposed VCO.

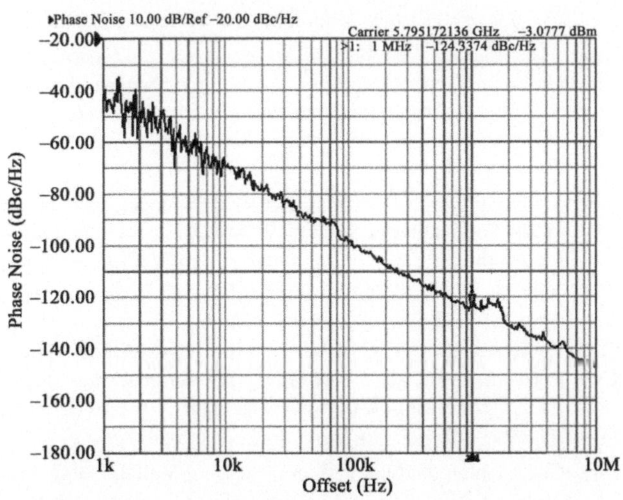

Figure 16. Measured phase noise of the proposed VCO.

figure of merit (*FOM*) of this proposed VCO is about –193.1 dBc/Hz at the carrier frequency of 5.8 GHz. The FOM_T is about –202 dBc/Hz.

The performance comparison between the proposed transformer feedback LC tank CMOS VCO with other previous reports are also summarized in **Table 2**. The comparison of FOM_T with phase noise of different VCOs are shown in **Figure 17**. The comparison of FOM_T with tuning range of different VCOs are shown in **Figure 18**. Our work exhibits excellent results. Due to the use of the transformer feedback technique, significant performance improvements are demonstrated in terms of output swing, frequency tuning range, and phase noise, making it extremely attractive for low-power and low-voltage RF applications.

4. Conclusion

A transformer feedback CMOS VCO is presented for low phase noise, low power and wideband operation. The VCO utilizes a gate to drain the transformer feedback configuration, and makes use of the quality factor improvement of the transformer-based LC tank. The proposed

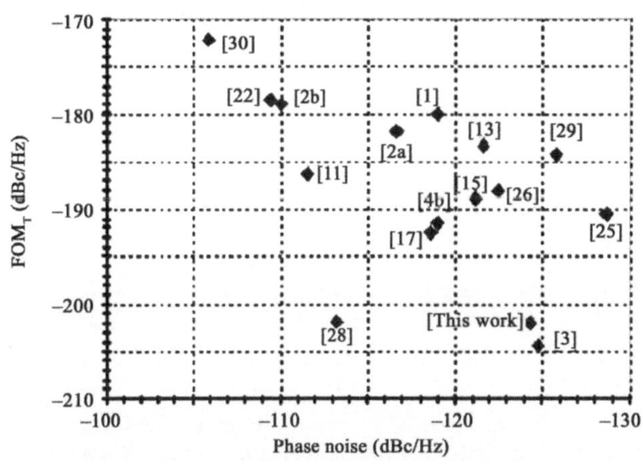

Figure 17. Comparison of FOM_T vs Phase noise.

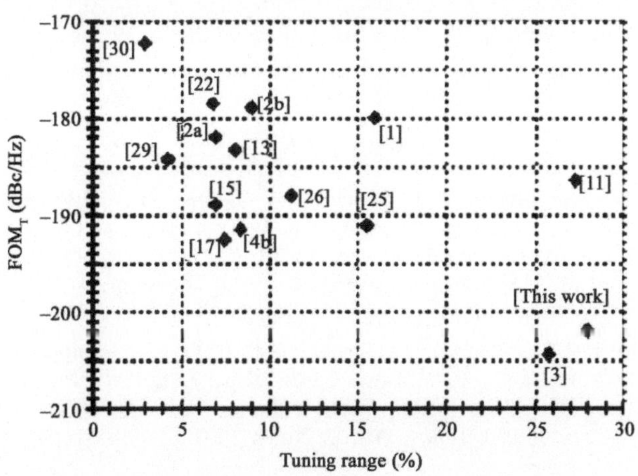

Figure 18. Comparsion of FOM_T vs Tuning range.

Table 2. Comparison of transformer feedback VCO performance.

Reference	Technology (μm)	f_0 (GHz)	V_{DD} (V)	Core Power Diss. (mW)	Phase noise (dBc/Hz)	Tunung range (%)	*FOM* (dBc/Hz)	FOM_T (dBc/Hz)
[7]a	0.18	1.38	0.35	1.46	−128.6	0	−189.8	N/A
[7]b	0.18	3.8	0.5	0.57	−119	8.4	−193	−191.5
[8]	0.18	8.5	0.7	4.7	−121.6	8.1	−190	−183.4
[9]	0.18	1.14	0.34	0.1	−121.2	7	−192	−188.9
[10]	0.18	5.45	0.6	6.9	−111.5	27.2	−177.8	−186.5
[15]	0.35	2.4	2.5	10	−121	26	−176	−180
[16]	0.13	5	1.2	1.8	−124.8	25.8	−196.2	−204.4
[18]	0.18	21.3	2.4	9.6	−105.9	3	−182.8	−172.3
[19]	0.18	16	1.8	6	−118.6	7.5	−195	−192.5
[20]	0.18	11.2	1.8	6.8	−109.4	6.84	−181.8	−178.5
[21]	0.35 SiGe	5.72	1.2	8.64	−125.8	4.3	−191.6	−184.3
[23]	0.18	17	1	5	−110	16.5	−187.6	−191.9
[24]a	0.18	5.12	0.6	3.9	−116.7	7	−185	−181.9
[24]b	0.18	5.13	1.8	2.59	−110	9	−180	−179
[25]	0.18	1.2	1.8	2.16	−128.7	15.6	−186.9	−190.7
[26]	0.18	4.5	1.5	6.75	−122.5	11.3	−187	−188
[27]	0.18	21.37	0.6	3.5	−109.8	5.1	−190.9	−185.1
[28]	0.13	5.8	1.5	6.1	−113.5	12.8	−180.1	−202.2
This work	0.18	5.8	0.8	4.4	−116.36 ~ −124.3	28	−185.2 ~ −193.1	−194.1 ~ −202

transformer feedback VCO is fabricated in 0.18-μm CMOS process with six metal layers. The measurement results show that the phase noise is –124.3 dBc/Hz at 1 MHz offset frequency and the output power is about –2.3 dBm. The measured tuning range is about 28% from 4.39 GHz to 5.8 GHz and the power dissipation is about 4.4 mW. The *FOM* and *FOMT* of the VCO is about –193.1 dBc/Hz and –202 dBc/Hz, respectively. The performance of our work is good for low voltage and low phase noise application of RF VCO.

REFERENCES

[1] A. Ravi, K. Soumyanath, R. E. Bishop, B. A. Bloechel and L. R. Carley, "An Optimally Transformer Coupled, 5 GHz Quadrature VCO in a 0.18-μm Digital CMOS Process," *Symposium on VLSI Circuits*, *Digest of Technical Papers*, June 2003, pp. 141-144.

[2] S. L. J. Gierkink, S. Levantino, R. C. Frye, C. Samori and V. Boccuzzi, "A Low-Phase-Noise 5-GHz CMOS Quadrature VCO Using Superharmonic Coupling," *IEEE Solid-State Circuits*, Vol. 38, No. 7, 2003, pp. 1148-1154.

[3] D. Baek, T. Song, E. Yoon and S. Hong, "8-GHz CMOS Quadrature VCO Using Transformer-Based LC Tank," *IEEE Microwave and Wireless Components Letters*, Vol. 13, No. 10, 2003, pp. 446-448.

[4] B. Çatlı and M. M. Hella, "A Dual Band, Wide Tuning Range CMOS Voltage Controlled Oscillator for Multi-Band Radio," *IEEE Radio Frequency Integrated Circuits Symposium* (*RFIC*), Honolulu, 3-5 June 2007, pp. 595-598.

[5] G. Cusmai, M. Repossi, G. Albasini, A. Mazzanti and F. Svelto, "A Magnetically Tuned Quadrature Oscillator," *IEEE Journal of Solid-State Circuits*, Vol. 42, No. 12, 2007, pp. 2870-2877.

[6] M. Demirkan, S. P. Bruss and R. R. Spencer, "Design of Wide Tuning-Range CMOS VCOs Using Switched Coupled-Inductors," *IEEE Journal of Solid-State Circuits*, Vol. 43, No. 5, 2008, pp. 1156-1163.

[7] K. Kwok and H. C. Luong, "Ultra-Low-Voltage High-Performance CMOS VCOs Using Transformer Feedback," *IEEE Solid-State Circuits*, Vol. 40, No. 3, 2005, pp. 652-660.

[8] C.-H. Chang and C.-Y. Yang, "A Low-Voltage High-Frequency CMOS LC-VCO Using a Transformer Feedback," *IEEE Radio Frequency Integrated Circuits* (*RFIC*) *Sym-*

posium, 17 April-17 June 2008, pp. 545-548.

[9] S.-L. Jang and C.-F. Lee, "A Low Voltage and Power LC VCO Implemented with Dynamic Threshold Voltage MOSFETS," *IEEE Microwave and Wireless Components Letters*, Vol. 17, No. 5, 2007, pp. 376-378.

[10] M.-T. Hsu and H.-J. Li, "Design of High Performance and Wide-Band VCO with Transformer Feedback," *APMC Microwave Conference*, Macau, 16-20 December 2008, pp. 1-4.

[11] M.-T. Hsu and Y.-H. Huang, "A Low Phase Noise and Low Power CMOS VCO with Transformer Feedback," *APMC Microwave Conference*, Macau, 7-10 December 2009, pp. 2280-2283.

[12] M.-T. Hsu, C.-T. Chiu and S.-H. Chen, "Low Power Design of CMOS 5-GHz Voltage-Controlled Oscillator from Narrowband to Wideband with Switching Capacitor Module," *IET Circuit Devices System*, Vol. 3, No. 3, 2009, pp. 99-105.

[13] J. Yang, C.-Y. Kim, D.-W. Kim and S. Hong, "Design of a 24-GHz CMOS VCO with an Asymmetric-Width Transformer," *IEEE Transactions on Circuits and Systems*—II: *Express Briefs*, Vol. 57, No. 3, 2010, pp. 173-177.

[14] P. Kinget, "Integrated GHz Voltage Controlled Oscillators," Kluwer, New York, 1999.

[15] D. Ham and A. Hajimiri, "Concepts and Methods in Optimization of Integrated LC VCOs," *IEEE Journal of Solid-State Circuit*, Vol. 36, No. 6, 2001, pp. 896-909.

[16] Y. Lee, S. Hyun and C. Kim, "Current Reuse Cross-Coupling CMOS VCO Using the Center-Tapped Transformer in LC Tank for Digitally Controlled Oscillator," *IEEE Radio Frequency Integrated Circuits* (*RFIC*) *Symposium*, Atlanta, 17 April-17 June 2008, pp. 549-552.

[17] C. H. Cao and K. O. Kenneth, "Millimeter-Wave Voltage-Controlled Oscillators in 0.13-μm CMOS Technology," *IEEE Solid-State Circuits*, Vol. 41, No. 6, 2006, pp. 1297-1304.

[18] C.-C. Li, T.-P. Wang, C.-C. Kuo, M.-C. Chuang and H. Wang, "A 21 GHz Complementary Transformer Coupled CMOS VCO," *IEEE Microwave and Wireless Components Letters*, Vol. 18, No. 4, 2008, pp. 278-280.

[19] C.-H. Chang and C.-Y. Yang, "A 0.18-μm CMOS 16-GHz Varactorless LC-VCO with 1.2-GHz Tuning Range," *IEEE Solid-State Circuits Conference*, Jeju, 12-14 November 2007, pp. 107-110.

[20] N.-J. Oh and S.-G. Lee, "11-GHz CMOS Differential VCO with Back-Gate Transformer Feedback," *IEEE Microwave and Wireless Components Letters*, Vol. 15, No. 11, 2005, pp. 733-735.

[21] J.-S. Syu, C. C. Meng and G.-W. Huang, "SiGe HBT Quadrature VCO Utilizing Trifilar Transformers," *IEEE Solid-State Circuits Conference*, 3-5 November 2008, pp. 465-468.

[22] A.-S. Porret, T. Melly, C. C. Enz and E. A. Vittoz, "Design of High-Q Varactors for Low-Power Wireless Applications Using a Standard CMOS Process," *IEEE Journal of Solid-State Circuits*, Vol. 35, No. 3, 2000, pp. 337-345.

[23] A. W. L. Ng and H. C. Luong, "A 1-V 17-GHz 5-mW CMOS Quadrature VCO Based on Transformer Coupling," *IEEE Solid-State Circuits*, Vol. 42, No. 9, 2007, pp. 1933-1941.

[24] Y.-H. Chuang, S.-L. Jang, S.-H. M. Lee, R.-H. Yen and J.-J. Jhao, "5-GHz Low Power Current-Reused Balanced CMOS Differential Armstrong VCOs," *IEEE Microwave and Wireless Components Letters*, Vol. 17, No. 2, 2007, pp. 139-141.

[25] S.-J. Yun, N. D. B. Yen, I. Y. Lee, J.-T. Lee and S.-G. Lee, "A Differentially-Tuned Voltage Controlled Oscillator Using Symmetric Transformer," *IEEE Microwave and Wireless Components Letters*, Vol. 18, No. 7, 2008, pp. 464-466.

[26] S.-H. Lee, Y.-H. Chuang, S.-L. Jang and C.-C. Chen, "Low-Phase Noise Hartley Differential CMOS Voltage Controlled Oscillator," *IEEE Microwave and Wireless Components Letters*, Vol. 17, No. 2, 2007, pp. 145-147.

[27] S.-L. Liu, K.-H. Chen, T. Chang and A. Chin, "A Low-Power K-Band CMOS VCO with Four-Coil Transformer Feedback," *IEEE Microwave and Wireless Components Letters*, Vol. 20, No. 8, 2010, pp. 459-461.

[28] Z. Safarian and H. Hashemi, "Wideband Multi-Mode CMOS VCO Design Using Coupled Inductors," *IEEE Transactions on Circuits and Systems I*, Vol. 56, No. 8, 2009, pp. 1830-1843.

[29] J. M. Lopez-Villegas, J. Samitier, C. Cane, P. Losantos and J. Bausells, "Improvement of the Quality Factor of RF Integrated Inductors by Layout Optimization," *IEEE Transactions on Microwave Theory and Techniques*, Vol. 48, No. 1, 2000, pp. 76-83.

[30] T. H. Lee and A. Hajimiri, "Oscillator Phase Noise: A Tutorial," *IEEE Journal of Solid-State Circuits*, Vol. 35, 2000, pp. 326-336.

Novel Optimized Cross-Layer Design with Maximum Weighted Capacity Based Resource Allocation for AMC/HARQ Wireless Networks

Reham A. El-Mayet[1], Hesham M. El-Badawy[2], Salwa H. Elramly[1]

[1]Electronic and Communication Department, Ain Shams University, Cairo, Egypt; [2]National Telecommunication Institute, Cairo, Egypt.

ABSTRACT

To provide quality-of service (QoS) guarantees for heterogeneous applications, most recent wireless communications technologies and standards combine the error-correcting capability of hybrid automatic repeat request (HARQ) schemes at the data link layer (DLL) with the adaptation ability of the adaptive modulation and coding (AMC) modes at the physical layer (PHY) layer. This paper aims to investigate the aggregated system capacity as well as the breakdown of this capacity for different ACM modes in each HARQ scheme. This investigation was done by using maximum weighted capacity (MWC) resource allocation at the PHY layer in conjunction with a novel packet error rate (PER)-based scheduling at the medium access control (MAC) layer. As a result, the dominant AMC mode corresponding to channel SNR was available.

Keywords: AMC; Cross-Layer Design; HARQ; Nakagami-m Fading Channel; QoS

1. Introduction

In order to enhance the throughput of the wireless communication system, link adaptation (LA) technologies are considered not only at the physical layer, but also at the upper-protocol-layers such as data link layer when designing the wireless networks. LA technologies use the instantaneous channel state information (CSI) to adaptively control the data transmission of wireless channel, maintain constant transmission power to reduce the interference of other users and satisfy different business' needs, and save resources to improve overall throughput for the system. In addition, the adaptive system can easily provide services with different qualities, such as higher information transmission rate, lower packet error rate, and higher spectral efficiency. LA technologies mainly include adaptive modulation and coding (AMC) at the physical layer (PHY) and hybrid automatic repeat request (HARQ) at the data link layer (DLL). The AMC schemes can be used to match transmission rates to time-varying channel, in order to obtain maximum throughput and improve spectral efficiency [1]. Cross-layer design (CLD) which combines AMC and HARQ is extensively studied in order to match the transmission rates to time-varying wireless communication channel conditions [2].

HARQ provides a good tackling technique to solve the tradeoff between channel coding and packet retransmission. It is a well suited solution to enforce the link quality at the medium access control (MAC) layer in wireless environments, and has been adopted in some towards-4G standards (3GPP LTE Release 8, for instance) [3]. It also can be considered as a combination of forward error correction (FEC) and automatic repeat request (ARQ) which is used to improve reliability [1].

A major concern in data communication is how to control transmission errors caused by channel noise so that error-free data can be delivered to the user. A solution to this problem is using HARQ schemes:

Type-I HARQ, the ARQ adopted in 3GPP (R99) is called as HARQ type I. It uses a fixed rate FEC code along with ARQ. In this scheme, both error detection and FEC bits are added to each packet prior to transmission, when the received code word is detected in error, two situations may arise. If the number of errors is within error-correcting capabilities of the code, the errors are corrected. Otherwise, the received coded data block is discarded and a retransmission is requested by the receiver, similar to standard ARQ [4]. The disadvantage of the type I HARQ scheme is that once the code rate is fixed, all parity bits for error correction are transmitted even if they are not all needed, thus reducing the channel use efficiency [5].

Type-II HARQ, it's an incremental redundancy (IR) HARQ. It considers the time selective property of the wireless channels. At first transmission, the data packet has no or little redundancy. If it fails, each retransmission includes additional redundancy bits from channel encoder; the redundancy is increased in the repeated transmission of the same packet. Thus, retransmission of type-II HARQ only uses the information bits that have not been transmitted in the previous transmissions. If all the channel coded information bits have been transmitted, then the channel code will be retransmitted. When the decoder has received the information packet, it combines all the information packets together to decode. The retransmission packet could not decode alone, it could be decoded only if it was combined together with the previous transmitted information packet [2].

Type-III HARQ, it also an IR scheme but it uses the concept of self-decodable codes; it can be combined with the error data stored and then decoded. The retransmission uses the puncture matrix which is complementary with the puncture matrix for the first transmission. Every puncture matrix has the same code rate and the same error-correction ability [2]. At the receiver, if a frame is detected to be in error, it is stored and a negative acknowledgement (NACK) is fed back. Upon arrival of the retransmission, the receiver attempts to decode the second transmission of the frame. If the decoding is successful, an ACK is fed back; otherwise the frame is stored and combines all the packets together to decode. The retransmission will continue until the decoding is successful or maximum number of transmission has reached [6].

HARQ in 3GPP Long Term Evolution (LTE), both LTE and LTE-Advanced (LTE-A) use HARQ. In LTE, HARQ reordering is done in the radio link control (RLC) layer. So, it operates into different modes: the unacknowledged mode (without ARQ) or in acknowledged mode (with active ARQ). In LTE-A, HARQ is very similar to HARQ in LTE, only small adaptations have been done to cope with the change in the supported bandwidth since LTE-A is supposed to support bandwidth up to 100 MHz [7].

In wireless networking, quality of service (QoS) plays a crucial role in performance measurement. There are four main schemes that are carried out for dynamic subcarrier and power allocation to different OFDM systems users, which allow a flexible multiuser access and enhancement of the multiuser diversity [2]: maximum capacity based resource allocation, proportional fairness based resource allocation, adaptive subcarrier and power allocation in OFDM based on maximizing utility and maximum weighted capacity (MWC) based resource allocation.

The AMC/HARQ CLD concept has been proposed in many literatures such as [1,2,4,6]. In [1], the performance of type I, II, III HARQ with AMC over Nakagami-m fading channel was analyzed, a closed form expression of PER and average spectral efficiency were derived and the conditional probability density function of signal-to-noise ratio (SNR) was calculated using Marcov model. In [2], the HARQ (type-II, type-III) was applied to the cross-layer framework in order to satisfy the prescribed delay and PER constraints at LL. In [4], Jaume Ramis had proposed an analytical link level queuing model to formulate a cross-layer design conceived as a constrained optimization problem to exploit the joint impact on QoS performance of both AMC at the PHY layer and HARQ-chase combining based-error control at the DLC-layer. Feijn Shi studied the joint design of type-III HARQ protocol at the data link layer and AMC scheme at the PHY layer through closed-form expressions for throughput, average delay and packet loss rate in [6]. Cross-layer scheduling techniques for multiuser OFDM system were introduced in [8,9] by Nan Zhou: in [8], a packet batch based cross-layer scheduling scheme was proposed. This scheme considers the differences between the batches in the same queue, so it is more flexible and efficient than conventional queue based scheduling. In [9], cross-layer resource allocations based on MWC technique and delay satisfaction (DS) scheduling scheme in the downlink multiuser OFDM system were proposed. The current paper will try to investigate the CLD and its enhancements for OFDM based wireless networks. Differently from the previously mentioned publications, the current paper introduces another vision for CLD in both uplink and downlink multiuser OFDM system by considering the packet error rate (PER) as well as the SNR constraints on the overall performance.

In this paper, based on the AMC/HARQ cross-layer design, the system capacity was investigated for different SNR of OFDM channels for different HARQ schemes. Also, a novel PER based data scheduling technique based on MWC resource allocation was proposed.

In Section 2, the system model is presented. Sub-carrier and power allocation are presented in Section 3. PER based data scheduling is presented in Section 4. Numerical results and analysis are shown in Section 5. Finally, Section 6 concludes this paper.

2. System Model

The point-to-point wireless packet communication system under consideration in this paper is illustrated in

Figure 1, showing the most important blocks involved in the transmission and reception processes for a single user. The HARQ protocol with AMC is used over the Nakagami-m fading channel, the channel gain is assumed to be constant during the transmission of one frame, and it varies during the transmission of the next frame. Also, the channel is assumed to support QoS-guaranteed traffic characterized by a packet error rate (PER) and a predetermined acceptable link layer PER P_{loss}. The processing unit at the data link layer is a packet of fixed size, and the processing unit at the physical layer is a frame made of a variable number of packets that depends on the transmission mode (TM) selected by the AMC scheme. The receiving buffer can accurately detect the received SNR and decide the next modulation scheme, then feedback channel state message to the source node. On the other hand, if the destination node cannot decode the received data packet correctly, it feeds back NACK signaling to the source node, otherwise, it feeds back acknowledge (ACK) signaling, and the transmission of all signaling messages are assumed to be error free, the maximum number of retransmissions is denoted $R_{\max}$.

2.1. Channel Modeling

This paper adopts The Nakagami-m channel model. .The probability density function (pdf) of the received SNR γ, is given by [2]:

$$P_\gamma(\gamma) = \frac{m^m \gamma^{m-1}}{\bar{\gamma}^m \Gamma(m)} e^{\left(-\frac{m\gamma}{\bar{\gamma}}\right)} \quad m \geq \frac{1}{2} \tag{1}$$

where, γ is the instantaneous received SNR, and $\bar{\gamma} := E\{\gamma\}$ is the average received SNR; m is the Nakagami fading parameter ($m \geq 1/2$). $\Gamma(m) := \int_0^\infty t^{m-1} e^{-t}$ is the gamma function.

2.2. Combining AMC with HARQ Schemes

Given γ in the previous channel model, the objective of AMC is to select a suitable TM to maximize the data rate while maintaining a prescribed PER P_{loss}.

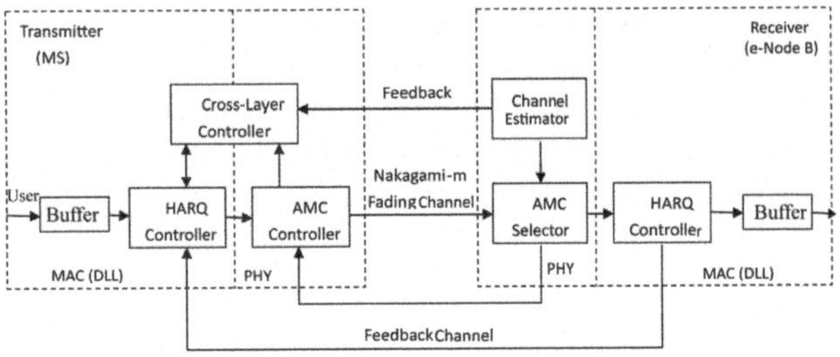

Figure 1. Single link between a single-antenna transmitter (MS) and a single-antenna receiver (e-Node B).

To combine the AMC at PHY and HARQ schemes at LL, only finite delays could be afforded. So, the maximum number of the transmission of one packet $R_{\max}$ could be determined by dividing the maximum allowable system delay over the round trip delay required for each transmission. If a packet is still erroneous after $R_{\max}$ transmissions, it will be dropped, packet loss is declared. So, acceptable P_{loss} should be predetermined [2].

To simplify the design, the following approximate instantaneous PER expression is used [1,2,6]:

$$\text{PER}_M(\gamma) \approx \begin{cases} 1, & \text{if } 0 < \gamma < \gamma_{pM} \\ a_M e^{(-g_M \gamma)}, & \text{if } \gamma \geq \gamma_{pM} \end{cases} \tag{2}$$

here, M is the mode index to identify the AMC mode. The parameters a_M, g_M and γ_{pM} are mode-dependent parameters, which are listed in **Tables 1** and **2** for packet length N_p = 1080 bits. These parameters are obtained by fitting Equation (2) to the exact PER [10]. This expression is used to approximate the PER, when the convolutional code is used. **Tables 1** and **2** show the values of these parameters for HARQ-type II and HARQ-type III respectively [2].

In the following **Tables 1** and **2**, the index r in parameters $a_{r,M}$ and $g_{r,M}$, represents the number of retransmission. r = 0, 1, 2, 3 represent the first, second, third and fourth transmission. M = 1 - 6 represent the AMC mode.

For the type-I HARQ, PER of each transmission is the same as the first transmission of type-II, III HARQ. So, the parameters $a_{0,M}$ and $g_{0,M}$ are the PER curve fitting parameters of type-I HARQ for each AMC mode. The expression of the PER for type-I HARQ is given by:

$$P_M = a_{0,M} e^{(-g_{0,M}\gamma)} \tag{3}$$

Table 1. Transmission modes for type II HARQ [2].

Type-II	Mode 1	Mode 2	Mode 3	Mode 4	Mode 4	Mode 5
Modulation	BPSK	QPSK	QPSK	16-QAM	16-QAM	64-QAM
Code Rate	1/2	1/2	3/4	3/4	3/4	3/4
R_n (bits/sym.)	0.5	1	1.5	3	3	4.5
$a_{0,n}$	1525.9	424.06	27.429	126.88	133.27	60.556
$g_{0,n}$	6.0354	2.6532	0.8483	0.4446	0.2430	0.0553
$a_{1,n}$	0.2608	163.34	1940	99.5411	359.118	172.666
$g_{1,n}$	3.5497	1.5291	2.4865	0.4986	0.4812	0.1242
$a_{2,n}$	13.134	401.49	149.355	178.966	224.796	56.311
$g_{2,n}$	7.8735	3.1401	2.4442	0.5944	1.5547	0.1485
$a_{3,n}$	1.0888	1807.2	46.2931	207.57	1064	40.7133
$g_{3,n}$	8.7707	3.9224	2.3718	0.6092	0.7082	0.1412

Table 2. Transmission modes for type III HARQ [2].

Type-III	Mode 1	Mode 2	Mode 3	Mode 4	Mode 4	Mode 5
Modulation	BPSK	QPSK	QPSK	16-QAM	16-QAM	64-QAM
Code Rate	1/2	1/2	3/4	3/4	3/4	3/4
R_n (bits/sym.)	0.5	1	1.5	3	3	4.5
$a_{0,n}$	1525.9	424.06	27.429	126.88	133.27	60.556
$g_{0,n}$	6.0354	2.6532	0.8483	0.4446	0.2430	0.0553
$a_{1,n}$	0.2608	0.96857	0.017374	1.2538	0.0273	9.1637
$g_{1,n}$	3.5497	1.5291	0.61528	0.4569	0.1341	0.14264
$a_{2,n}$	13.134	1.4037	877.64	1.8815	0.5573	0.0762
$g_{2,n}$	7.8735	2.6295	3.4533	0.5538	0.4107	0.0909
$a_{3,n}$	1.0888	0.07827	0.11309	1.0951	1.4994	0.5364
$g_{3,n}$	8.7707	2.0908	1.1894	0.5096	0.5258	0.1286

Based on the parameters in the **Tables 1** and **2**, the PER curves for the type-II, III HARQ can be calculated from:

$$\begin{aligned} P_{0,M} &= a_{0,M} e^{(-g_{0,M}\gamma)} \\ P_{R_{\max}-1,M} &= a_{R_{\max}-1,M} e^{(-g_{R_{\max}-1,M}\gamma)} \end{aligned} \tag{4}$$

$P_{0,M}$ is the PER for the first transmission, $P_{R_{\max}-1,M}$ is the PER for the $R_{\max}$ th transmission, the index M is the AMC mode.

Equation (5) must be valid, in order to satisfy that the PER is lower than the packet loss probability P_{loss} after the $R_{\max}$ transmissions at LL:

$$\begin{aligned} &\sqrt[R_{\max}]{a_{0,M} \cdots a_{R_{\max}-1,M}}\, e^{\left[\frac{(g_{0,M}+\cdots+g_{R_{\max}-1,M})\gamma}{R_{\max}}\right]} \\ &\le P_{\text{loss}}^{\left(\frac{1}{R_{\max}}\right)} := P_{\text{target}}\left(R_{\max}\right) \end{aligned} \tag{5}$$

Then, the average PER for each mode is given by:

$$\overline{\text{PER}}_M\left(\gamma, R_{\max}\right) = \overline{a}_M e^{(-\overline{g}_M\gamma)} \tag{6}$$

where,

$$\begin{aligned} \overline{a}_M\left(R_{\max}\right) &= \sqrt[R_{\max}]{a_{0,M} \cdots a_{R_{\max}-1,M}} \\ \overline{g}_M\left(R_{\max}\right) &= \frac{\left(g_{0,M} + \cdots + g_{R_{\max}-1,M}\right)\gamma}{R_{\max}} \end{aligned} \tag{7}$$

For the type-I HARQ, in order to satisfy the PER requirement at LL, $P_M^{R_{\max}} \le P_{\text{loss}}^{1/R_{\max}}$, then we get

$$P_M \le P_{\text{loss}}^{(1/R_{\max})} := P_{\text{target}} \tag{8}$$

2.3. Different AMC Mode Interval Calculation

The parameter $P_{\text{target}}(R_{\max})$ is a key parameter transferred from LL to PHY. $P_{\text{target}}(R_{\max})$ transfers the PER constraint P_{loss} and the time delay requirement $R_{\max}$ to PHY, in order to control the determination of the intervals for different modulation and coding schemes. The intervals are dynamic because of two reasons: first, $P_{\text{target}}(R_{\max})$ important factor to determine the intervals and it is affected by P_{loss} and $R_{\max}$; second, PER curve in PHY for each transmission mode $\overline{\text{PER}}_M\left(\gamma, R_{\max}\right)$ are affected by $R_{\max}$. For type-I HARQ, P_M is not affected by $R_{\max}$. For type-II, and type-III HARQ, P_M decreases as $R_{\max}$ increases. The intervals of different AMC modes can be calculated from the expression [2,6]:

$$\begin{aligned} \gamma_M &= 1/\overline{g}_M \ln\left(\overline{a}_M / P_{\text{target}}\right), M = 1,2,\cdots,M^* \\ \gamma_M &= +\infty \end{aligned} \tag{9}$$

By substituting by $\overline{a}_M\left(R_{\max}\right)$, $\overline{g}_n\left(R_{\max}\right)$ for (type-II & III) and P_{target} into the expression (9), we could figure out the intervals of type-I, II, III HARQ for different transmission modes, when the maximum number of transmission is $R_{\max}$ and the maximum PER probability is P_{loss}.

2.4. ACM Model Selection

At PHY layer, when SNR $\gamma \in [\gamma_M, \gamma_{M+1})$, the mode M modulation and coding will be chosen. So if we put Equation (1) into the expression below, then we will choose mode M with probability [2,6]:

$$\begin{aligned} \pi_i &= P_\gamma\left(M\right) = \int_{\gamma_M}^{\gamma_{M+1}} P_\gamma\left(\gamma\right) \mathrm{d}\gamma \\ &= \frac{\Gamma\left(m, \frac{m\gamma_M}{\gamma}\right) - \Gamma\left(m, \frac{m\gamma_{M+1}}{\gamma}\right)}{\Gamma(m)} \end{aligned} \tag{10}$$

where, $\Gamma(m,x) := \int_x^\infty t^{m-1} e^{-t} \mathrm{d}t$ is complementary in complete Gamma function.

3. Subcarrier and Power Allocation for the Downlink

By considering a downlink OFDM system with U users, it's assumed that each subcarrier is occupied by only one user [11]. With cross-layer optimization, the QoS information is transferred from the traffic controller at the MAC layer to the subcarrier, and power controller at the PHY layer for resource allocation. Then, the resource allocation results are fed-back to the traffic controller in the base station for scheduling of the data to be sent out in each slot [8].

In consistence with [9], this paper assumes a total bandwidth of B shared by N subcarriers, and the OFDM signaling is time slotted. Each slot has duration of T_{slot}.

Let, Ω_u denotes the index set of subcarriers allocated to user u $(u = 1, \cdots, U)$. Let $p_{u,n}$ be the power allocated to user u on subcarrier n; ($n \in \Omega_u$). $h_{u,n}$ is the corresponding channel gain. N_o is the power spectral density of Additive White Gaussian Noise (AWGN). By assuming perfect channel estimation, the achievable data rate of user u on subcarrier n is expressed as [9]:

$$R_{u,n} = \frac{B}{N}\log_2\left(1 + p_{u,n}\varphi_{u,n}\right) \tag{11}$$

where,

$$\varphi_{u,n} = \frac{|h_{u,n}|^2}{N_o B/N} \tag{12}$$

is the channel-to-noise power ratio for user u on subcarrier n.

Therefore, the total data rate of user u is given by:

$$R_u = \sum_{n\in\Omega_u} R_{u,n} \tag{13}$$

A resource allocation scheme had been proposed in [2], to maximize the weighted sum of all users' capacity J, *i.e.*, to maximize

$$J = \sum_{u=1}^{U} W_u R_u \tag{14}$$

Subject to: $p_{u,n} \geq 0, \sum_{u=1}^{U}\sum_{n\in\Omega_u} p_{u,n} \leq P_{\text{total}}$

$\Omega_i \cap \Omega_j = \varnothing$ $(i \neq j)$ and,

$\Omega_1 \cup \Omega_2 \cup \cdots \cup \Omega_U \subseteq \{1, 2, \cdots, N\}$,

where, W_u denotes the weight for user u, which is evaluated based on the cross-layer optimization criteria. So it indicates the QoS information for user u, and is obtained from the result of data scheduling at the MAC layer as will be investigated in Section 4 and P_{total} denotes the total power.

The optimization process is divided into two stages namely, subcarrier allocation and power allocation.

3.1. Optimal MWC Based Subcarrier Allocation

For simplicity, by assuming uniform power allocation across all subcarriers, *i.e.*, each subcarrier is allocated a power $p = P_{\text{total}}/N$. Optimal subcarrier allocation leads to the maximum cost function, which is denoted by $J_{\max}$. If an arbitrary subcarrier n allocated to user u $(u = 1, \cdots, U)$ with optimal subcarrier allocation is now reassigned to user s, let J' denote the resulting cost function. The difference between the two cost functions is given by [9]:

$$J_{\max} - J' = W_u R_{u,n} - W_s R_{s,n} \geq 0 \tag{15}$$

Substituting (11) into (15), we have

$$\frac{W_u}{W_s} \geq \frac{\log_2\left(1 + p\varphi_{s,n}\right)}{\log_2\left(1 + p\varphi_{u,n}\right)} \tag{16}$$

which implies that, with optimal subcarrier allocation, subcarrier n should be allocated to user u rather than users if Equation (16) is satisfied. However, it is prohibitively complex to perform optimal subcarrier allocation with a large number of subcarriers. Therefore, a suboptimal scheme is desired.

- *Suboptimal MWC based subcarrier allocation*

Intuitively, to maximize the cost function in Equation (14), a larger weight demands a higher data rate. Let R_u/W_u denote the Rate-to-Weight Ratio (RWR) and assuming uniform power allocation across all subcarriers. Then, the following suboptimal subcarrier allocation scheme is employed, where the user with the lowest RWR is allowed to pick subcarriers in each of iteration [9]:

1) Initialization:

a) Set R_u =0, $\Omega_u = \varnothing$ for all u $(u = 1, \cdots, U)$, sort W_u in the descending order, and let $L = \{1, 2, \cdots, N\}$ denote the set of unallocated subcarriers.

For u = 1 to U:

b) If $\varphi_{u,m} \geq \varphi_{u,n}$ $(m, n \in L;\ m \neq n)$, assign subcarrier m to user u, *i.e.*, add subcarrier m to Ω_u. Remove subcarrier m from L. Update R_u according to Equation (5).

2) Find the minimum R_u/W_u $(k = 1, \cdots, U)$, and repeat 1 b) for the corresponding user u.

3) Repeat 2) until $L = \varnothing$.

3.2. Optimal MWC Based Power Allocation

Following subcarrier allocation, the optimal power allocation for each user can be obtained by using the Lagrange multiplier, *i.e.*, (14) can be rewritten as:

$$J = \sum_{u=1}^{U} W_u R_u + \lambda\left(\sum_{u=1}^{U}\sum_{n\in\Omega_u} p_{u,n} - P_{\text{Total}}\right) \tag{17}$$

Subject to $\sum_{u=1}^{U}\sum_{n\in\Omega_u} p_{u,n} = P_{\text{Total}}$ and $p_{u,n} \geq 0$. Letting $\partial J/\partial p_{u,n} = 0$, the optimal solution for $p_{u,n}$ is given by:

$$p_{u,n} = \max\left\{\frac{W_{u,n}}{\sum_{u=1}^{U}\left(W_i \text{ size of } (\Omega_i)\right)}\left[P_{\text{Total}} + \sum_{i=1}^{U}\sum_{q\in\Omega_j}\frac{1}{\varphi_{i,q}}\right] - \frac{1}{\varphi_{u,n}}, 0\right\} \tag{18}$$

Consider users j and u ($j, u \in \{1, \cdots, U\}$), and subcarrier $m \in \Omega_j$ and $n \in \Omega_u$ are two arbitrary subcarriers allocated to users j and u, respectively. By $\partial J/\partial p_{u,n} = \partial J/\partial p_{i,m} = 0$, we can drive:

$$\frac{W_i}{W_j} = \frac{\varphi_{u,n}\left(1 + p_{j,m}\varphi_{j,m}\right)}{\varphi_{j,m}\left(1 + p_{u,n}\varphi_{u,n}\right)} \tag{19}$$

1) If $W_j/W_u = 1$, from (19) it can be derived that:

$$p_{j,m} - p_{u,n} = \frac{\varphi_{j,m} - \varphi_{u,n}}{\varphi_{j,m}\varphi_{u,n}} \quad (20)$$

Equation (20) implies that with equal weight, the subcarrier with a better channel gain is allocated more power, which is the same result as water-filling [12].

2) If $W_j/W_u > 1$, we have:

$$p_{j,m} - p_{u,n} > \frac{\varphi_{j,m} - \varphi_{u,n}}{\varphi_{j,m}\varphi_{u,n}} \quad (21)$$

Equation (21) implies that the subcarrier corresponding to the user with a higher weight is allocated more power than the case using water-filling.

4. Packet Error Rate Based Data Scheduling

After receiving the resource allocation results from the PHY layer, which indicates the amount of data allowed for each user, the MAC layer performs scheduling for each batch of data to be sent out. This paper proposes a novel scheduling scheme, which assigns the weight depending on the average PER in each ACM mode, after the intervals of different ACM modes are calculated from Equation (9).

The weight W_u denotes the weight of the batch of user u corresponding to the ACM mode M_i, which is given by:

$$W_u = \begin{cases} \frac{1}{U} & \text{for PER} < \delta \text{ PER max} \\ \frac{1}{T_i} & \text{for PER} \geq \delta \text{ PER max} \end{cases} \quad (22)$$

where, T_i is the total number of users in ACM mode i, and $\delta \leq 1$.

Equation (22) gives higher priority to the user whose batch of data has the maximum PER.

To give the higher priority to the user whose batch of data has the minimum PER, the following equation may be used:

$$W_u = \begin{cases} \frac{1}{U} & \text{for PER} \geq \delta \text{ PER max} \\ \frac{1}{T_i} & \text{for PER} < \delta \text{ PER max} \end{cases} \quad (23)$$

5. Numerical Results and Analysis

In order to investigate the proposed cross-layer optimization scheme, numerical results will be deduced for HARQ schemes over a Nakagami-m fading channel. The data unit is assumed to be one frame, and each frame contains a fixed number of symbols. The receiving buffer can accurately detect the received γ and decides the next modulation scheme, then feedback channel state message to the source node; on the other hand, if the destination node cannot decode the received data packet correctly, it feeds back NACK signaling.

5.1. SNR Margins for Different HARQ Modes

In order that, the different modulation and coding schemes is assessed in the case of having only the AWGN. So, **Figure 2** presents the obtained probability of PER against the apparent SNR for different AMC modes.

Figures 2 and **3** show the case of having different SNR margins to sustain the PER at a certain range. The obtained results in **Figures 2** and **3** are in consistence with those published in many references as [1,2].

The retransmission technique depends on the used HARQ scheme. Also, it is assumed that the transmissions of all signaling messages are assumed to be error free, maximum number of retransmissions, $R_{\max} = 4$ and the ACM modes $M = 6$.

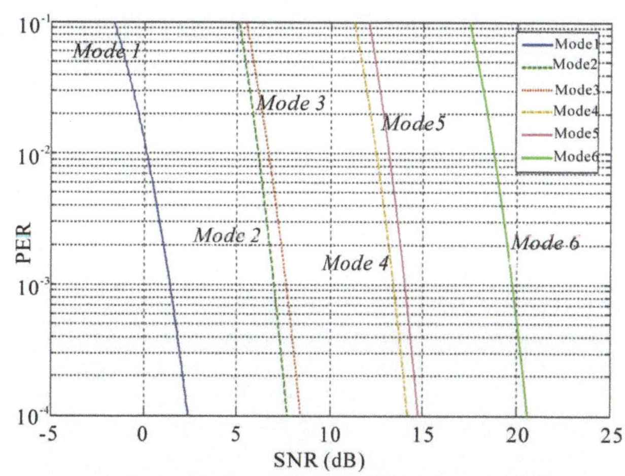

Figure 2. Different ACM modes for type II HARQ.

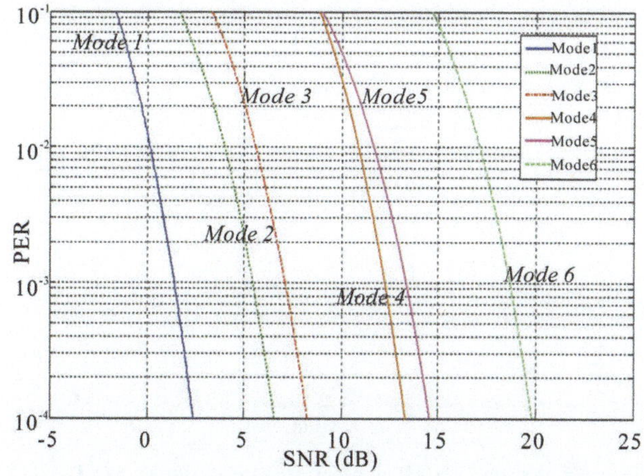

Figure 3. Different ACM modes for type III HARQ.

5.2. Effect of Proposed PER Scheduling Criteria on the System Capacity

This paper employs the proposed MWC based resource allocation at the PHY layer and the PER based scheduling at the MAC layer for consistency. It considers a LTE system listed in [13], with $U = 100$ users, a total transmit power of $P_{\text{total}} = 1$ W, a total bandwidth of $B = 10$ MHz, subcarrier width of 15 KHz, physical source block code of 180 KHz bandwidth and 50 available Physical Resource Block (PRB). The power spectral density of AWGN is $N_0 = -80$ dBW/Hz [9] and δ in Equations (22) and (23) is taken to be 0.8.

Figures 4 and **5** show the obtained throughput for each mode by deploying the minimum PER using Equation (23) after running different number of iterations with different HARQ techniques. Whereas, **Figures 6** and **7** show the obtained throughput for each mode by deploying the maximum PER using Equation (22) after running different number of iterations with different HARQ techniques.

So, it may be noticed that, the PER maximum as well as the minimum margins are closely estimating the validation range for each AMC mode separately. This may be explained as a result of having good and stable CLD optimization for different AMC modes with different HARQ.

5.3. Effect of Service Requests on HARQ-Type III

Figures 8-10, are investigating the effect of having more customers talking on the same Enhanced e-NodeB site and contenting upon the available throughput.

So, these figures show that by increasing the number of customers, it's noticed that there's a steady increase in the required SNR to achieve the same throughput. This may be explained by that the system will be more starving to the power in order to overwhelm the existing interference which arises from the existing customers.

5.4. LTE System Capacity with Proposed CLD Based PER Scheduling

In **Figures 11-14**, it is shown that for LTE system parameters, HARQ type III will have much better performance (aggregated throughput) than that of HARQ type II. This

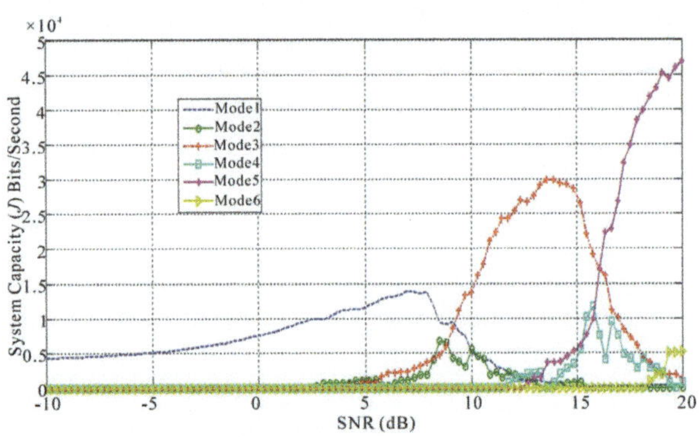

Figure 4. Average SNR against system capacity (*J*) HARQ type II-minimum PER.

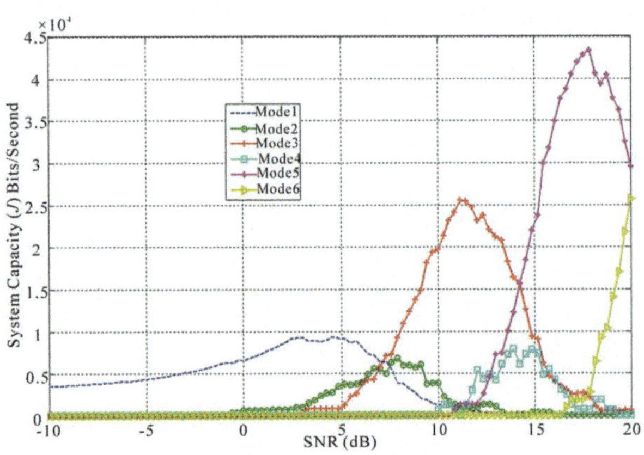

Figure 5. Average SNR against system capacity (*J*) HARQ type III-minimum PER.

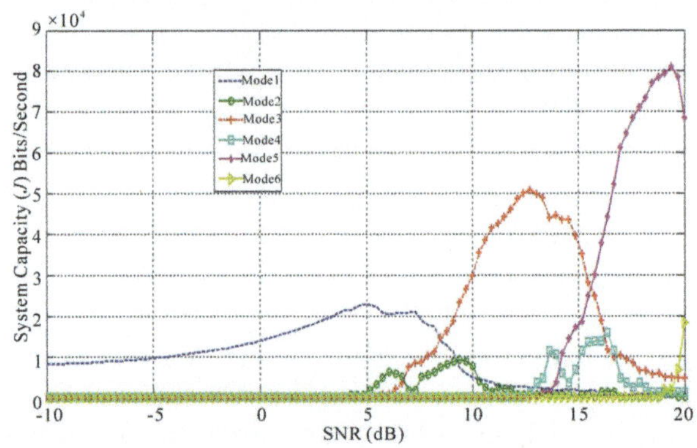

Figure 6. Average SNR against system capacity (*J*) HARQ type II-maximum PER.

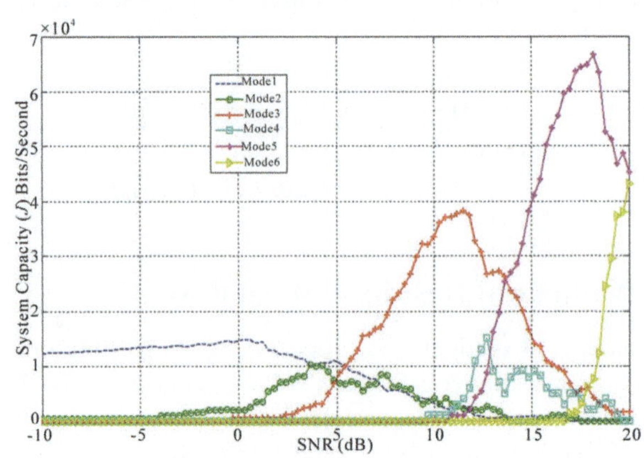

Figure 7. Average SNR against system capacity (*J*) HARQ type III-maximum PE.

may be explained by that the HARQ type III will have higher coding gain so, the obtained throughput is improved and can be achieved in lower SNR (compared to HARQ type II required SNR).

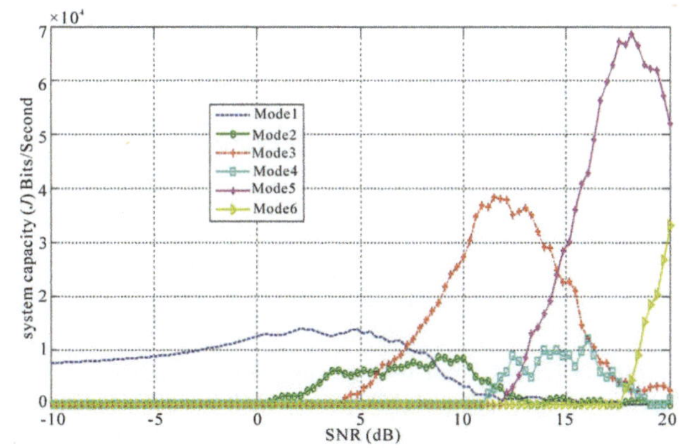

Figure 8. Average SNR against system capacity (*J*) K = 100-HARQ type III.

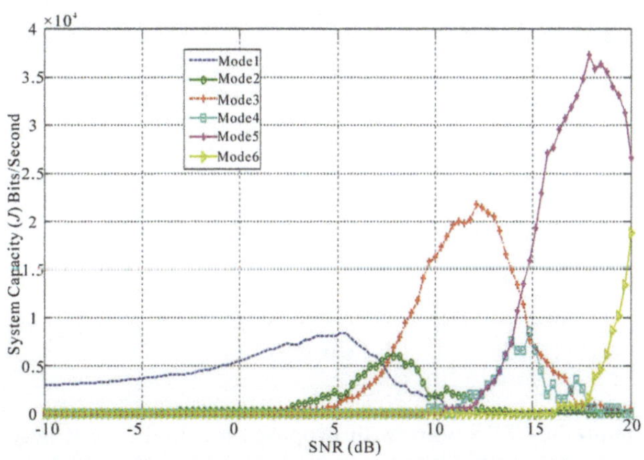

Figure 9. Average SNR against system capacity (*J*) K = 200-HARQ type III.

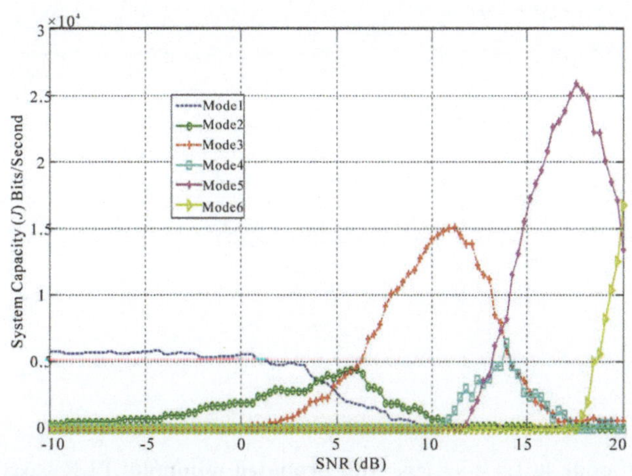

Figure 10. Average SNR against system capacity (*J*) K = 300-HARQ type III.

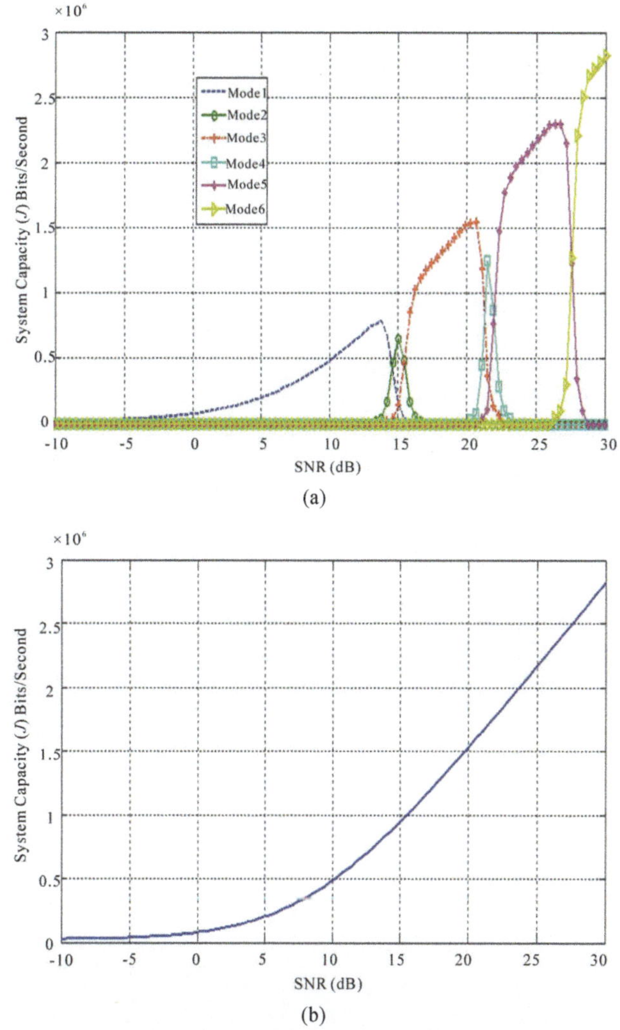

Figure 11. (a) Average SNR against system capacity (*J*) for each mode in LTE system with proposed maximum PER based scheduling-HARQ type II; (b) Average SNR against aggregated system capacity (*J*) in LTE system with proposed maximum PER based scheduling-HARQ type II.

6. Conclusion

A novel optimized cross-layer design (CLD) for AMC/HARQ wireless networks was presented. In this paper, MWC resource allocation at PHY layer is used. In addition, the new PER based data scheduling is proposed in order to investigate the aggregated throughput, after developing AMC/HARQ schemes. Also, this paper introduces an analysis and complete investigation for breaking down of different AMC modes in the same HARQ scheme. The concurrent paper aims to put in the spot the relation between the dominant operational ACM modes in conjunction with the available SNR, which will be useful for practical implementation and deployment for the cross-layer design of certain operator or service provider.

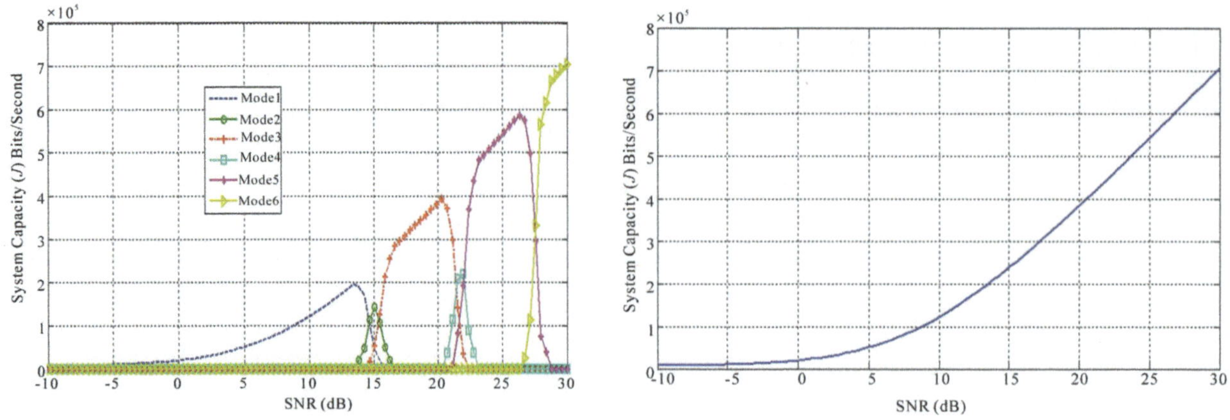

(a) (b)

Figure 12. (a) Average SNR against system capacity (*J*) for each mode in LTE system with proposed minimum PER based scheduling-HARQ type II; (b) Average SNR against aggregated system capacity (*J*) in LTE system with proposed minimum PER based scheduling-HARQ type II.

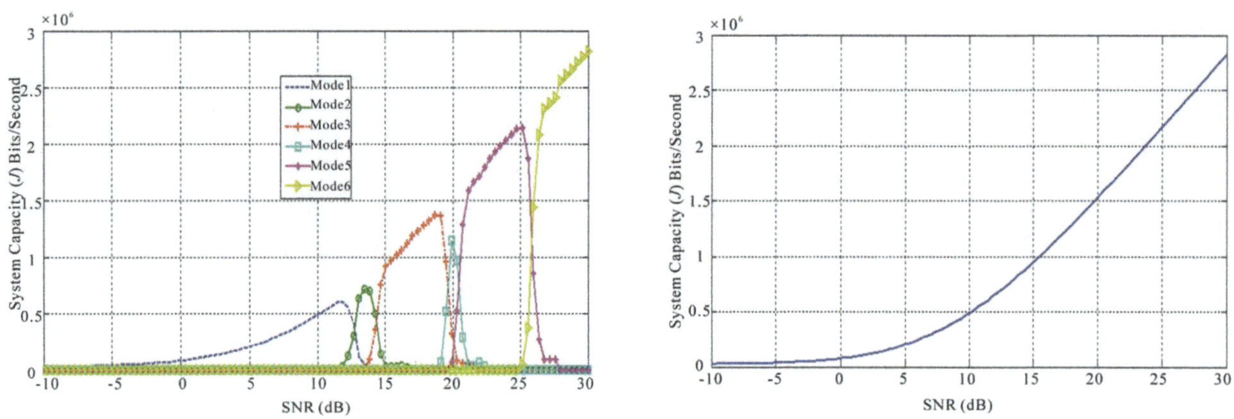

(a) (b)

Figure 13. (a) Average SNR against system capacity (*J*) for each mode in LTE system with proposed maximum PER based scheduling-HARQ type III; (b) Average SNR against aggregated system capacity (*J*) for each mode in LTE system with proposed maximum PER based scheduling-HARQ type III.

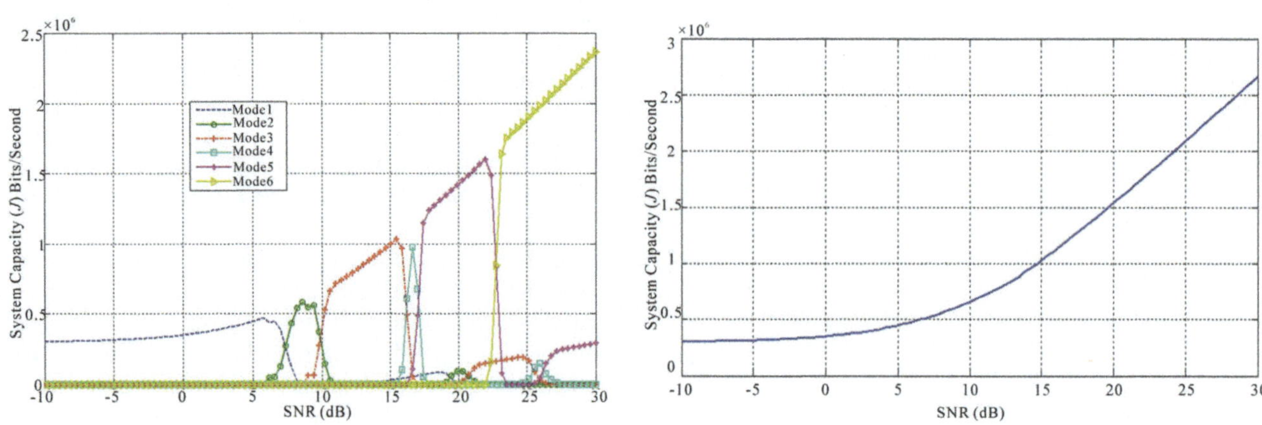

(a) (b)

Figure 14. (a) Average SNR against system capacity (*J*) for each mode in LTE system with proposed minimum PER based scheduling-HARQ type III; (b) Average SNR against aggregated system capacity in LTE system with proposed minimum PER based scheduling-HARQ type III.

REFERENCES

[1] C.-Q. Dai, Y. Rao, L. Jiang, Q.-B. Chen and Q. Huang, "Cross-Layer Design of Combining HARQ with Adaptive Modulation and Coding for Nakagami-m Fading Channels," *Journal of Communications, Academy Publisher*, Vol. 7, No. 6, 2012, pp. 458-463.

[2] C. C. Li, L. L. Xu and D. F. Yuan, "Performance Analysis and Comparison of HARQ Schemes in Cross Layer Design," *Fourth International Conference on Communications and Networking in China*, Ji'nan, 26-28 August 2009, pp. 1-5.

[3] S. Marcille, P. Ciblat and C. J. Le Martret, "A Cross-Layer HARQ scheme Robust to Imperfect Feedback," 2012 *Conference Record of the Forty Sixth Asilomar Conference on Signals, Systems and Computers*, 4-7 November 2012, pp. 143-147.

[4] J. Ramis, G. Femenias, F. Riera-Palou and L. Carrasco, "Cross-Layer Optimization of Adaptive Multi-Rate Wireless Networks Using Truncated Chase Combining HARQ," *Global Telecommunications Conference*, Palma de Mallorca, 6-10 December 2010, pp. 1-6.

[5] S. Kallel and D. Haccoun, "Generalized Type II Hybrid ARQ Scheme Using Punctured Convolutional Coding," *IEEE Transactions on Communications*, Vol. 38, No. 11, 1990, pp. 1938-1946.

[6] F. Shi, D. Yuan and L. Xu, "Performance Analysis of Type-III HARQ Scheme in Cross-Layer Design for Quos-Guaranteed Traffic," *International Workshop on Cross Layer Design*, Ji'nan, 20-21 September 2007, pp. 133-137.

[7] A. M. Cipriano, P. Gagneuri, G. Vivier and S. Sezginer, "Overview of ARQ and HARQ in Beyond 3G System," *IEEE 21st International Symposium on Personal, Indoor and Mobile Radio Communications Workshops*, Colombes, 26-30 September 2010, pp. 424-429.

[8] N. Zhou, X. Zhu, Y. Huang and H. Lin, "Novel Batch Dependant Cross-Layer Scheduling for Multiuser OFDM Systems," *IEEE International Conference*, Beijing, 19-23 May 2008, pp. 3878-3882.

[9] N. Zhou, X. Zhu and Y. Huang, "Cross-Layer Optimization with Guaranteed QoS for Wireless Multiuser OFDM Systems," *IEEE 18th International Symposium on Personal, Indoor and Mobile Radio Communications*, Athens, 3-7 September 2007, pp. 1-5.

[10] X. Wang, Q. Liu and G. B. Giannakis, "Analyzing and Optimizing Adaptive Modulation Coding Jointly with ARQ for QoS-Guaranteed Traffic," *IEEE Transactions on Vehicular Technology*, Vol. 56, No. 2, 2007, pp. 710-720.

[11] Z. Shen, J. G. Andrews and B. L. Evans, "Adaptive Resource Allocation in Multiuser OFDM Systems with Proportional Rate Constraints," *IEEE Transactions on Wireless Communications*, Vol. 4, No. 6, 2005, pp. 2726-2737.

[12] T. M. Cover and J. A. Thomas, "Elements of Information Theory," Wiley, New York, 1991.

[13] "3GPP Long Term Evolution: System Overview, Product Development and Test Challenges," *Agilent Technology*, 2009.
http://cp.litcraturc.agilcnt.com/litwcb/pdf/5989-8139EN.pdf

11

Spectrum Sensing and AM-FM Decomposition through Synchrosqueezing

K. Vandhana, P. V. S. Sowmya, P. Roshni, K. Divya, S. Ashwin, K. A. Narayanankutty

Department of Electronics and Communication Engineering, Amrita Vishwa Vidyapeetham, Coimbatore, India.

ABSTRACT

In this paper we have accomplished one of the tasks of cognitive radio *i.e.* dynamic spectrum sensing by using wavelet based Synchrosqueezing transform [1], a novel technique, which was proposed to analyze a signal in time-frequency plane. The distinctive feature of this transform compared to other techniques is that it enables us to decompose amplitude and frequency modulated signals and allows individual reconstruction of these components. The objective is also to separate the occupied band into amplitude modulated and frequency modulated bands.

Keywords: Cognitive Radio; Synchrosqueezing; AM-FM; Spectrum Sensing; Time-Frequency Reassignment; Empirical Mode Decomposition

1. Introduction

The tremendous growth of wireless data use in our day-to-day life piloted the problem of spectrum scarcity. Recent developments in the line of wireless networks have resulted in huge leap which diminishes underutilization of spectrum. Cognitive radio (CR) [2] originates as an alluring solution to this issue. Yesteryears had witnessed fixed allocation of bands but modern methods in cognitive radio have resulted in dynamic allocation of the unused licensed bands. Prime concern of cognitive radio is spectrum sensing.

Spectrum Sensing

Spectrum sensing in cognitive radio ensures the efficient utilization of spectral holes in the licensed frequency band. This feature enables a secondary (unlicensed) user to avail the spectrum holes without causing interference to a primary (licensed) user. It has got additional functions which find the bandwidth, waveform, modulation and carrier frequency of the signal that occupied the spectrum. There are three different approaches in which spectrum sensing can be done namely:

1) Transmitter detection: Detects the existence of a signal in a particular spectrum. e.g. Energy based sensing, matched filtering.

2) Cooperative detection: Detects primary users through the information obtained from various cognitive radios. e.g. Centralized sensing, Distributed sensing.

3) Interference based detection: Interference temperature level restricts the coexistence of secondary and primary users in a CR system.

The second half of the paper handles the decomposition of amplitude modulated and frequency modulated signals. Previous works done [3,4] on AM-FM decomposition deal with separating AM and FM harmonics presented in audio signals which contribute to applications like audio coding, singer segregation etc. On the other hand, we have realized AM or FM modulated schemes used by the primary users and also their corresponding bandwidths.

The overall essence of this paper is to appraise the concept of cognitive radio's spectrum sensing, and specifically, on the establishment of an innovative Synchrosqueezing transform on this particular spectrum. This fetches us a more appropriate conclusion that shows the distinction in AM and FM signal, from which we can deduce the range of bandwidth of AM and FM signals. The contents are arranged as follows. Section 2 deals with brief history in spectrum sensing, Section 3 deals with Synchrosqueezing transform and its algorithm, Section 4 shows the results of spectrum sensing and AM-FM decomposition, Section 5 includes discussion of the scope of the concept and conclusion.

2. Spectrum Sensing through Ages [5]

Spectrum sensing can be done on the basis of several methods. Let us take a peep on the methods through the ages. Initially Energy based Detection was popular as it required no information about primary users. The detection was performed by comparing the output of energy detector with a previously set threshold value. Energy based detection method faces problems in deciding the threshold, its incapability to discriminate between primary user and noise. This method is not favorable when the noise is nonstationary. Cyclostationarity paved the way in utilizing the features like periodicity of received signals. Channel fading influences cyclostationarity approach on a large scale making it prone to offsets and losing cyclostationary features of received signals. In matched filter it obtains an optimal solution for signal detection where the transmitted signal is known. This method requires information regarding operating frequency, bandwidth, modulation type etc. But implementing receivers for all those signal types make it complex and impractical. Hilbert Huang Transform is a combination of EMD [6] (Empirical Mode Decomposition) and HSA (Hilbert Spectral Analysis) for decomposing a signal and obtaining its instantaneous frequency for non-stationary and nonlinear data. Major limitation in this method is that it cannot distinguish different components in narrow band signals. To overcome this limitation a new concept of Synchrosqueezing Transform (SST) method of time-frequency analysis, where it can differentiate closely spaced components in frequency domain. The following section gives us an insight to the SST.

3. Synchrosqueezing Transform [1,7]

Analysis of time series data primarily aims at separating meaningful components, and knowing the structure and to extrapolate for predictions. Nonstationary real time signals that are observed in the application of any scientific work are analyzed using time-frequency transforms such as the short-time Fourier transform (STFT), the Wigner-Ville distribution [8], and the continuous wavelet transform (CWT) [9]. To comprehend more clearly about the signal, the classical Empirical mode decomposition is used. Although EMD helps us to understand this concept, and able to deal with nonstationary signals, its basic mathematical explanations are fuzzy. This is where Synchrosqueezing comes in handy, as Synchrosqueezing algorithm extracts and explains the constituent oscillatory components obtained by the decomposition of the signal. This transform comes under the clan of time-frequency reassignment (TFR) techniques, but its uniqueness is that reassignment is done in frequency domain unlike other TFR techniques basically used in time domain.

Given any signal, $s(t)$, which is always a sum of amplitude modulated and frequency modulated signals. *i.e.*,

$$s(t)=\sum_{k=1}^{K} s_k(t) \tag{1}$$

where,

$$s_k(t)=A_k(t)\cos(\phi_k(t)) \tag{2}$$

is an intrinsic mode type function (IMF) with an amplitude $A_k(t)$ and frequency $\Phi'(t)$. These IMFs are well separated in frequency domain and are obtained by means of Synchrosqueezing.

Algorithm

a) Compute the CWT of the signal $s(t)$.

$$W(a,b)=\int s(t)\left(1/\sqrt{a}\right)\overline{\psi((t-b)/a)}\,\mathrm{d}t \tag{3}$$

where $\psi(t)$ is mother wavelet chosen, a is scaling factor of mother wavelet, b is the translating factor in time axis.

b) In order to convert time-scale plane to time-frequency plane, candidate instantaneous frequency is obtained by,

$$\omega_s(a,b)=-i\left(W_s(a,b)\right)^{-1}(\partial/\partial b)W_s(a,b) \tag{4}$$

where $\omega_s(a,b)$ is the instantaneous frequency.

c) The obtained instantaneous frequencies are divided into different frequency regions such that $\omega_l-\omega_{l-1}=\Delta\omega$, where ω_l is the center frequency. Synchrosqueezed transform is obtained by assigning the net contributions of $\omega_s(a,b)$ in these frequency regions to its corresponding center frequencies.

$$T_f(\omega_l,b) \\ =(\Delta\omega)^{-1}\sum_{a_k:|\omega(a_k,b)-\omega l|\le\Delta\omega/2} W_s(a_k,b)a_k^{-3/2}(\Delta a)_k \tag{5}$$

where a_k's are discrete values of a for which the transform values are calculated and $a_k-a_{k-1}=(\Delta a)_k$.

Hence time-scale plane is converted to time-frequency plane by this reassignment technique thereby enhancing the resolution of TF plane which facilitates the easy extraction of the IMF components.

The primary requirement in demodulation of a signal is to get amplitude, phase and instantaneous frequency information. Synchrosqueezing takes advantage of its reassignment principle to overcome the noise interference of signal (non-linear and non-stationary). In the reassignment mode, Synchrosqueezing uses superposition of a class of functions that slowly varies with respect to the instantaneous frequencies.

Reference [10] summarizes a bi-dimensional version of Synchrosqueezing transform on images, this enables to decompose its AM-FM harmonics. This paper focuses on separation of AM-FM components of 1-D signal. Decomposition can be done in either of the two categories where AM and FM modulated signals co-exist or exist in different bands.

4. Results

4.1. Spectrum Sensing

In this paper we have taken a wideband signal of bandwidth 2.5 GHz ranging from 750 MHz to 3.25 GHz with a noise power of −3 db as input signal shown in **Figure 1**. Applying Synchrosqueezing transform on this signal yields a sharpened TF plot as shown in **Figure 2**. It is evident from this figure that Synchrosqueezing transform gives us finite thread like instantaneous frequency curves in spite of the noise presented in the signal. **Figure 3** depicts the presence and absence of frequencies in the signal as 1 representing the former and 0 representing the latter.

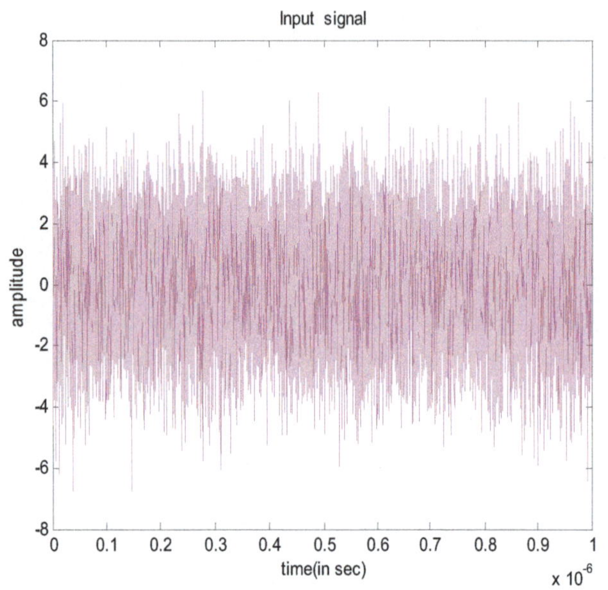

Figure 1. Input signal with noise power of −3 dB.

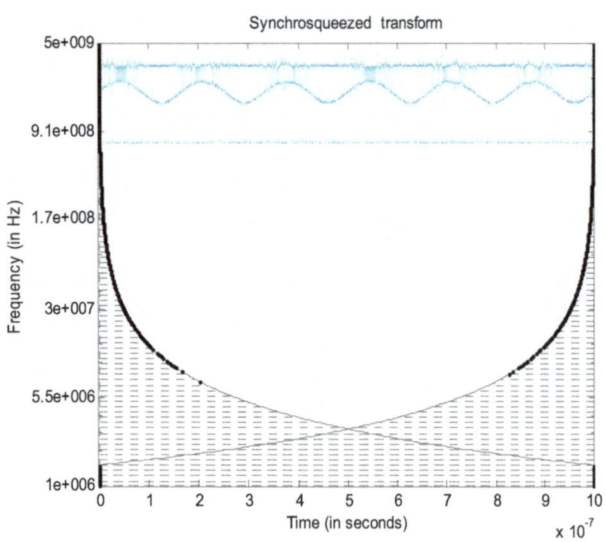

Figure 2. Synchrosqueezed transform of the input signal.

4.2. AM-FM Decomposition

1) Non Overlapping Case:

The input signal taken here is same as that of spectrum sensing (**Figure 1**) which consists of AM and FM band. In **Figure 3** the wider band corresponds to FM modulated components and the other two bands correspond to AM modulated components. Extraxcted curves (**Figure 4**) from **Figure 2** portray the IMFs presented in the signal. **Figures 4(a)** and **(c)** represent AM components and **Figure 4(b)** represents FM component in the signal. **Figure 5** shows the occupancy of FM band (1.5 - 2.4 GHz) and **Figure 6** shows the occupancy of AM band (750 MHz & 3.25 GHz).

2) Overlapping Case:

This is the worst case scenario of AM-FM decomposi-

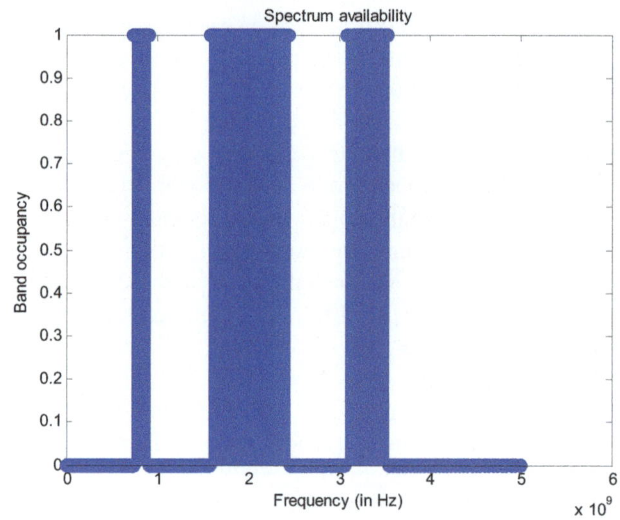

Figure 3. Spectrum occupied of the input for efficient usage.

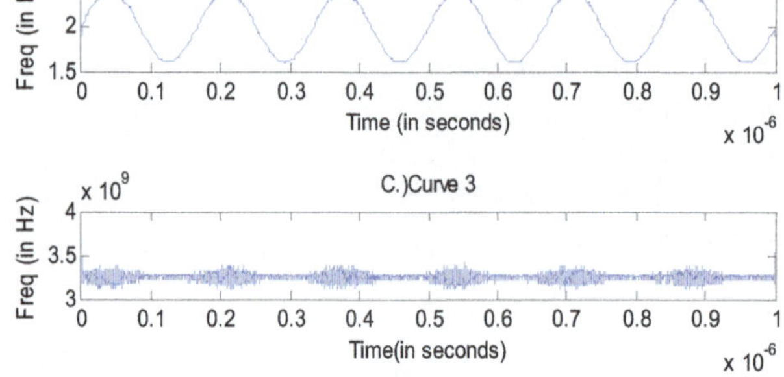

Figure 4. Extracted curves from the synchrosqueezed transform (Figure 2).

tion. Despite the hindrance, we can still distinguish AM and FM components. The bandwidth of the input signal (**Figure 7**) is 8 GHz ranging from 2.8 GHz to 10.8 GHz. The results of this case are shown in **Figures 8-11**.

5. Conclusion

The crux of this paper reveals us an interesting develop ment in time-frequency reassignment technique, which is in Synchrosqueezing. The involvement of Synchrosqueezing in spectrum sensing delivers us with sharpened TF image of the signal. Co-existence of FM and AM signals in a specific domain can enhance the value of spectral utility. The paper also reveals the present status of AM-FM decomposition where it will be beneficial for future references and for more complex modulation

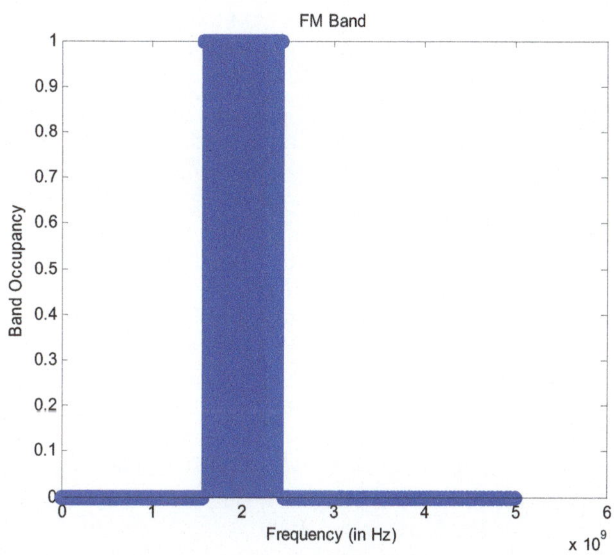

Figure 5. Bandwidth occupied by FM.

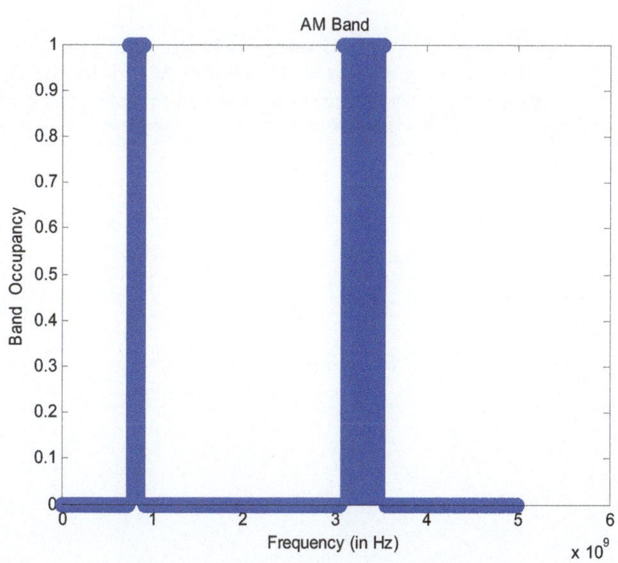

Figure 6. Bandwidth occupied by AM.

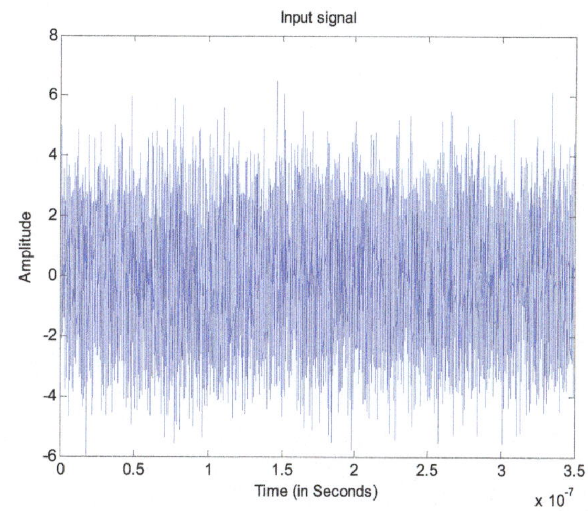

Figure 7. Input signal with noise power −3 dB (for overlapping case).

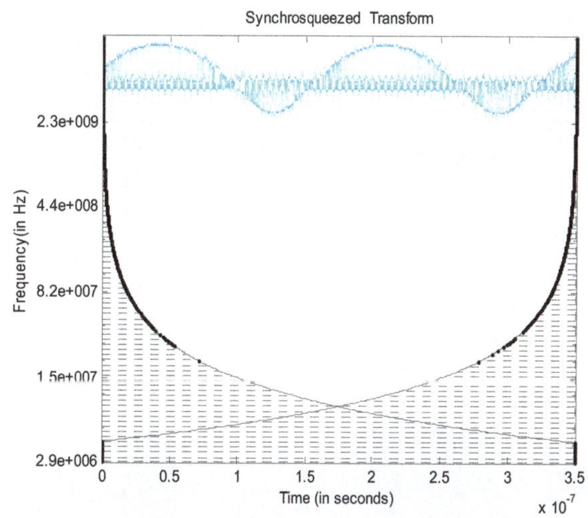

Figure 8. Synchrosqueezed transform of the input signal (Figure 7).

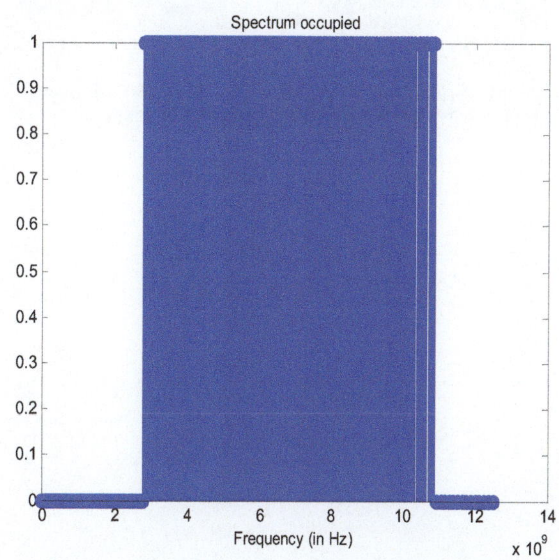

Figure 9. Spectrum occupied by input signal.

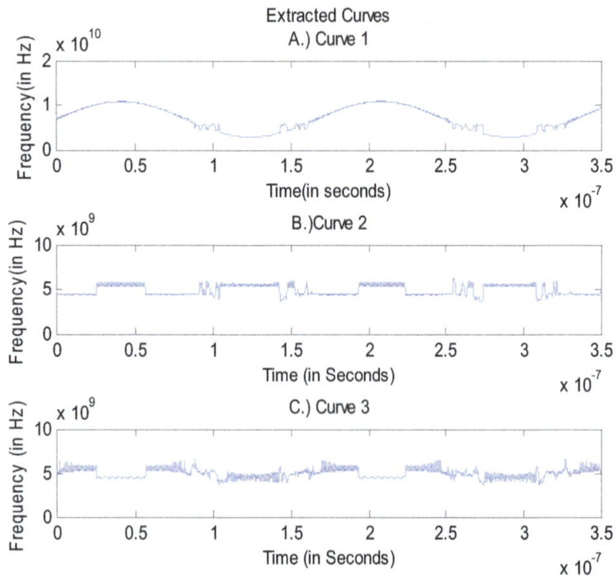

Figure 10. Extracted curves from synchrosqueezed transform.

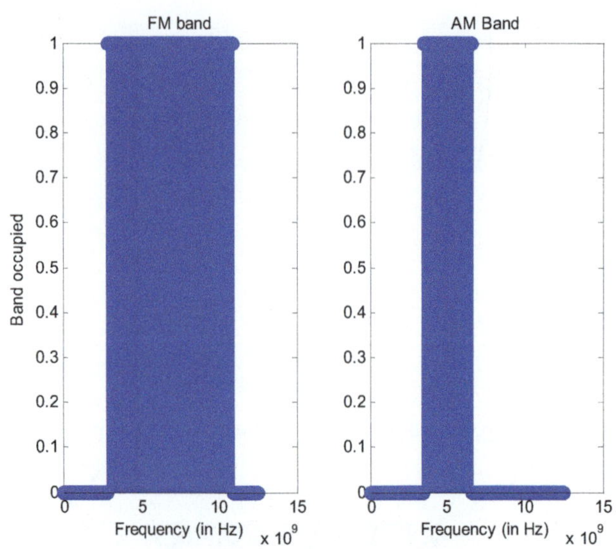

Figure 11. Spectrum occupied by FM (left) of band 2.8 - 10.8 GHz and AM (right) of band 4.5 - 5.5 GHz.

techniques. This also can act as evidentiary support for futuristic cognitive radio ideas in areas of research and development.

REFERENCES

[1] I. Daubechies, J. F. Lu1 and H.-T. Wu, "Synchrosqueezed Wavelet Transforms: An Empirical Mode Decomposition-Like Tool," 25 July 2010.

[2] S. Haykin and L. Fellow, "Cognitive Radio: Brain Empowered Wireless Communications," *IEEE Journal*, Vol. 23, No. 2, 2005.

[3] Gianfelici, F. Biagetti, G. Crippa and P. Turchetti, "AM-FM Decomposition of Speech Signals: An Asymptotically Exact Approach Based on the Iterated Hilbert Transform," *IEEE Conference Publication*, July 2005, pp. 333-338.

[4] T. Backstrom, "Parametric AM/FM Decomposition for Speech and Audio Coding," *IEEE Conference Publication*, October 2009, pp. 333-336.

[5] T. Yucek and H. Arslan, "A Survey of Spectrum Sensing Algorithms for Cognitive Radio Applications," *IEEE Communication Surveys & Tutorials*, Vol. 11, No. 1, 2009.

[6] S. Meignen and V. Perrier, "A New Formulation for Empirical Mode Decomposition Based on Constraint Optimization," 2007. http://www-ljk.imag.fr/membres/Valerie.Perrier/PUBLI/EMD-optim07.pdf

[7] G. Thakur, E. Brevdo, N. S. Fuˇckar and H.-T. Wu, "The Synchrosqueezing Algorithm for Time-Varying Spectral Analysis: Robustness Properties and New Paleoclimate Applications," 2012. arXiv:1105.0010v2[math.CA]

[8] P. Flandrin, "Time-Frequency/Time-Scale Analysis," Academic Press, San Diego, 1999.

[9] A. Graps, "An Introduction to Wavelets," *IEEE Computational Science and Engineering*, Vol. 2, No. 2, 1995.

[10] M. Clausel, T. Oberlin1 and V. Perrier, "The Monogenic Synchrosqueezed Wavelet Transform: A Tool for the Decomposition/Demodulation of AM-FM Images," 2012.

Time-Reversal UWB Positioning Beacon for Railway Application

Bouna Fall[1,2,3,4], Fouzia Elbahhar[1,2], Marc Heddebaut[1,2], Atika Rivenq[1,3,4]

[1]Université Lille Nord de France, Lille, France; [2]Institut Français des Sciences et Technologies des Transports de L'aménagement et des Réseaux (IFSTTAR), Laboratoire Electronique, Ondes et Signaux Pour les Transports (LEOST), Villeneuve d'Ascq, France; [3]Département Opto-Acousto-Electronique (DOAE), Université de Valenciennes et du Hainaut-Cambrésis (UVHC), Institut D'électronique, de Microélectronique et de Nanotechnologie (IEMN), Valenciennes, France; [4]Centre National de la Recherche Scientifique (CNRS), Unité Mixte de Recherche (UMR), Villeneuve d'Ascq, France.

ABSTRACT

This paper studies a new positioning beacon for railway transport using Ultra Wideband (UWB) radio and Time Reversal (TR) techniques. UWB radio has the potential to offer a good level of performance in terms of localization accuracy. Time Reversal channel pre-filtering facilitates signal detection and also helps increasing the received energy in the targeted area. In this paper, we evaluate the characteristics of TR technique in terms of temporal focusing. The theoretical and simulation results for Power Delay Profile, equivalent channel model and focusing gain of TR-UWB are given. We analyze the contribution of Time Reversal associated with UWB technology to enhance the localization resolution. The IEEE 802.15.3a channel models are used to evaluate the performance of this system. In terms of localization error, the theoretical and simulation results show that TR-UWB technique delivers improved performance over the UWB localization approach.

Keywords: UWB; Time Reversal; Focusing Gain; Channel Model; Power Delay Profile; TDOA; Localization Error

1. Introduction

Guided urban automated transportation systems are progressing significantly nowadays, highlighting many benefits. User's security and accessibility to these guided transport systems constitute a major issue dealt with many teams. In order to obtain an efficient and safe control and command system of the trains, it is essential to ensure an adequate exchange of information between vehicles and infrastructure and to determine accurately and constantly the absolute localization of the train.

The track to train data exchange is usually called CBTC for Communication Based Train Control. It can use many different radio techniques [1]. The localization system is based on proprioceptive sensors (phonic wheels, Doppler radars…) embarked in the train. This on-board system is coupled to the use of beacons located at ground, on the track, between the rails. These beacons are kilometer markers. They are used to compensate the drift of the localization information computed using the proprioceptive sensors alone, when the train moves. The beacons provide absolute localization information whenever the train passes over them. They are adequately spaced along the track in order to limit the localization drift to a tolerable safe value. In many railway systems, these beacons constitute the only equipment remaining between the rails [2]. Therefore, it could be interesting to remove this last equipment from the track, for example to facilitate track maintenance. In our studied case, the beacon would be situated on the side of the track, a few meters away, therefore not interfering with track maintenance operations.

In this paper, we propose a new solution allows to increase the positioning performances in terms of localization errors and to reduce the cost of track maintenance operations, thus constituting a new beacon realization. This new solution is based on the Ultra Wideband Impulse-Radio technology (UWB-IR) associated to the Time Reversal (TR) technique.

Otherwise, the combination of these techniques could be applied in various areas of the localization, such as the detection and localization of persons-through barriers or to find victims of accidents, especially in the mountains or in mines [3]. Through-the-Wall sensing takes also advantage of TR techniques and this application is very useful in safety and peace-keeping applications.

In this context, the proposed association between UWB and TR techniques for the precise localization of trains is illustrated in **Figure 1**. It shows the particular case of a railway tunnel. The beacons are geo-referenced. TR-UWB beacons are installed on the side of the track. The TR-UWB beacons are kilometer markers. They can also exchange local CBTC information over the UWB radio link. When arriving in the range of the UWB communication, the train computes its absolute localization to the beacons using time of flight information. Localization algorithms such as TDOA (Time Difference of Arrival) and TOA (Time of Arrival) can be used. Moreover, multiple UWB transmitters can be located in a beacon to enhance focusing and availability. The local Channel State Information (CSI) between any beacon transmitter and the virtual optimal beacon localization between the rails is identified a single time during the initial installation. This information is then introduced as pre-filtering data in the different UWB transmitters. Therefore, focusing is obtained in the required position, along the track, potentially improving the absolute localization process [4].

This study evaluates the temporal focusing performance of the time reversal (TR) method using the IEEE 802.15.3a channel models. For these evaluations, theoretical and simulation methods are used. Since the railway localization problem can be considered as a one dimension (1D) problem, we analyse the contribution of TR on the quality of localization in 1D.

This work also aims to verify whether the combination of UWB and TR techniques allows obtaining the decimetre required localization level of precision necessary to the railway application.

This paper is organized as follows: the second part presents the UWB and TR techniques. The third part presents the propagation channel modeling using the IEEE 802.15.3a channel models. The fourth section provides the analytical results of the TR-UWB system in terms of temporal focusing including focusing gain and Power Delay Profiles. The fifth section provides a comparative study between analytical and simulation results of the TR-UWB system in terms of temporal focusing. The sixth section presents the principle of TR-UWB localization and the performance of the proposed solution in terms of localization error. Finally, the conclusion summarizes the results and suggests future work.

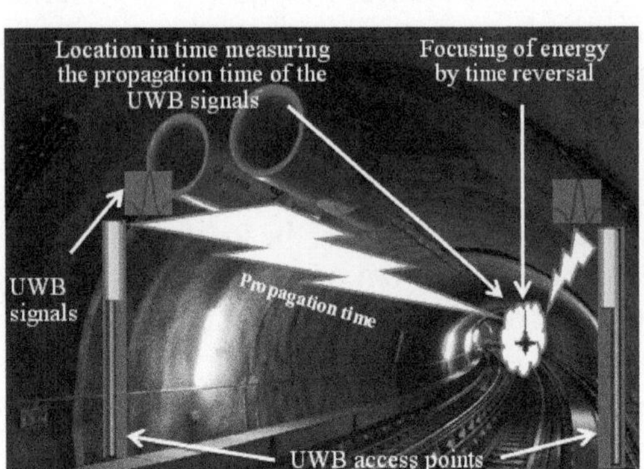

Figure 1. TR-UWB proposed localization system.

2. UWB and TR Techniques for Railway Applications

2.1. Ultra Wideband (UWB) Technology

UWB radio is typically defined as a wireless transmission scheme with a bandwidth over 500 MHz, or occupying 20% or more of the carrier frequency. There are different UWB approaches. In our study, we use an Impulse Radio UWB-IR system. This model involves the transmission of very short pulses occupying a very wide spectrum (**Figure 2**). The Gaussian waveform and its derivative are usually used. **Figure 3** shows the second derivative of Gaussian function that we will use.

A direct benefit of UWB concerns the use of low-cost, small size radio interface circuits [5].

UWB radio technology also offers some potential for railway applications [6]:

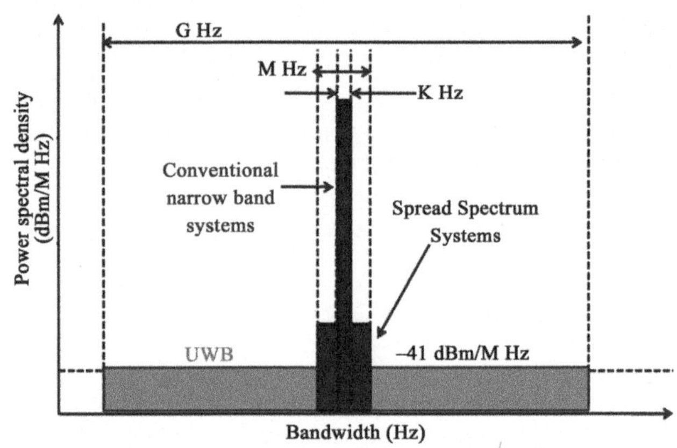

Figure 2. Comparison of the UWB spectrum and other wireless radio systems.

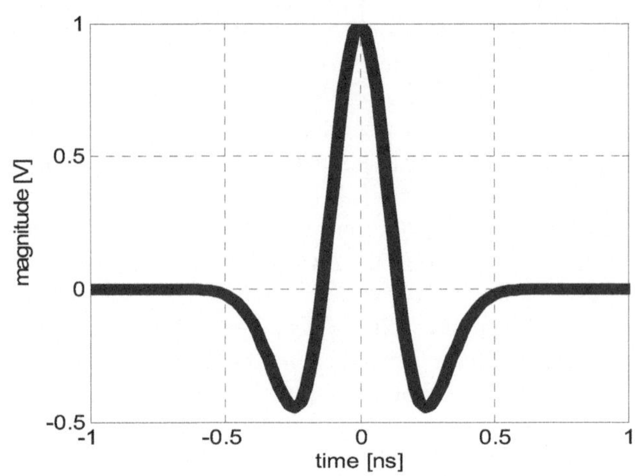

Figure 3. Transmitted pulse (second derivative of Gaussian function).

- It provides potentially high transmission data-rate, using a very large bandwidth;
- It offers high resolution train location because of the fine temporal resolution of the transmitted pulses;
- It adds ability to detect obstacles (radar) due to the impulsive nature of the signals. This capability is essential to detect obstacles in front of the train;
- Availability and robustness to multi-path are inherent to the large frequency bandwidth.

Localization in indoor environments such as tunnels is the subject of two major sources of error, the first being the lack of line of sight between the transmitter and the receiver and the second being the excessive presence of multipath. With the introduction of UWB in wireless communications, it seemed that this technology can provide improvements. However the studies have also raised some major problems like the complexity of the signal processing at the reception. Acquisition of UWB signals is critical, and its fundamental limits [7,8].

In order to solve all the problems facing the UWB, several studies have been conducted associating UWB and time reversal technique [9].

2.2. Principle and Characteristics of Time Reversal (TR) Technique

Classically, Time Reversal has been applied to acoustics and underwater systems [10,11]. It is closely related to retro-directive array in microwave [12,13] and phase conjugation in optics. More recently, it has also been studied for broadband especially for UWB communications [14]. The first Time Reversal experiment using electromagnetic waves in the 2.45 GHz band was reported in [15]. This contribution suggests that the techniques developed for ultrasound might also be used for the electromagnetic case. It is an interesting challenge because in many real environments (buildings or cities), microwaves, using wavelengths between 5 and 30 cm, are scattered off objects such as walls, desks, vehicles and so on, which produces a multitude of paths from the transmitter to the receiver. In such situations, a time reversal system should be able not only to compensate the multipath effect, but also to improve some radio communication parameters, thanks to the many reflections/reverberations that occur.

We propose to use the properties of time reversal [16] for localization systems.

The principle of the proposed TR-UWB system uses three steps. Firstly, we select the signal we want to transmit. In our case, it is the second derivative of the Gaussian function; it is an ultra short pulse. Secondly, the channel impulse response is measured between the transmitter (Tx) and the receiver (Rx) and the channel state information is loaded into Tx. Thirdly, the selected signal and the impulse response are reversed in time and transmitted by Tx in the propagation channel, up to Rx. This TR-UWB principle, represented in **Figure 4**, can be mathematically described by noting $s(t)$ the transmitted pulse, $h(t)$ the complex impulse response of the channel and $h^*(-t)$ the complex conjugate of the time reversed version of $h(t)$. We note $y(t)$ the received signal without TR and $y_{rt}(t)$, the received signal with TR at the receiver. Their expressions are given by:

$$y(t) = s(t) \otimes h(t) + n(t) \quad (1)$$

$$y_{rt}(t) = s(t) \otimes h^*(-t) \otimes h(t) + n(t) \quad (2)$$

where $\otimes$ represents the convolution operation and $n(t)$ is the Gaussian noise.

From Equation (2), we deduce the equivalent impulse response $heq(t)$ which corresponds to the autocorrelation function of the channel [17]:

$$heq(t) = h^*(-t) \otimes h(t) \quad (3)$$

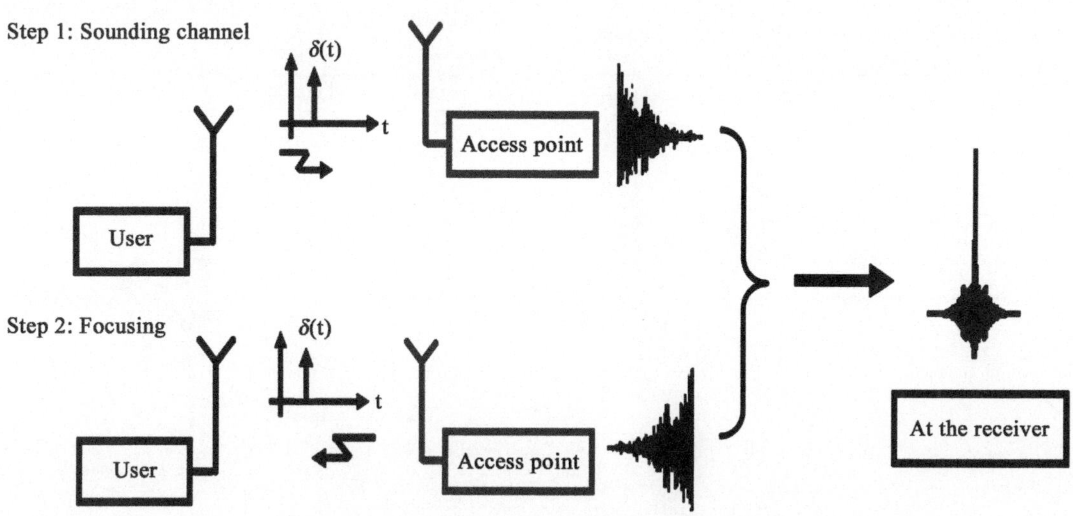

Figure 4. Principle of time reversal technique.

The autocorrelation function is used to evaluate two main characteristics associated to time reversal, *i.e.*, the temporal focusing and the spatial focusing. These characteristics are very beneficial to the UWB system [17, 18]. To study the temporal focusing, we evaluate the Focusing Gain (FG), which is defined as the ratio of the strongest peak in TR received to the strongest peak received by a conventional UWB system [19]. It can be written as:

$$FG_{[dB]} = 20\log_{10}\left(\frac{\max\left(\left|y_{rt}(t)\right|\right)}{\max\left(\left|y(t)\right|\right)}\right) \quad (4)$$

Higher FG could potentially translate into higher communication range and higher precision of localization for a localization system as compared to a classical UWB system.

3. IEEE 802.15.3a Channel Model

In this part, we describe the IEEE 802.15.3a channel model. The objective of using this channel model is to characterize the channel impulse response [20,21]. The time reversal would benefit from the complexity of the propagation environment. If the environment is increasingly complex, best will be the focus of energy [22].

The IEEE 802.15.3a model was developed from around 10 contributions, all referring to distinct experimental measurements, performed in indoor residential or office environments [23,24].

In order to reflect the phenomenon of ray clustering that was observed in several measurement campaigns, the model is based on the Saleh-Valenzuela formalism. Parameters are provided to characterize the clusters and ray arrival rates (Λ and λ), as well as the inter- and intraluster exponential decay constants (Γ and γ). Four sets of parameters are provided to model the four following channel types:

- The channel model CM 1 corresponds to a distance of 0 - 4 m in a LOS situation;
- The channel model CM 2 corresponds to a distance of 0 - 4 m in an NLOS situation;
- The channel model CM 3 corresponds to a distance of 4 - 10 m in an NLOS situation;
- The channel model CM 4 corresponds to an NLOS situation with a large delay spread τ_{rms} = 25 ns.

For our study we use the three first scenarios CM1, CM2 and CM3. CM4, generated to fit a large 25-ns RMS delay spread was considered not relevant to our application. Some important characteristics of these scenarios based upon a 167-ps sampling time are shown in **Table 1**.

This comprehensive model is a reference for the study of UWB systems. It can be applied in indoor environments and short range conditions. We use these models for performance evaluation in terms of temporal focusing and error localization. **Figure 5** presents the impulse response of channel models CM1, CM2 and CM3. The impulse response complexity increases from CM1 to CM3.

4. Analytical Study of Time Reversal Characteristics

In this section, we develop an analytical study of the temporal focusing of time reversal. The study of the temporal focusing is based on the characterization of the propagation channel. For each considered channel model (CM1, CM2, CM3), we determine the equivalent impulse response, the Power Delay Profile and Focusing Gain.

Throughout the study, we denote by:

- $s(t)$: transmitted pulse (derivative Gaussian pulse);
- $h(t)$: Channel Impulse Response CIR;
- $h^*(-t)$: conjugated and reversed CIR.

Table 1. IEEE 802.15.3a UWB channel model characteristics.

Channel	CM1	CM2	CM3
Mean excess delay (ns)	5.05	10.38	14.18
RMS delay (ns)	5.28	8.03	14.28
Distance (m)	0 - 4	0 - 4	4 - 10
Scenario	LOS	NLOS	NLOS
No of significant paths (85%)	20.8	33.9	64.7

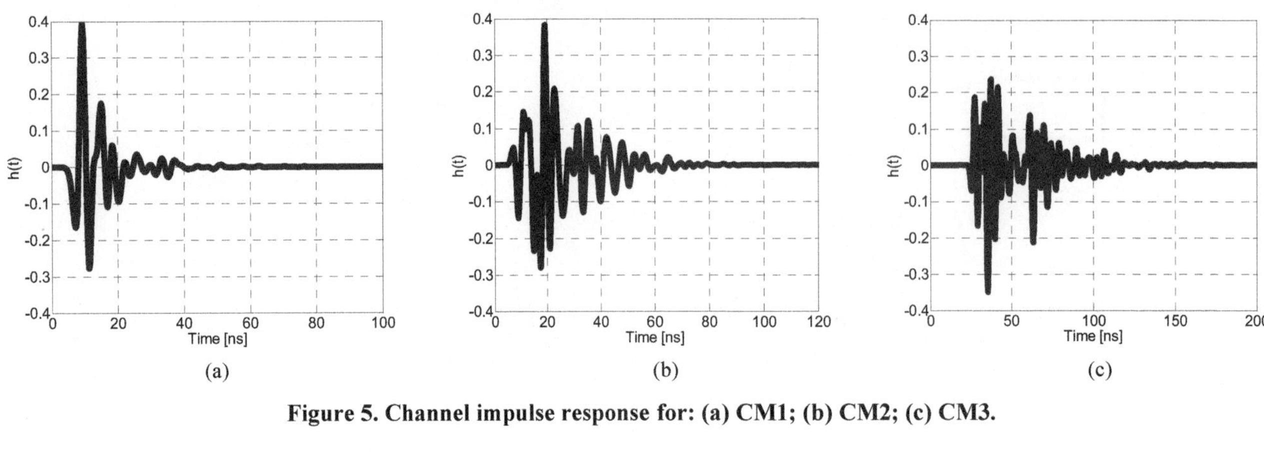

Figure 5. Channel impulse response for: (a) CM1; (b) CM2; (c) CM3.

The analytical study in the case of the IEEE 802.15.3a model is done using the statistical moments of interferences that affect the performance of a TR-UWB receiver [25].

For this channel model, we can define CIR by:

$$h(t)=\sum_{i=0}^{\infty}\alpha_i s(t-t_i) \quad (5)$$

The expected value of the equivalent CIR is: [10]

$$E\{heq(t)\}=E\left\{\int_A\sum_{i=0}^{\infty}\alpha_i^2 s(\tau-t_i)s(\tau+t-t_i)\mathrm{d}\tau\right\} \quad (6)$$

Using the Auto-Correlation Function (ACF) for prototype pulse defined by Equation (7):

$$\phi_s(t)=\int_A s(\tau-t_i)s(\tau-t_i+t)\mathrm{d}\tau \quad (7)$$

The Equation (6) becomes:

$$E\{heq(t)\}=E\left(\sum_{i=0}^{\infty}\alpha_i^2\right)\phi_s(t) \quad (8)$$

Computing the average energy of the CIR in a generic time window $W=[a,b]$ given:

$$E\left\{\sum_{i\in I_W}\alpha_i^2\right\}=\int_W P_g(t)\mathrm{d}t \quad (9)$$

where I_W is the random set of multipath components within W and $P_g(t)$ referred to Average Power Delay Profile (APDP).

A continuous-time in W is given by a variance of the CIR energy function [26]:

$$Var\left\{\sum_{i\in I_W}\alpha_i^2\right\}=\int_W R_g(t)\mathrm{d}t \quad (10)$$

where $R_g(t)$ is the kurtosis delay profile.

Using the equivalence (9) in (8), we obtain:

$$E\{heq(t)\}=\phi_s(t)\int_A P_g(t)\mathrm{d}t \quad (11a)$$

We observe, for $t>T_s$ $\phi_s(t)=0$ therefore

$$E\{heq(t)\}\approx 0.$$

The Average Power Delay Profile $P_g(t)$ is described by a model with exponential decay which is characterized by the average received energy E_g and the *rms* delay spread τ_{rms} of the channel.

$$P_g(t)=\frac{E_g}{\tau_{rms}}\exp(-t/\tau_{rms}) \quad (12)$$

For determining $R_g(t)$ we choose a uniform Poisson arrival process with an arrival rate of $\bar{\lambda}$ ray/s and Nakagami m distributed ray amplitude a simplified version of the Saleh-Valenzuela channel model is [27]:

$$R_g(t)=P_g^2(t)\left(1+\frac{1}{m}\right)\frac{1}{\bar{\lambda}} \quad (13)$$

In this case the Power Delay Profile for the equivalent CIR is given by:

$$PDP_{\text{TR-UWB}}(t)=E\left\{|heq(t)|^2\right\} \quad (14)$$

It becomes:

$$PDP_{\text{TR-UWB}}(t)=E\left\{\iint_A h(\tau)h(\mu)h(t+\tau)h(\mu+t)\mathrm{d}\tau\mathrm{d}\mu\right\} \quad (15)$$

The equivalences (11a) and (15) are developed in Appendix, and their final expressions are given by the Equations (11b) and (16):

$$E\{heq(t)\}=E_g\phi_s(t) \quad (11b)$$

$$\begin{aligned}&PDP_{\text{TR-UWB}}(t)\\&=E_g^2\phi_s^2(t)+\frac{E_g^2}{2\tau_{rms}}\\&\times\left[(1+\Phi_{\phi s}(t))c_1\exp(-t/\tau_{rms})+\left(1+\frac{1}{m}\right)\frac{1}{\bar{\lambda}}\phi_s^2(t)\right]\end{aligned} \quad (16)$$

The focusing gain presented in (4) can be written in this case as:

$$FG_{[dB]}=10\log_{10}\left(\frac{\max(PDP_{\text{TR-UWB}}(t))}{\max(PDP_{\text{UWB-IR}}(t))}\right) \quad (17)$$

with

$$PDP(t)=P_g(t)=\frac{E_g}{\tau_{rms}}\exp(-t/\tau_{rms}) \quad (18)$$

The correlation function peak is located at zero time shift, therefore:

$$\max(PDP(t))=\frac{E_g}{\tau_{rms}} \quad (19)$$

$$\max(PDP_{\text{TR-UWB}}(t))=E_g^2(\tau_{rms}+c_1+c_2)/\tau_{rms} \quad (20)$$

where $c_2=\left(1+\frac{1}{m}\right)\frac{1}{2\bar{\lambda}}$

Then, (17) becomes:

$$FG_{dB}=10\log_{10}(E_g(\tau_{rms}+c_1+c_2)) \quad (21)$$

5. Analytical and Simulation Results

In this section, we compare the simulation and the analytical results in UWB and TR-UWB cases using IEEE 802.15.3a channel models. Thereafter, we study the contribution of Time Reversal in UWB in terms of temporal focusing.

Tables 2 gives general input parameters for analytical and simulation steps.

We performed a comparative study of the *m* value appearing in the Nakagami distribution and we considered the four following cases [28]:

$m = 0.5$: corresponding to a Gaussian situation;

$m = 1$: is the Rayleigh law, i.e. deep fading;

$m = 1.5$: corresponding to a severe fading;

$m = 4$: corresponding to a low fading.

This study showed that the value of m has a negligible impact on system performance TR-UWB. We agreed to choose the value $m = 1.5$ in the remainder of our studies.

Figures 6(a)-(f) show analytical and simulation results for TR-UWB system, in case of IEEE 802.15.3a channel models. The average over 1000 runs of channels is performed for each channel model (CM1, CM2 and CM3). These results correspond to the equivalent channel impulse response $heq(t)$ and Power Delay Profile $PDP_{\text{TR-UWB}}(t)$ for transmitted a second Gaussian derivative pulse. We can observe then, the simulation results are close to analytical results.

Table 3 gives a comparison between analytical and simulation results for IEEE 802.15.3a channel models in term of FG value. This study shows a slight difference between the values of FG to the analytical study and the study by simulation. This can be explained by the sets of input parameters made.

By expanding by simulation the study interval, we can now compare PDP-UWB and PDP-TR-UWB for the 3 considered channel models. **Figures 7-9** provide a comparison between PDP_{UWB} and $PDP_{\text{TR-UWB}}$. Considering PDP_{TR}, we obtain a very effective temporal focusing as well as an increase of the amplitude of the power over PDP_{UWB} alone. This translates into the values of FG presented in **Table 4**.

Table 2. Input parameters for IEEE 802.15.3a channel models.

Channel models	CM1	CM2	CM3
E_g (normalized)	1	1	1
$\bar{\lambda}$ Ray arrival rate [1/ns]:	2.5	0.5	2.1
τ_{rms} RMS delay [ns]:	5.28	8.03	14.28
The *m* value (Nakagami-m)	1.5	1.5	1.5

Table 3. Analytical and simulation results for IEEE 802.15.3a channel models in term of FG Value.

Channel models	CM1	CM2	CM3
$FG_{\text{analytical}}$ [dB]	7.38	8.20	10.91
$FG_{\text{simulation}}$ [dB]	7.31	8.11	10.80

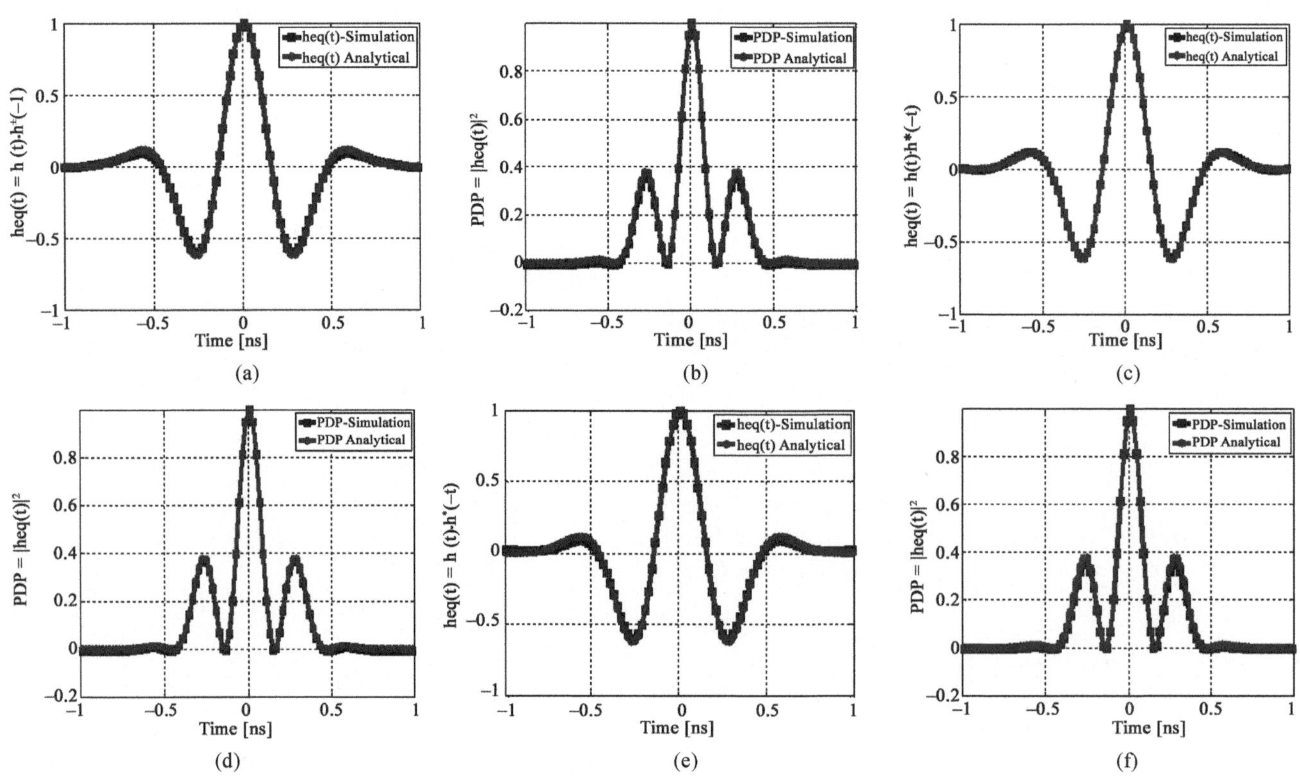

Figure 6. Analytical and simulation results for TR-UWB correspond respectively to the equivalent channel impulse response *heq(t)* and Power Delay Profile $PDP_{\text{TR-UWB}}(t)$: (a), (b) for CM1; (c), (d) for CM2; (e), (f) for CM3.

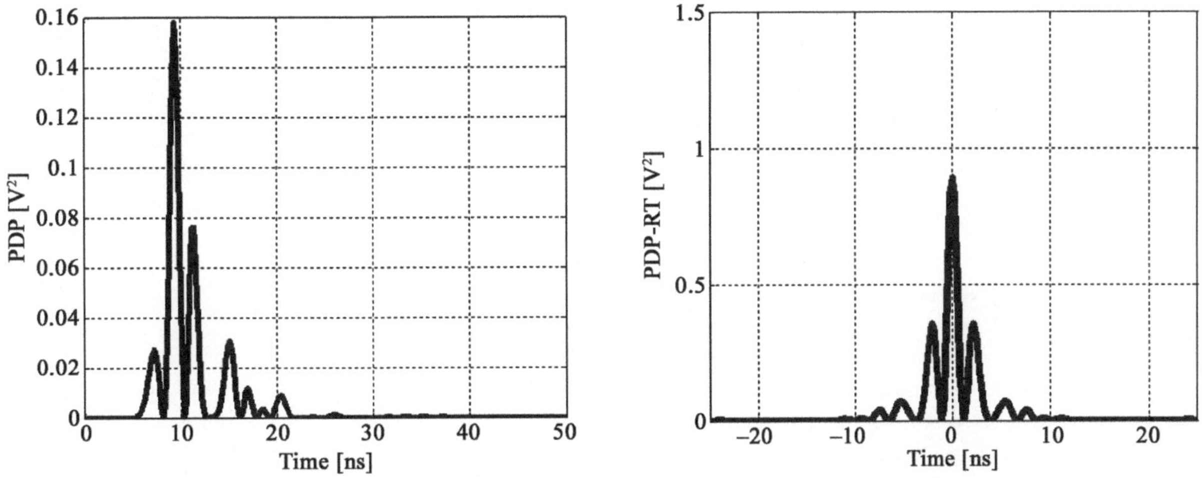

(a) (b)

Figure 7. (a) PDP_{UWB}; (b) $PDP_{TR\text{-}UWB}$ for CM1.

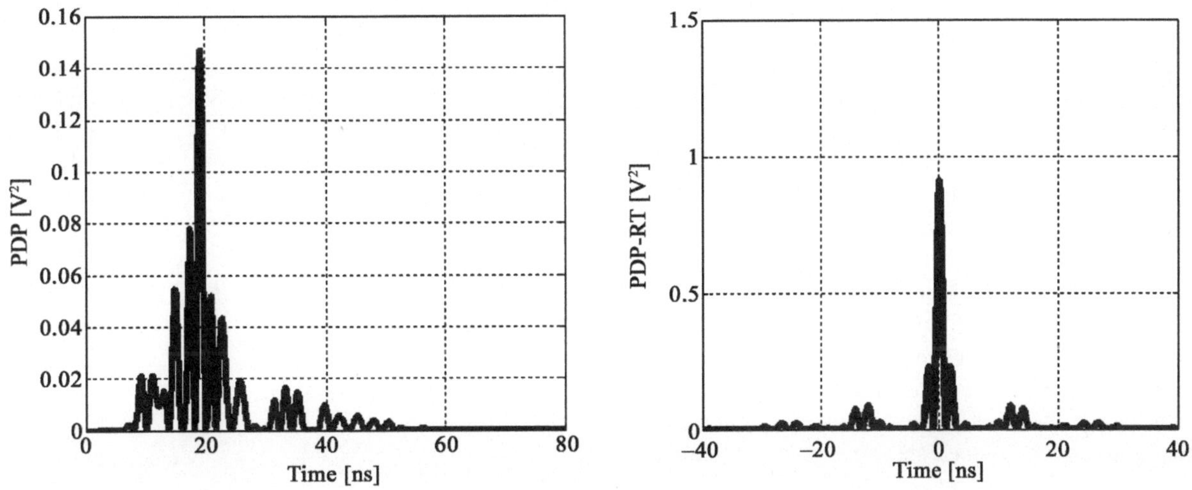

(a) (b)

Figure 8. (a) PDP_{UWB}; (b) $PDP_{TR\text{-}UWB}$ for CM2.

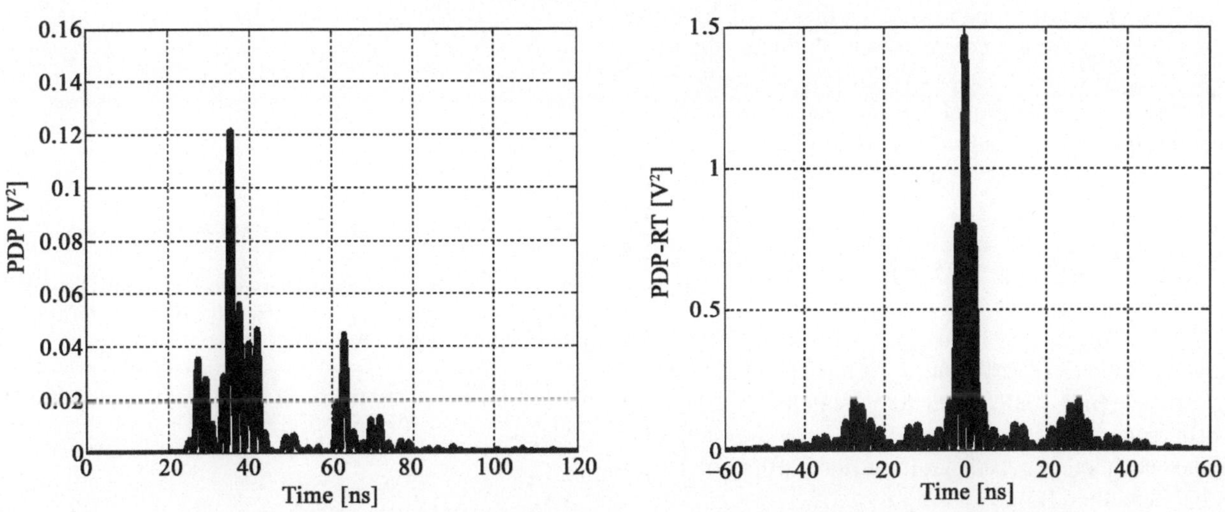

(a) (b)

Figure 9. (a) PDP_{UWB}; (b) $PDP_{TR\text{-}UWB}$ for CM3.

From CM1 to CM3 FG increases, due in particular to the stronger multipath. Indeed, TR takes advantage of the complexity of the channel.

This would be very beneficial for the purpose of locating in confined environments such as tunnels.

6. TR-UWB-IR for Localization

In this section, we carry on this study by evaluating the localization error. We still consider the two preceding UWB alone and TR-UWB techniques. The goal is to estimate the contribution of TR to the location in terms of localization error, as a function of the propagation environment position of a mobile in a 2D plane. To locate the mobile, each base station sends its own signal (recorded and reversed in time in the case of TR). The UWB signals are modulated by an antipodal modulation and coded by a Gold code [29,30], one code per base station. Processing is performed at the mobile (R_x) to determine its position relative to the base stations. The localization error is given by the difference between the calculated position and the actual position of the mobile.

The mobile receives the signals from each base station and performs an adequate signal processing to determine its position, relative to the base stations. Using the TDOA technique, the signal received at the mobile is processed to retrieve the position of the latter [31]. The localization technique (TDOA) is combined with the Chan algorithm to calculate the position of the mobile [32,33]. This technique is illustrated in **Figure 10**, where S_1, S_2 and S_3 represent the three base stations of known coordinates, M is the mobile to be located. The difference distance between mobile and the ith base station is given by Equation (32):

$$R_i = \sqrt{(X_i - x)^2 + (Y_i - y)^2} \qquad (22)$$

where, (x, y) are the unknown coordinates of mobile position, (X_i, Y_i) are the coordinates of the base stations. Considering as a reference station S_1 (the reference station is the nearest station of the mobile, in the studied case), the difference distance between reference station S_1 and other stations is given by:

$$\begin{aligned} R_{i,1} &= c \cdot d_{i,1} = R_i - R_1 \\ &= \sqrt{(X_i - x)^2 + (Y_i - y)^2} - \sqrt{(X_1 - x)^2 + (Y_1 - y)^2} \end{aligned} \qquad (23)$$

where, c is the celerity of light, $d_{i,1}$ is the TDOA estimate between reference station and the ith station. Calculating the differences in distance allows us to define a system of nonlinear equation of hyperbolas (Equation (24)), resolvable by Chan's method [29].

$$R_{i,1}^2 + 2R_{i,1} = X_i^2 + Y_i^2 - 2X_{i,1}x - 2Y_{i,1}y + x^2 + y^2 \qquad (24)$$

For a three base station system, Chan's method produces two TDOA to determine the coordinates (x, y) of mobile:

$$\begin{bmatrix} x \\ y \end{bmatrix} = -\begin{bmatrix} X_{2,1} & Y_{2,1} \\ X_{3,1} & Y_{3,1} \end{bmatrix}^{-1} \cdot \left\{ \left[\begin{bmatrix} R_{2,1} \\ R_{3,1} \end{bmatrix} . R_1 + \frac{1}{2} \begin{bmatrix} R_{2,1}^2 - K_2 + K_1 \\ R_{3,1}^2 - K_3 + K_1 \end{bmatrix} \right] \right\} \qquad (25)$$

where:

$$K_1 = X_1^2 + Y_1^2;\ K_2 = X_2^2 + Y_2^2;\ K_3 = X_3^2 + Y_3^2;$$
$$R_{2,1} = cd_{2,1};\ R_{3,1} = cd_{3,1}.$$

Our comparison is based on the computation of the Root Mean Square Error of localization between the conventional UWB system and the proposed TR-UWB system. An Additive White Gaussian Noise (AWGN) is also injected in the CM1, CM2 and CM3 models. A large number of iterations are used (1000 iterations). **Table 5** presents a comparative study between the UWB system alone and the TR-UWB system; the computed focusing gain and localization errors are indicated using a Signal to Noise Ratio (SNR) of 8 dB. We obtain that the combination of UWB and *TR* greatly reduces the localization error. Indeed, in the particular case treated, the localization error is 7.45 cm in the case of CM3 for UWB system,

Table 4. Focusing gain for CM1 CM2 and CM3.

Channel models	CM1	CM2	CM3
FG [dB]	7.31	8.11	10.80

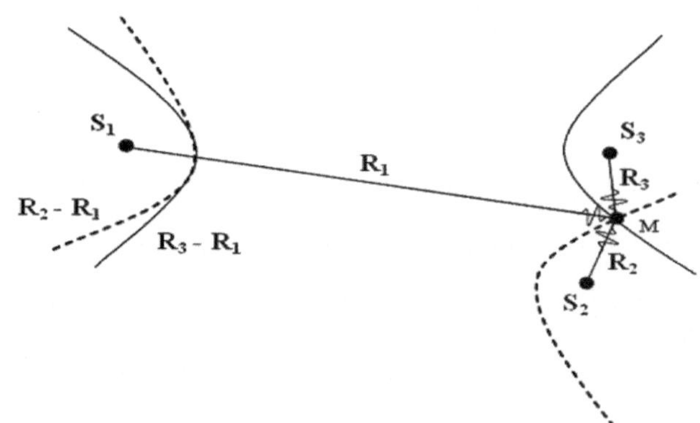

Figure 10. 2-D TDOA localization process.

Table 5. Comparative study between UWB and TR-UWB in terms of focusing gain and error localization (CM1, CM2 and CM3, SNR = 8 dB).

	$FG_{[dB]}$	$RMSE_{[cm]}$ for UWB	$RMSE_{[cm]}$ for TR-UWB
CM1	7.31	2.08	~= 0
CM2	8.11	2.66	~= 0
CM3	10.80	7.45	1.12

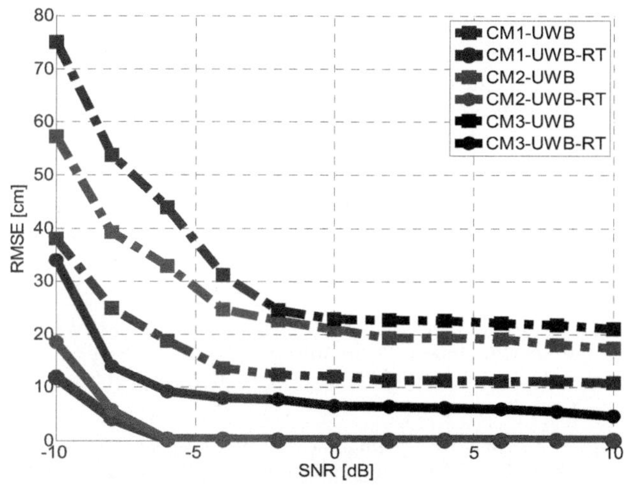

Figure 11. Comparison UWB/TR-UWB, RMSE vs. SNR for channel models CM1, CM2 and CM3.

whereas it is only 1.12 cm for TR-UWB system. This remark also applies to CM1 and CM2.

Different SNR versus were also tested. **Figure 11** presents the Root Mean Square Error of localization (RMSE) versus SNR for a large number of CM1 to CM3 realizations. Over the 2 to 10 dB computed range, a stable, considerable improvement is obtained.

7. Conclusion

In this contribution, we presented results obtained working on the association of time reversal and UWB impulse radio for a localization application. The IEEE 802.15.3a channel models are used to perform these studies. The analytical and simulation results have shown that this particular combination can reduce the error localization thanks to the focusing characteristics of time reversal. In future work, we will study different waveforms and modulation types to identify the most appropriate ones. This study will also be conducted with other channel models as well as different single transmitter and multiple transmitter configurations. To validate the analytical and simulation results, experimentations will be soon conducted in an anechoic chamber and a real environment.

REFERENCES

[1] H. Saghir, M. Heddebaut, F. Elbahhar, J. M. Rouvaen and A. Rivenq, "Train-to-Wayside Wireless Communication in Tunnel Using Ultra-Wideband and Time Reversal," *Transportation Research Part C: Emerging Technologies*, Vol. 17, No. 1, 2009, pp. 81-97.

[2] M. N. Benslimen, "Recherche de Procedure de Caractérisation de l'Environnement Elecromagnétique Ferroviaire Adaptées au Contexte des Systèmes de Communications Embarqués," Ph.D. Thesis, Lille University, Lille, 2009.

[3] N. Maaref, P. Millot, X. Ferriéres, C. Pichot and O. Picon, "Electromagnetic Imaging Methode Based on Time Reversal Processing Applied to through the Wall Target Localization," *Progress in Electromagnetics Research M*, Vol. 1, 2008, pp. 59-67.

[4] B. Fall, M. F. Elbahhar, M. Heddebaut and A. Rivenq, "Time Reversal and UWB Techniques for Positioning System in Transport Applications," *International Symposium on Signal, Image, Video and Communications (ISIVC)*, Valenciennes, 4-6 July 2012, pp. 1-4.

[5] M. L. Welbom, "System Considerations for Ultra-Wide-Band Wireless Networks," *IEEE: Radio and Wireless Conference RAWCON*, Boston, 19-22 August 2001, pp. 5-8.

[6] H. Saghir, M. Heddebaut, F. Elbahhar, A. Rivenq and J. M. Rouvaen, "Time Reversal UWB Wireless Communication-Based Train Control in Tunnel," *Journal of Communications*, Vol. 4, No. 4, 2009, pp. 248-256.

[7] W. Suwansantisuk, M. Z. Win and L. A. Shepp, "On the Performance of Wide-Bandwidth Signal Acquisition in Dense Multipath Channels," *IEEE Transactions on Vehicular Technology*, Vol. 54, No. 5, 2005, pp. 1584-1594.

[8] W. Suwansantisuk and M. Z. Win, "Multipath Aided Rapid Acquisition: Optimal Search Strategies," *IEEE Transactions on Information Theory*, Vol. 53, 2006, pp. 174-193.

[9] M. Fink, "Ondes et Renversement du Temps," *Bulletin de L'union des Professeurs de Physique et de Chimie*, 2005, pp. 25-31.

[10] M. Fink, "Time Reversal Waves and Super Resolution," *Journal of Physics: Conference Series* 124. *4th AIP International Conference and the 1st Congress of the IPIA*, 2008.

[11] A. Derode, P. Roux and M. Fink, "Robust Acoustic Time Reversal with High Order Multiple Scattering," *Physical Review Letters*, Vol. 75, No. 23, 1995, pp. 4206-4209.

[12] L. G. Van Atta, "Electromagnetic Reflector," US Patent No. 2908002, 1959.

[13] B. E. Henty and D. D. Stancil, "Multipath-Enabled Super-Resolution for RF and Microwave Communication Using Phase-Conjugate Arrays," *Physical Review Letters*, Vol. 93, No. 24, 2004, pp. 1-4.

[14] T. Strohmer, M. Emami, J. Hansen, G. Papanicolaou and A. Paulraj, "Application of Time-Reversal with MMSE Equalizer to UWB Communications," *Proceedings of Globecom Conference*, Dallas, December 2004, pp. 3123-3127.

[15] G. Lerosey, J. de Rosny, A. Tourin, A. Derode, G. Montaldo and M. Fink, "Time Reversal of Electromagnetic Waves," *Physical Review Letters*, Vol. 92, No. 19, 2004, pp. 1-3.

[16] X. Liu, B.-Z. Wang, S. Xiao and J. Deng, "Performance of Impulse Radio UWB Communication Based on Time Reversal Technique," *Progress in Electromagnetics Research, PIER* 79, Vol. 79, No. 11, 2008, pp. 401-413.

[17] D. Abassi-Moghadam and D. T. Vakili, "Channel Characterization of Time Reversal UWB Communication Systems," *Wiley International Journal of Communication Systems*, Vol. 65, No. 9-10, 2010, pp. 601-614.

[18] C. Zhou, "Time Reversal Ultra-Wideband (UWB) Multiple Input Multiple Output (MIMO) Based on Measured Spatial Channels," *IEEE Transactions on Vehicular Technology*, Vol. 58, No. 6, 2009, pp. 2884-8898.

[19] C. Zhou, "Impulsive Radio Propagation and Time Reversal MIMO System for UWB Wireless Communications," Ph.D. Thesis, Faculty of the Graduated School, Tennessee Technological University, Cookeville, 2008.

[20] P. Pagani, F. T. Talom, P. Pajusco and B. Uguen, "Ultra Wide Band Radio Propagation Channels, A Practical Approach," ISTE Ltd. and John Wiley & Sons, Inc., Chichester, 2008, 239 p.

[21] J. A. Gubner K. Hao, "The IEEE 802.15.3a UWB Channel Model as a Two-Dimensional Augmented Cluster Process," *IEEE Transaction on Information Theory*, 2006, pp. 1-9.

[22] I. H. Naqvi, "Application of Time Reversal (TR) Technique to Ultra-Wide Band (UWB) and Multi-Antenna (MIMO) Communication Systems," Ph.D. Thesis, Institute National des Sciences Appliquées de Rennes, 2009.

[23] A. Malish, "Ultra Wideband Propagation Channels—Theory, Measurement and Modeling," *IEEE Transactions on Vehicular Technology*, Vol. 54, No. 5, 2005, pp. 1528-1545.

[24] J. A. Gubner and K. Hao, "Analysis of the IEEE 802.15.3a UWB Channel Model," *IEEE Journal on Selected Area in Communications*, 2005, pp. 1-8.

[25] Y. Chen and N. C. Beaulieu, "Moment Based Interference Analysis of CM1, CM2, CM3 and CM4 UZB Systems," *IEEE GLOBECOM Proceedings*, 2007, Washington, pp. 3858-3862.

[26] W. Klaus and P. Marco, "Statistical Analysis of UWB Channel Correlation Functions," *IEEE Transactions on Vehicular Technology*, Vol. 57, No. 3, 2008, pp. 1359-1373.

[27] A. Adel, M. Saleh, A. Reinaldo and A. Valenzuela "A Statistical Model for Indoor Multipath Propagation," *IEEE Journal Selected Areas in Communications*, Vol. 5, No. 2, 1987, pp. 128-137.

[28] D. P. Dhiraj, "On the Simulations of Correlated Nakagami-m Fading Channels Using Sum of Sinusoids Method," Ph.D. Thesis, University of Missouri, Columbia, 2006.

[29] F. Elbahhar, B. Fall, M. Heddebaut, A. Rivenq and J. M. Rouvaen, "Indoor Positioning System Based on the UWB Technique," *IPIN* 2011, Guimarães, 21-23 September 2011, pp. 1-4.

[30] A. Elabed, F. Elbahhar, Y. Elhillali, A. Rivenq and R. Elassali, "UWB Communication System Based on Bipolar PPM with Orthogonal Waveforms," *Wireless Engineering and Technology*, Vol. 3, No. 3, 2012, pp. 181-188.

[31] M. Ghavami, L. V. Michael and R. Kohno, "Ultra Wide Band Signals and Systems in Communication Engineering," Wiley, London, 2007.

[32] R. I. Reza, "Data Fusion for Improved TOA/TDOA Position Determination in Wireless Systems," Master of Science in Electrical Engineering, Faculty of the Virginia Polytechnic Institute, 2000.

[33] X. Wang, Z. Wang and B. O'Dea, "A TOA-Based Location Algorithm Reducing the Errors Due to Non-Line-of-Sight (NLOS) Propagation," *IEEE Transaction on Vehicular Technology*, Vol. 52, No. 1, 2003, pp. 112-116.

Appendix

The Analytical Study of the PDP of Time Reversal

This appendix computes the equivalence (18) that is given by:

$$PDP_{\text{TR-UWB}}(t)=E\left\{\iint_A h(\tau)h(\mu)h(t+\tau)h(\mu+t)\mathrm{d}\tau\mathrm{d}\mu\right\} \tag{26}$$

Substituting $h(t)=\sum_i \alpha_i s(t-t_i)$ and expanding all the products. We obtain by noticing that $E\{\alpha_i^3\alpha_j\}=0$:

$$\begin{aligned}&PDP_{TR-UWB}(t)\\&=E\Bigg\{\int_A\int_A\sum_i\alpha_i^2 s(\tau-t_i)\\&\times\Big[s(\tau+t-t_i)\sum_k\alpha_k^2 s(\mu-t_k)s(\mu+t-t_k)\\&+s(\mu-t_i)\sum_{j,j\neq i}\alpha_j^2 s(\tau+t-t_j)s(\mu+t-t_j)\\&+s(\mu+t-t_i)\sum_{j,j\neq i}\alpha_j^2 s(\tau+t-t_j)s(\mu-t_j)\mathrm{d}\tau\mathrm{d}\mu\Bigg\}\end{aligned} \tag{27}$$

This equation is then composed of three terms named I, II and III:

$$\begin{aligned}I&=E\left\{\sum_i\alpha_i^2\int_A s(\tau-t_i)s(\tau+t-t_i)\mathrm{d}\tau\times\sum_k\alpha_k^2\int_A s(\mu-t_k)s(\mu+t-t_k)\mathrm{d}u\right\}\\&\approx E\left\{\sum_i\alpha_i^2\sum_k\alpha_k^2\right\}\phi_s^2(t)\end{aligned} \tag{28}$$

Knowing that:

$$\begin{aligned}E\{\sum\alpha_i\}^2&=Var\{\sum\alpha_i^2\}+\left(E\{\sum\alpha_i^2\}\right)^2\\&=\int_A R_g(\tau)\mathrm{d}\tau+\iint_A P_g(\tau)P_g(\mu)\mathrm{d}\tau\mathrm{d}\mu\end{aligned} \tag{29}$$

Then (28) becomes:

$$I=\phi_s^2(t)\left\{\int_A R_g(\tau)\mathrm{d}\tau+\iint_A P_g(\tau)P_g(\mu)\mathrm{d}\tau\mathrm{d}\mu\right\} \tag{30}$$

Using the property $E\{\alpha_i^2\alpha_j^2\}=E\{\alpha_i^2\}E\{\alpha_j^2\}$ the second term II becomes:

$$\begin{aligned}II&=\iint_A E\left\{\sum_i\alpha_i^2 s(\tau-t_i)s(\mu-t_i)\right\}\\&\times E\left\{\sum_{j,j\neq i}\alpha_j^2 s(\tau+t-t_j)s(\mu+t-t_j)\right\}\mathrm{d}\tau\mathrm{d}\mu\end{aligned} \tag{31}$$

By following the change of variable: $\mu=\tau+\xi$ we get:

$$\begin{aligned}&E\left\{\sum_i\alpha_i^2 s(\tau-t_i)s(\tau+\xi-t_i)\right\}\\&=E\{h(\tau)h(\tau+\xi)\}\end{aligned} \tag{32}$$

and

$$\begin{aligned}&E\left\{\sum_{j,j\neq i}\alpha_j^2 s(\tau+t-t_j)s(\tau+\xi+t-t_j)\right\}\\&\approx E\{h(\tau+t)h(\tau+t+\xi)\}\end{aligned} \tag{33}$$

According to [16]

$$E\{h(\tau)h(\tau+\xi)\}=P_g(\tau)\phi_s(\xi) \tag{34}$$

Therefore, referring to (4), (31) becomes:

$$II\approx\int_A P_g(\tau)P_g(\tau+t)\int_{A'}\phi_s^2(\xi)\mathrm{d}\xi\mathrm{d}\tau \tag{35}$$

where $A'=[a-\tau,a-\tau+T_l]$

It also defines that: $\int_{A'}\phi_s^2(\xi)\mathrm{d}\xi=c_1$ is the autocorrelation energy of pulse, then *II* becomes:

$$II\approx c_1\int_A P_g(\tau)P_g(\tau+t)\mathrm{d}\tau \tag{36}$$

The third tem *III* gives:

$$\begin{aligned}III&=\iint_A E\left\{\sum_i\alpha_i^2 s(\tau-t_i)s(\mu+t-t_i)\right\}\\&\times E\left\{\sum_{j,j\neq i}\alpha_j^2 s(\tau+t-t_j)s(\mu-t_j)\right\}\mathrm{d}\tau\mathrm{d}\mu\end{aligned} \tag{37}$$

$$\begin{aligned}III&=\int_A\int_{A'}E\{h(\tau)h(\tau+t+\xi)\}\\&E\{h(\tau+t)h(\tau+\xi)\}\mathrm{d}\tau\mathrm{d}\xi\\&\approx\int_A P_g(\tau)P_g(\tau+t)\int_{A'}\phi_s(\xi+t)\phi_s(\xi-t)\mathrm{d}\xi\mathrm{d}\tau\end{aligned} \tag{38}$$

Also, arguing with the short support of $\phi_s(\xi)$, we can approximate the inner integral by

$$c_1\Phi_{\phi s}(t)=\int_{A'}\phi_s(\xi+t)\phi_s(\xi-t)\mathrm{d}\xi\mathrm{d}\tau$$

Where $\phi_s(t)$ is the pulse auto-correlation function and $\Phi_{\phi s}(t)$ is a normalized time-compressed auto-correlation function of $\phi_s(t)$, ($\Phi_{\phi s}(0)=1$). Therefore *III* can be written as:

$$III\approx c_1\Phi_{\phi s}(t)\int_A P_g(\tau)P_g(\tau+t)\mathrm{d}\tau \tag{39}$$

Using this development, (27) becomes:

$$\begin{aligned}&PDP_{TR-UWB}(t)\\&\approx\phi_s^2(t)\left\{\int_A R_g(\tau)\mathrm{d}\tau+\iint_A P_g(\tau)P_g(\mu)\mathrm{d}\tau\mathrm{d}\mu\right\}\\&+c_1\left(1+\Phi_{\phi s}(t)\right)\int_A P_g(\tau)P_g(\tau+t)\mathrm{d}\tau\end{aligned} \tag{40}$$

Substituting (12) and (13) in (11a) and (15) for observation interval $A=[a,b]=[0,\infty[$ we obtain:

$$\begin{aligned} &E\{heq(t)\} \\ &= E_g\left[\exp(-\tau/\tau_{rms})-\exp(-\tau/\tau_{rms})\right]_b^a \phi_s(t) \\ &= E_g\left[\exp(-\tau/\tau_{rms})-\exp(-\tau/\tau_{rms})\right]_\infty^0 \phi_s(t) \\ &= E_g\phi_s(t) \end{aligned} \quad (41)$$

$$\begin{aligned} &PDP_{\text{TR-UWB}}(t) \\ &= \phi_s^2(t)E_g^2\left[\exp(-a/\tau_{rms})-\exp(-b/\tau_{rms})\right]^2 \\ &\quad +\frac{E_g^2}{2\tau_{rms}}\left[\exp(-2a/\tau_{rms})-\exp(-2b/\tau_{rms})\right] \\ &\quad \times\left[(1+\Phi_{\phi s}(t))c_1\exp(-t/\tau_{rms})+\left(1+\frac{1}{m}\right)\frac{1}{\lambda}\phi_s^2(t)\right] \end{aligned} \quad (42)$$

$$\begin{aligned} &PDP_{\text{TR-UWB}}(t) \\ &= E_g^2\phi_s^2(t)+\frac{E_g^2}{2\tau_{rms}}\times\left[(1+\Phi_{\phi s}(t))c_1\exp(-t/\tau_{rms})\right. \\ &\quad \left.+\left(1+\frac{1}{m}\right)\frac{1}{\lambda}\phi_s^2(t)\right] \end{aligned} \quad (43)$$

13

Design and Development of a 2 × 1 Array of Slotted Microstrip Line Fed Shorted Patch Antenna for DCS Mobile Communication System

G. A. Bidkar[1], R. M. Vani[2], P. V. Hungund[1]

[1]Department of PG Studies and Research in Applied Electronics, Gulbarga University, Gulbarga, India; [2]University Science Instrumentation Centre, Gulbarga University, Gulbarga, India.

ABSTRACT

Compact microstrip antennas have recently received much attention due to the increasing demand of small antennas for personal communication equipment. The problem of achieving a wide impedance bandwidth for compact microstrip antennas is becoming an important topic in microstrip antenna design. In this paper the design and development of a 2 × 1 array of a low cost slotted microstrip line fed shorted patch antenna (MFSPA) has been presented. Both the shorted patch and microstrip line feed network have air substrate. The material cost is thus reduced to a minimum. The array consists of two adjacent patches fed, using a simple microstrip T network. The impedance bandwidth of nearly 40%, covering the bandwidth requirement of 1750 MHz band is obtained. Also the antenna exhibits dual band operation. The cross polarization radiation in H-Plane observed with a single element antenna has been reduced considerably with 2 × 1 array. A peak antenna gain of 9.2 dBi is obtained with a small variation of 0.8 dBi. From the results obtained it is clear that the antenna array studied has a low cost fabrication and is suitable for applications in DCS mobile communication base station.

Keywords: Microstrip Antenna; Compact Antenna; Shorted Patch; Microstrip Line Fed

1. Introduction

With the ever increasing demand for mobile communication and emergence of many systems, it is important to design broadband antennas to cover wide frequency range [1]. The design of an efficient wideband small size antenna for recent wireless applications is a major challenge. Microstrip patch antennas have found extensive application in wireless communication systems owing to their advantages such as low profile, conformability, low cost fabrication and ease of integration with feed networks [2]. However, conventional microstrip patch antennas suffer from narrow bandwidth. This poses design challenges for broad band applications [3]. There are well-known methods to increase the bandwidth of antennas including increase of the substrate thickness, the use of dielectric substrate of low dielectric constant, the use of various impedance matching and feeding techniques and use of slot antenna geometry [4,5].

In this paper, design, fabrication and study of the prototype of 2 × 1 array of low cost slotted microstrip line fed shorted patch antenna has been presented. The radiating patch is shorted to ground with the help of shorting plates of suitable width. Both microstrip feed line and shorted patch have air substrate of different heights. The single element MFSPA shown in **Figure 1**, offers a good broadside radiation pattern in the H-plane. But relatively greater cross polarization radiation is observed in the H-plane patterns. The Auto-CAD diagram of a single element antenna is as shown in **Figure 2**. An effort is made to reduce the cross polarization radiation by designing a 2 × 1 array of elements as shown in **Figure 3**. In the array configuration, two adjacent patches are fed, using a simple microstrip T network having half guided wavelength difference in length between its two output feed lines.

2. The Structure and Geometry of Single Element and 2 × 1 array of MFSPA

The artwork of the proposed antenna array is carried out by software, Auto-CAD 2004 for achieving better accuracy. The antenna is fabricated on a low cost glass epoxy substrate material of thickness h = 1.6 mm and dielectric constant 4.2. The radiating patch has a length L and width W. The patches are supported by plastic posts above the ground plane. The distance of the radiating patch to the ground plane is "h". The radiating patch is short circuited to the ground plane by using identical shorting plates of width "d" placed at two ends of one of the patch's radiating edges.

At the centre of the patch edge with shorting plates, a 50 ohm microstrip feed line is used to directly feed the radiating patch. The strip of the feed line has a width W_f and is connected to the radiating patch at the patch's shorted edge by conducting strip of the same width W_f. The air substrates of shorted patch and feed line have heights "h" and "t" respectively. By selecting suitable value of "h" a wide impedance bandwidth can be obtained. Rectangular slots of size 2 mm × 10 mm inclined at an angle of 45 degrees on the top left and bottom right corners are placed.

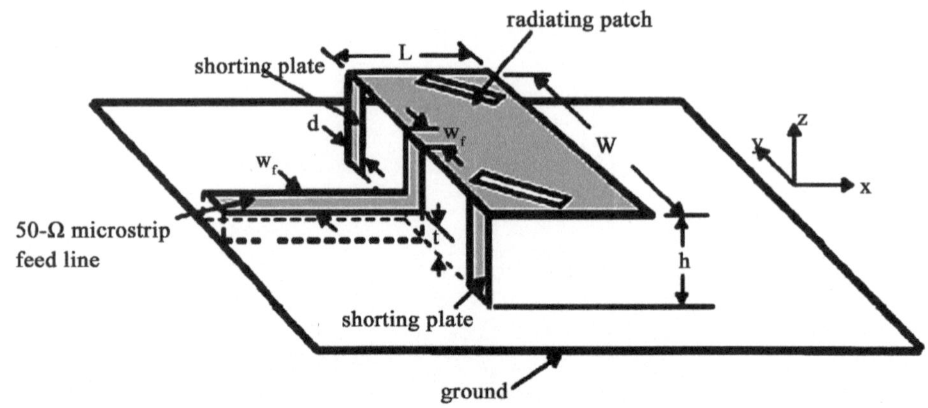

Antenna Parameters: L = Length of the patch; W = Width of the patch; W_f = Width of the 50 Ω feed line; h = Height of the radiating patch above the substrate; t = Height of the feed line above the substrate; d = Width of the shorting plates.

Figure 1. Basic geometry of MFSPA with inclined slots.

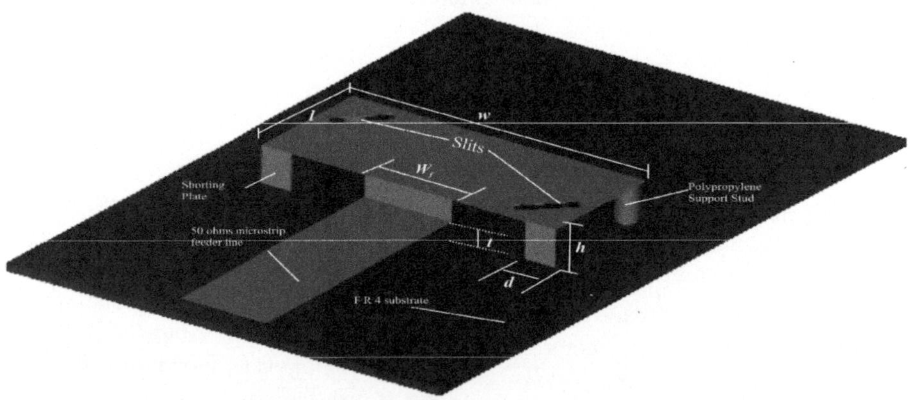

Figure 2. Auto-CAD diagram of MFSPA.

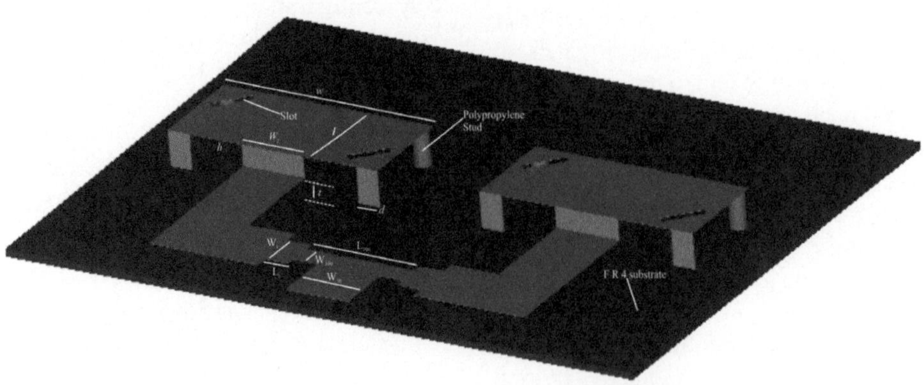

Figure 3. 2 × 1 array of slotted MFSPA.

In the 2 × 1 array of MFSPA the two adjacent patches are placed on the substrate at a distance of 7.1 cm between their centers *i.e.* $\lambda/2.3$, approximately equal to half wavelength. It consists of a 50 Ω microstrip line of length L_{50} and width W_{50} which is connected to a 100 Ω line of length L_{100} and width W_{100} to form a two way power divider. A matching quarter wave transformer of length Lt and width Wt is connected between 100 Ω feed line and mid point of the radiating elements in order to ensure perfect impedance matching. The array geometry and feed arrangement is shown in **Figure 3**. **Table 1** gives the design parameters of 2 × 1 array.

3. Results and Discussions

Measurements for the return loss and the radiation patterns of the antennas are made with the Vector Network Analyzer (ROHDE & Schwarz, German made SVK Model-7.8651). **Figure 4** shows the measured return loss versus frequency plot of the single element antenna. The experimental study of single element antenna has shown that it offers an impedance bandwidth of 25.2% at a resonant frequency of 1905 MHz as shown in **Figure 4** and the gain of 6.4 dBi. **Figure 5** shows the radiation patterns of a single element antenna.

From **Figure 6** it is clearly seen that an impedance bandwidth of 40% covering the bandwidth requirement of 1750 MHz band (1200 MHz - 1900 MHz) is obtained by 2 × 1 array configuration. The return loss curve also shows the dual band operation of the antenna. The antenna resonates at the second frequency f_2 (2350 MHz) offering a bandwidth of 36% (2000 MHz - 2850 MHz). Typical measured radiation patterns for the antenna are presented in **Figure 7**. Good broadside patterns are obtained in the H plane. A relatively lesser cross polarization radiation in H plane is observed for the antenna array as compared to that of a single element antenna.

Table 1. Design parameters.

Design Parameter	Dimension in mm	Design Parameter	Dimension in mm
L	23.5	L_{50}	12.5
w	54	W_{50}	11
h	8	L_{100}	28
t	3.2	W_{100}	5
W_f	16	w_t	10
d	5.5	l_t	6

The distance "D" between the adjacent patches is D = 7.1 cm ($\lambda_0/2.3$); Ground plane size: 100 × 150 mm^2.

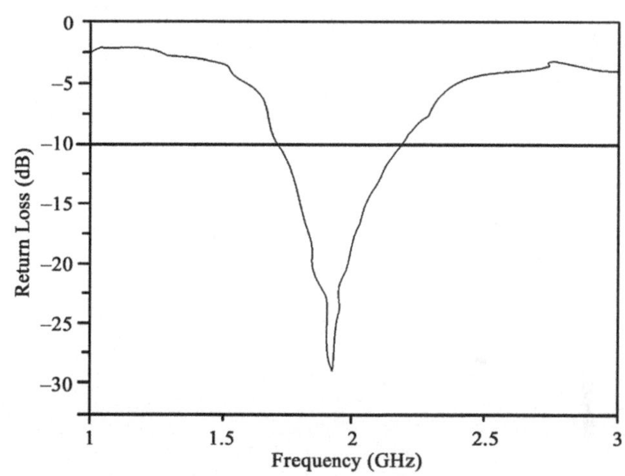

Figure 4. Return loss vs frequency plot of single element slotted MFSPA.

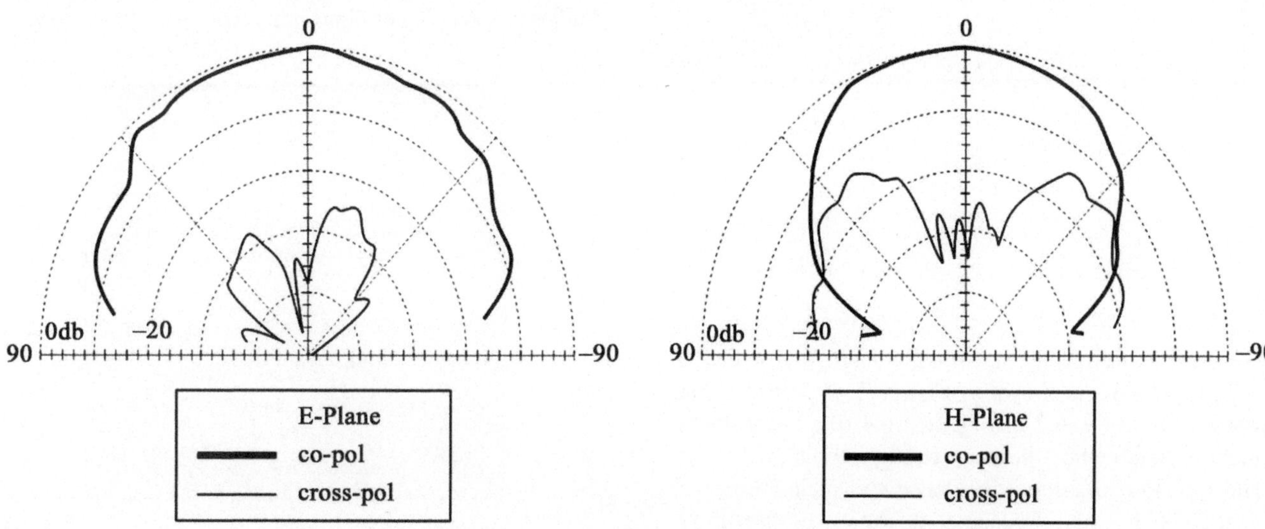

Figure 5. Radiation patterns of single element MFSPA with inclined slots.

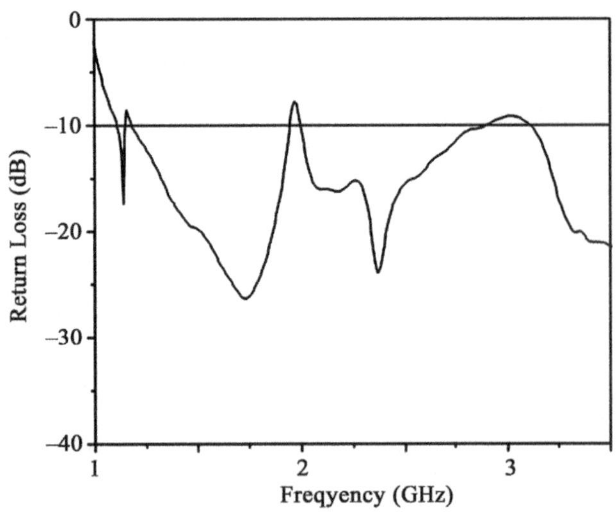

Figure 6. Return loss vs frequency plot of the 2 × 1 array of slotted MFSPA.

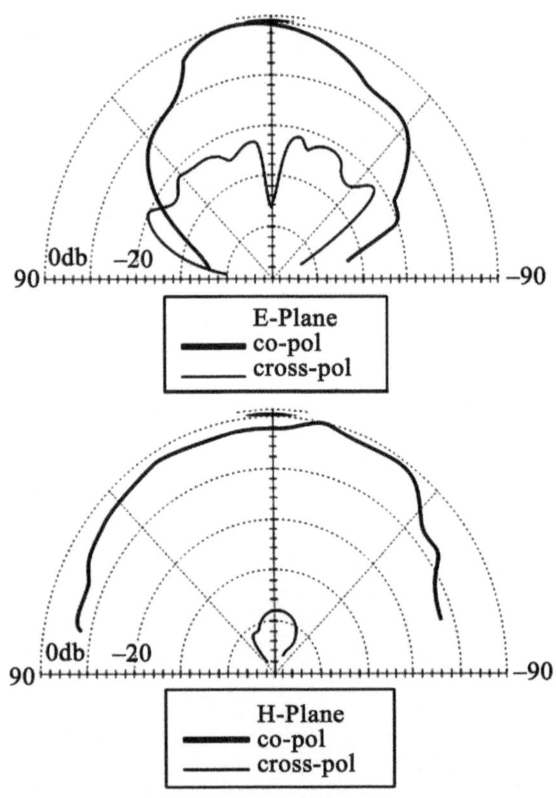

Figure 7. Radiation patterns of 2 × 1 Array of MFSPA with inclined slots.

The gain vs frequency plot for 2 × 1 array of MFSPA with inclined slots is shown in **Figure 8**. A considerable increase in gain *i.e.* 9.2 dBi against 6.4 dBi for a single element is achieved by the array configuration.

The variation of input impedance shown in **Figure 9** also depicts the dual frequency operation of the array configuration.

A typical MFSPA is simulated using IE3D software and is shown in **Figure 10**. The return loss vs frequency curve for the same is shown in **Figure 11**.

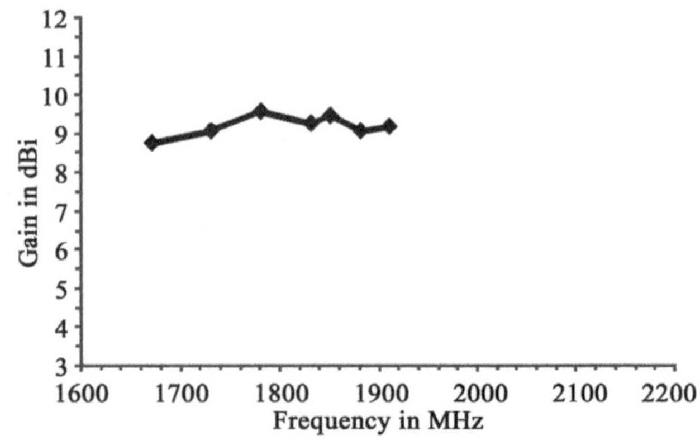

Figure 8. Gain vs Frequency plot of 2 × 1 Array.

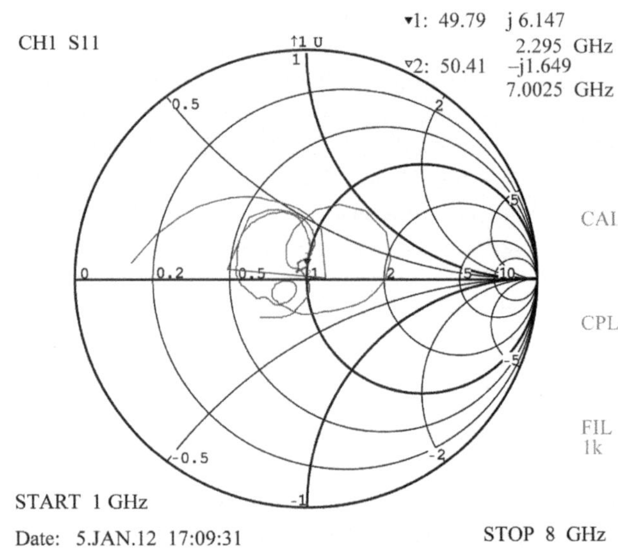

Figure 9. Variation of input impedance with frequency.

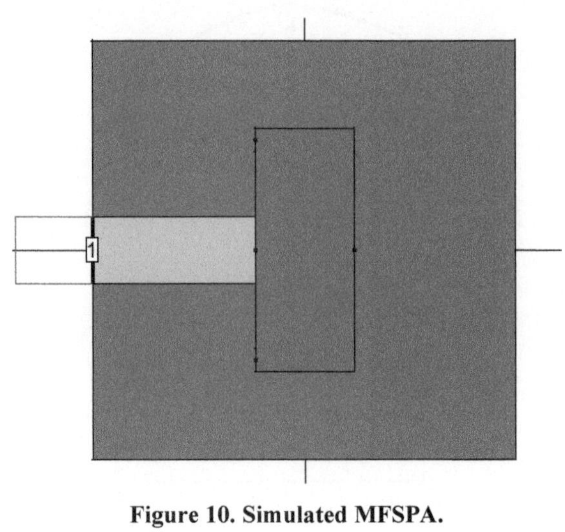

Figure 10. Simulated MFSPA.

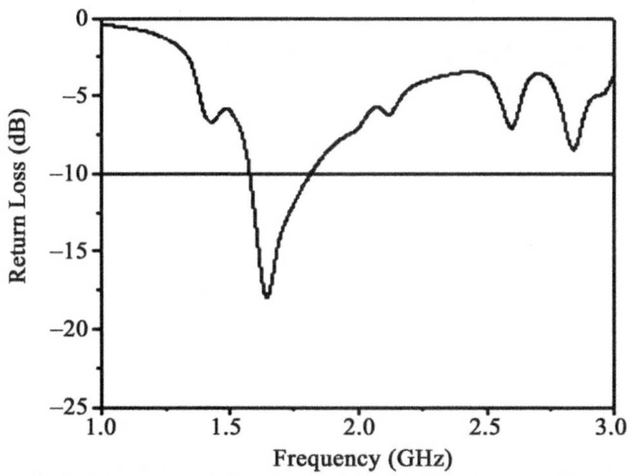

Figure 11. Return loss vs frequency plot of simulated MFSPA.

4. Conclusion

The prototype of a 2 × 1 array of MFSPA has been developed and its performance has been studied. The performance of the same is compared with that of the single element antenna. The proposed design of the 2 × 1 array provides an enhanced impedance bandwidth of 40% and 36% at the two bands with respect to the centre frequencies of 1750 MHz and 2350 MHz. The prototype of a single element antenna presents bandwidth of 25.2% at a centre frequency of 1950 MHz. Both the antennas show good broadside radiation patterns. Substantial reduction in cross polarization radiation is observed with antenna array configuration. The results show that the proposed antenna may conveniently be used for DCS base station applications and also for the other wireless applications due to its dual band operation.

REFERENCES

[1] K.-L. Wang, "Compact and Broadband Microstrip Antennas," John Wiley and Sons, Inc., Chichester, 2002.

[2] K. Fujimoto and J. R. James, "Mobile Antenna System Handbook," 2nd Edition, Artech House Inc., 2001.

[3] R. G. Vaughan and J. B. Anderson, "Antenna Diversity in Mobile Communication," *IEEE Transactions on Antennas and Propagation*, Vol. 49, 1987, pp. 954-960.

[4] S.-Y. Lin and K.-C. Huang, "A Compact Microstrip Antenna for GPS and DCS Application," *IEEE Transactions on Antennas and Propagation*, Vol. 53, No. 3, 2005, pp. 1227-1229.

[5] S. W. Su and J. H. Chou, "Low Cost Flat Metal Plate Dipole Antenna for 2.4/5-GHz WLAN Operation," *Microwave and Optical Technology Letters*, Vol. 50, No. 6, 2008, pp. 1686-1687.

14

Mixers of Ultra-High Gain from 5.0 to 18.0 GHz

Hsin-Chia Yang[1], Jui-Ming Tsai[2], Cheng-Huang Tsao[1], Mu-Chun Wang[1], Sungching Chi[1], Tsin-Yuan Chang[2]

[1]Department of Electronics Engineering, Ming Hsin University of Science and Technology, Hsin-Chu City, Chinese Taipei; [2]Department of Electrical Engineering, National Tsing-Hua University, Hsin-Chu City, Chinese Taipei.

ABSTRACT

Mixers in the communication system provide the possibility of encoding and decoding radio-frequency EM waves with signals through the help of local oscillators. A mixer with capability of high conversion gain, good isolation, and good linearity is comparably appreciated. Extensively wide ranges of frequencies, from 5.0 to 18.0 GHz, are to be examined addressing the promising functions of mixers in this study. A TSMC 0.18 μm CMOS model implanted in Agilent ADS is used for the circuit designs. Generated from Gilbert Cell Mixer, the modified circuits take advantage of extra active and passive devices to optimize the conversion gains. Characteristics of high conversion gain over 20 dB or even higher (as high as 29.842 dB at −40 mW RF power at working frequency 6 GHz) and low noise figures (NF) are shown.

Keywords: Gilbert Cell Mixer; P1dB; Linearity; Noise Figure (NF); Conversion Gain; Isolation

1. Introduction

Mixers are indispensable components in transceiver system (transmitter and receiver) in the communication system, providing the possibility of encoding and decoding radio-frequency EM waves with signals through the help of local oscillators. As quoted as "Mixer", a mixer mixes two input signals with different frequencies and yields an output signal with a specific frequency. A mixer with capability of high conversion gain, good isolation, and good linearity is comparably appreciated.

Traditional Gilbert Cell [1] in **Figure 1** composing of six NMOS devices mainly generates two differential currents, namely, i_{D1} and i_{D2}, flowing through M1 and M2 with the common trans-conductance (g_m). In one aspect, the two currents associate the two inputs, one from the local oscillator imposing on the two sets, (M3, M4) and (M5, M6), and the other from RF signals impacting on M1 and M2, as expressed as follows:

$$\begin{cases} i_{D1} \cong \dfrac{C_{ox}^{(1)}\mu_N W}{L}\left(v_G - v_{th}\right)v_{DS1} \\ i_{D2} \cong \dfrac{C_{ox}^{(1)}\mu_N W}{L}\left(v_G - v_{th}\right)v_{DS2} \end{cases} \quad (1)$$

with $g_{m1} = \dfrac{\partial i_D}{\partial v_G} = g_{m2}$. In Equation (1), $C_{ox}^{(1)}$ is oxide capacitance (F/m^2), μ_N is the mobility (m^2/sV), and L is the channel length. Because of the multiplication of $v_G = v_{Go}\cos\left(\omega_{LO}t\right)$ and $v_{DS} = v_{DSo}\cos\left(\omega_{RF}t\right)$ in Equation (1), the output IF_P signals are resulted, as derived in Equation (2), achieving raising and lowering of frequencies, *i.e.*, $\left(\omega_{RF} \pm \omega_{LO}\right)$.

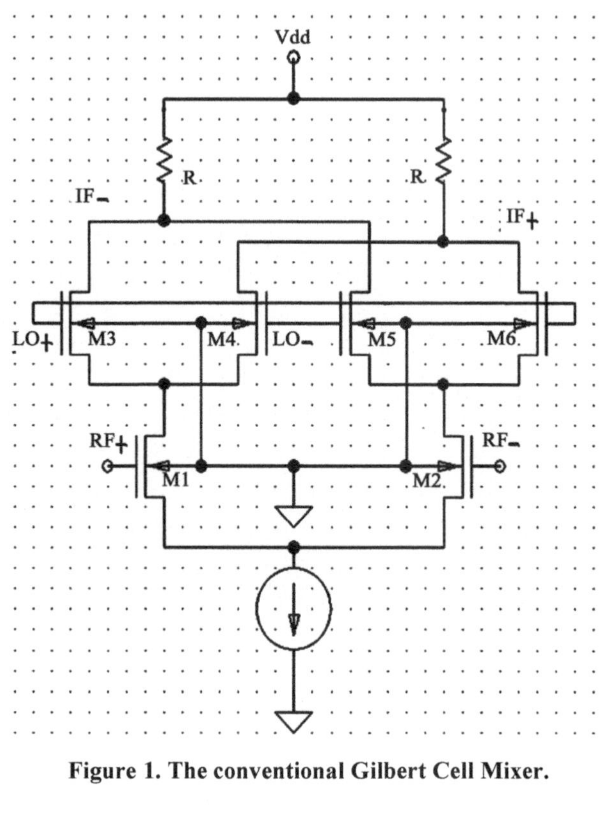

Figure 1. The conventional Gilbert Cell Mixer.

$$\begin{aligned} v_G v_D &= v_{Go} v_{DSo} \cos(\omega_{LO} t)\cos(\omega_{RF} t) \\ &= \frac{v_{Go} v_{DSo}}{2} \cos\left[(\omega_{RF} + \omega_{LO})t\right] \\ &+ \cos\left[(\omega_{RF} - \omega_{LO})t\right] \end{aligned} \quad (2)$$

In this paper, a powerful mixer using an extensively generated circuit based on Gilbert Cell with sufficient current sources in the downstream is proposed as shown in **Figure 2**. Somehow, to enhance the stability and the amplification, the symmetrical topology is used by duplicating the unit in **Figure 3** on both sides of **Figure 2**. Advanced Design System (ADS) by Agilent implemented with a TSMC 0.18 micron model is to be used to evaluate the circuit design.

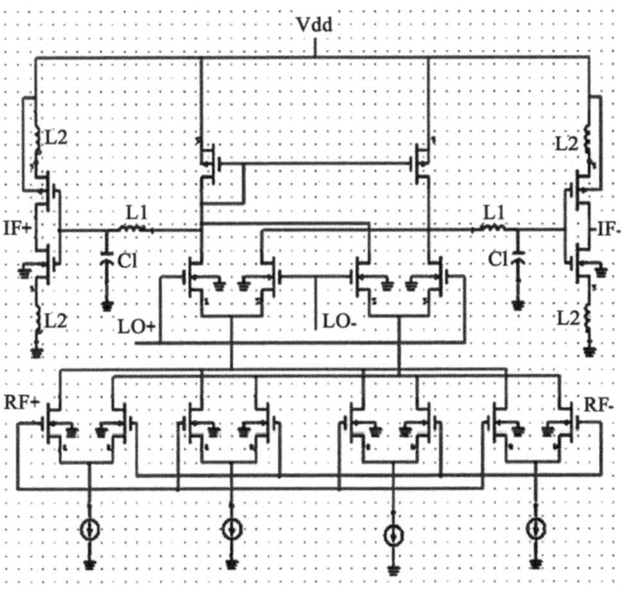

Figure 2. Four-current-source Gilber Cell Mixer with two inverters enhancing the amplifying effects.

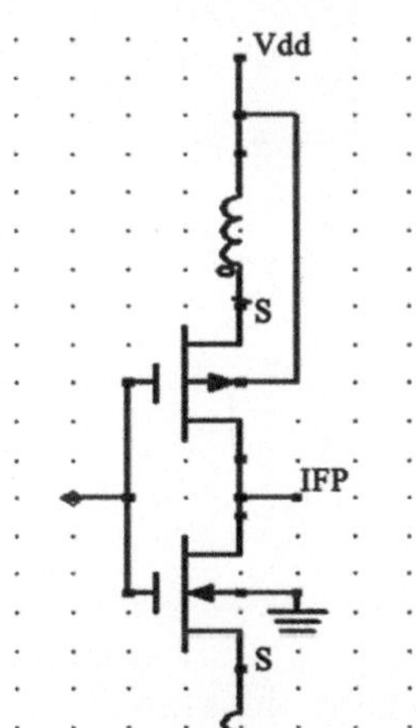

Figure 3. The inverter-like CMOS units.

2. Mixer Circuit Design

For the sake of larger current flow, the circuit in **Figure 2** generated from the conventional Gilbert Cell Mixer is to be optimized using four I_{bias}'s instead of only one I_{bias}. To avoid the unnecessary power consuming, passive resistors are never considered. With the introduction of inductors and capacitors (L1, and C1), the noise figures are expected to be lowered. And the isolations at higher frequencies ranged from 6 GHz to 18 GHz are supposed to be better because of the two sets of inverter-like units as shown in **Figure 3**. Each set is inserted with two inductors (L2), one in the upstream and the other in the downstream, to keep the radio frequency signals from leaking. The whole set in **Figure 3** is thus operated to raise the conversion gains as well. The symmetrical topology is to be used and the current sources are adjusted step by step. The number of mirror units is to be imposed and the number of symmetrical NMOS devices where LO and RF are applied is carefully examined and determined.

3. Results and Analysis

In general, the conversion gains are normally high above 20 dB from 6 GHz to 18 GHz as listed in **Table 1**, in which the maximum conversion gain can be even as high as almost 30 dB. As shown in **Figure 4(a)**, the exciting results are strongly appreciated at 15 GHz, such as the conversion gains = 23.256 dB at RF_power = –40 mW, 17.881 dB at RF_power = –20 mW, and even over 2.5 dB at RF power = 0 mW. This circuit executes not only the function of a mixer but also enhance the conversion gain, which is quite encouraging and convincing. Two reasons are speculated: 1) the four current sources require higher current to flow through the whole component; and 2) the two inserted inductors in series with the inverter-like CMOS devices keep the high frequency signals from dumping to the ground [2-7]. **Figure 4(b)** also demonstrates the surprisingly good linearity at such a high frequency from power –40 mW to power –20 mW, extending to about power –15 mW at P1dB point. In addition, the isolations among LO (Terminal 1), RF (Terminal 2), and IF (Terminal 3) are to be found in **Figure 5**, in which S12 = –15.173 dB, S23 = –28.902 dB, and S13 = –24.783 dB and all are satisfying. Unfortunately, the noise figure NF = 8.400 dB in **Figure 6** is not sufficiently low as expected. Nevertheless, it is still controllable and is expected to be lowered in the near future.

4. Discussions

The last two columns of **Table 1** show the encouraging conversion gains up to 18 GHz at RF_power = –40 mW and at RF_power = –20 mW. As compared to the previous work in [8], the conversion gains can be even

Table 1. Comparisons of the previous work and this work.

	Freq. (GHz)	Previous Work [3] Gain (dB) At RF_Power = –40 mW	Previous Work [3] Gain (dB) At RF_Power = –20 mW	This Work Gain (dB) At RF_Power = –40 mW	This Work Gain (dB) At RF_Power = –20 mW
TSMC 0.18 m Process	6.0	21.153	18.884	29.841	20.623
	7.0	22.583	19.167	28.640	20.524
	8.0	22.142	18.852	27.757	19.856
	9.0	21.599	18.349	27.220	19.726
	10.0	20.954	17.745	26.616	19.482
	11.0	NA	NA	26.042	19.231
	12.0	NA	NA	25.471	18.950
	13.0	NA	NA	24.854	18.577
	14.0	NA	NA	24.052	17.954
	15.0	NA	NA	23.256	17.881
	16.0	NA	NA	21.925	18.146
	17.0	NA	NA	20.987	17.711
	18.0	NA	NA	20.100	17.280

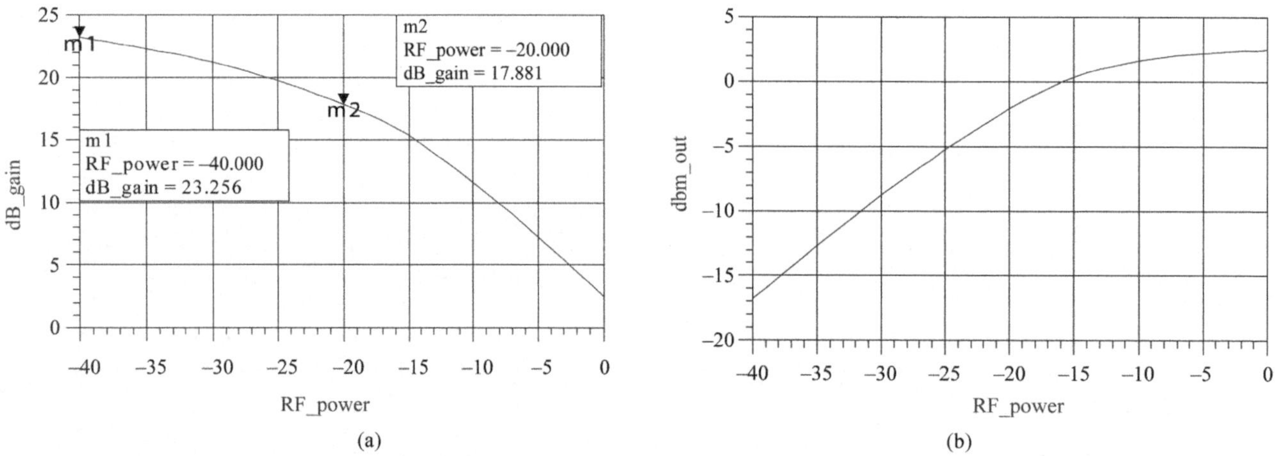

Figure 4. (a) The conversion gain; and (b) the linearity.

noticeably better from 6 GHz to 10 GHz. It is due to careful tuning on the passive devices and the stable symmetrical topology, which can be seen in the mirror feature. In addition, the inverter-like CMOS devices with inductors in series play the roles of keeping the high frequency signals away from dumping to the ground, and thus contribute significantly to the conversion gains. The current sources were kept the same at 2 mA for its satisfactory status in this structure compared to the previous design [8]. Somehow, the currents of the four current sources and the voltage of the voltage source are expected to be lower in the continuous work. In the mean time, noise figures are to be lowered in the near future, too, even at such high working frequencies.

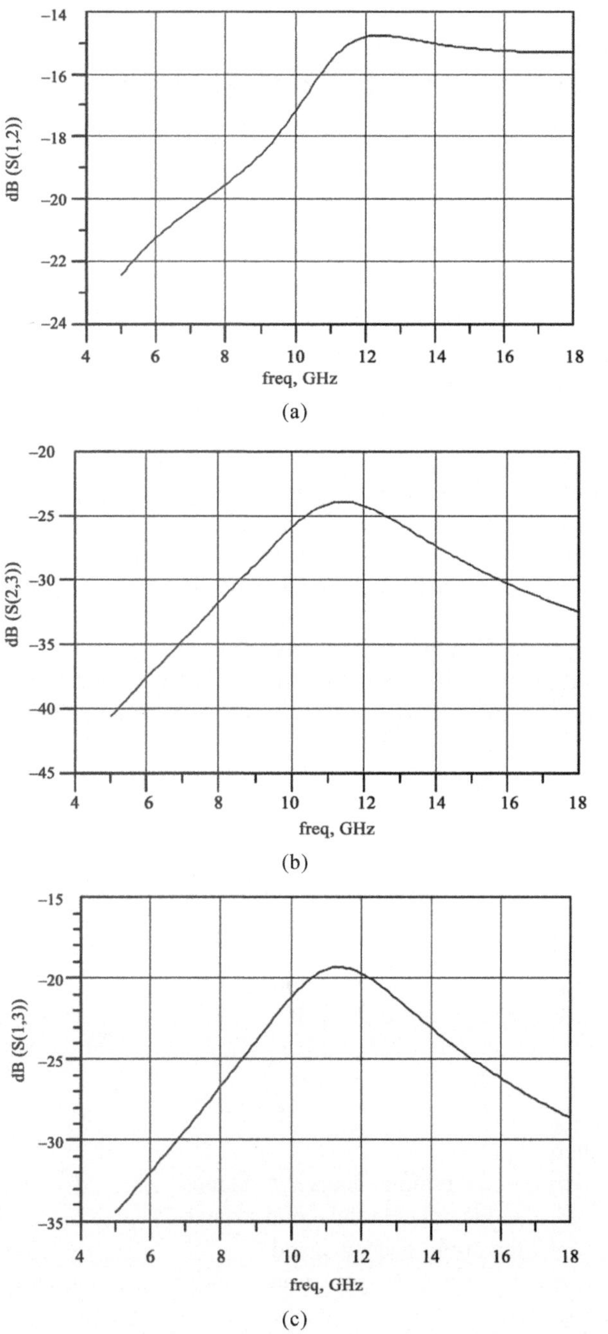

Figure 5. The isolation among LO (Terminal 1), RF, (Terminal 2) and IF (Terminal 3) at 15 GHz (a) LO-RF (b) RF-IF (c) LO-IF.

5. Conclusion

A Gilbert-Cell Mixer based mixer is introduced with two inverter-like CMOS units in series with two inductors. The conversion gains are found to be significantly improved. The good isolation, acceptable noise figure, and linearity are to be checked. The encouraging data shows the promising applicable circuit designs. One would continue to develop the total solution for the whole component and even the whole system.

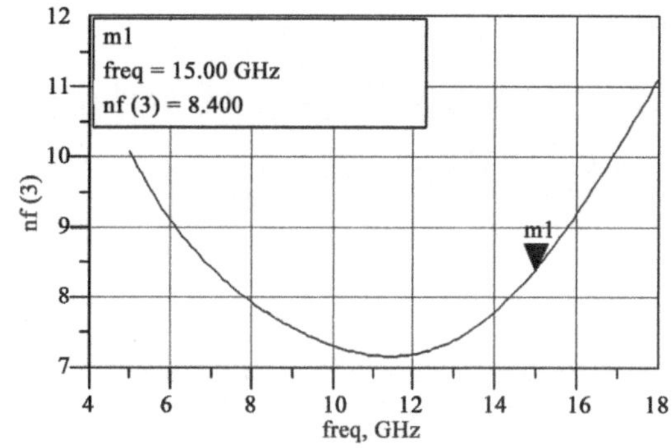

Figure 6. Noise figure at IF at 15 GHz.

6. Acknowledgements

The authors would like to thank Yi-Chang Cheng for the support. Special thanks are due to all the teaching assistants, Fang-Fang Chen and Ven-Chung Yu.

REFERENCES

[1] M. T. Terrovitis and R. G. Meyer, "Noise in Current-Commutating CMOS Mixers," *IEEE Journal of Solid-State Circuits*, Vol. 34, No. 6, 1999, pp. 772-783.

[2] K. Choi, D. H. Shin and C. P. Yue, "A 1.2-V, 5.8-mW, Ultra-Wideband Folded Mixer in 0.13-µm CMOS," *IEEE Radio Frequency Integrated Circuits (RFIC) Symposium*, Honolulu, 3-5 June 2007, pp. 489-492.

[3] H. Qi, J. Chen, H. Pan and J. Meng, "A 1.2 V High Linearity Mixer Design," *IEEE ICEMI*, Xi'an, 16-18 August 2007, pp. 274-277.

[4] P. J. Sullivan, B. A. Xavier and W. H. Ku, "Low Voltage Performance of a Microwave CMOS Gilbert Cell Mixer," *Digital Object Identifier*, Vol. 32, No. 7, 1997, pp. 1151-1155

[5] T.-P. Wang, C.-C. Chang, R.-C. Liu, M.-D. Tsai, K.-J. Sun, Y.-T. Chang, L.-H. Lu and H. Wang, "A Low-Power Oscillator Mixer in 0.18-m CMOS Technology," *Digital Object Identifier*, Vol. 41, No. 1, 2006, pp. 88-95.

[6] E. E. Bautista, B. Bastani, J. Heck and Member IEEE, "A High IIP2 Down Conversion Mixer Using Dynamic Matching," *IEEE Journal of Solid-State Circuits*, Vol. 35, No. 12, 2000, pp. 1934-1941.

[7] A. Q. Safarian, A. Yazdi and P. Heydari, "Design and Analysis of an Ultrawide-Band Distributed CMOS Mixer," *IEEE Transactions on Very Large Scale Integration (VLSI) Systems*, Vol. 13, No. 5, 2005, pp. 618-629.

[8] H.-C. Yang, M.-D. Chang, M.-C. Wang, S.-C. Chi and C.-T. Wang, "3.5 GHz to 10.0 GHz Mixers of High Gains and Good Isolations," *IEEE 7th International Conference on Wireless Communications, Neworking and Mobile Computing, (WiCOM)*, Wuhan, 23-25 September 2011.

Design of a CMOS Optical Receiver Front-End Using 0.18 µm Technology

Ajay Shukla, Radheshyam Gamad, Rohan Raikwar

Department of Electronics & Instrumentation Engineering, S. G. S. Institute of Technology and Science, Indore, India.

ABSTRACT

This paper reports design of a CMOS optical receiver front-end using 0.18 µm technology. Design process is current associated with photodiode using trans-impedance amplifier (TIA) for wide bandwidth, high gain, low input referred noise and wide dynamic range. The Automated Gain Control (AGC) voltage is used to provide variable gain for multi-level signals. This design is simulated in 0.18 µm UMC technology for the performance analysis. The best simulation results are reported the maximum TIA gain of 67.26 dBΩ at 0 V AGC followed by a post amplifier gain of 86.70 dBΩ. The bandwidth range is 7.03 GHz to 11.5 GHz corresponding to 0 - 3 V AGC respectively. The input referred noise level value is 43.86 pA/√Hz up to 10 GHz frequency. In addition authors have obtained the common mode rejection ratio (CMRR) is 72.42 dB and rectified group delay is 144.48 ps. Verification of the design, reported results are compared with earlier published work and improvements obtained in the present results.

Keywords: Component; Formatting; Style; Styling; Insert

1. Introduction

Optical receivers find applications in laptop computers, cellular phones, digital cameras, computer peripherals, personal digital assistants (PDAs), and many other consumer electronics equipped with a short-distance communication port. A commonly used topology is the trans-impedance (TIA) amplifier, whose relative low input impedance and wide bandwidth is well suited for the application [1]. A wide dynamic range is required for infrared wireless optical receivers in order to accommodate variable link distances, 0 ~ 100 cm. In the design of fixed-gain trans-impedance feedback amplifiers for infrared wireless receivers, there is a direct trade-off between input noise current and the input current overload level via the value of the shunt feedback resistor employed. Therefore, in order to enlarge the dynamic range, various means have been adopted to vary the gain of the trans-impedance amplifiers in response to the input signal levels. Variable-gain trans-impedance feedback amplifiers, however, are prone to instability [2]. TIA needs an Automated Gain Control (AGC) circuit to control the gain of the TIA in order to keep the different symbol levels equally spaced for multilevel signaling [3]. The TIA uses an AC coupled, differential version of the low-voltage topology outlined in [4], which is modified common-gate architecture. It is AC coupled to the photo detector (PD) using vertical parallel plate interconnects capacitors. **Figure 1** shows the block representation of the receiver.

For open loop TIA the input resistance of this amplifier can be determined by [5]:

$$Rin = \frac{rds\_1 + Rd}{1 + (gm\_1 + gmb\_1)rds\_1} \tag{1}$$

where,

rds_1 = the drain to source resistance

gm_1 = the device trans-conductance

Rd = the drain resistance and

gmb_1 = the back-gate trans-conductance due to the body effect

For long channel devices operating in the saturation region, the value of *rds* is large and it can be reduced by this relation:

$$Rin = \frac{1}{gm + gmb} \tag{2}$$

This is an important result because the bandwidth is independent of the trans-impedance gain set by RD. The unfortunate downside of open loop TIA is that the noise current produced by the load resistance RD and the bias transistor are directly referred to the input with a unity factor therefore closed loop TIA are more preferable because the feedback resistor can be increased independently to the supply voltage since no bias current flows through it. The trans-impedance gain of an ideal inverting voltage amplifier with a feedback resistor *Rfb* can be given as [1]:

$$Zt = \frac{A.Rfb}{(A+1)\left(1 + jw\frac{Rfb.Cpd}{A+1}\right)} \quad (3)$$

where,

A = the open loop voltage gain of the amplifier.

Cpd = the photodiode capacitance.

If the voltages gain "A" of the amplifier is sufficiently high; the trans-impedance is approximately equal to Rfb in the amplifier's pass band. Assuming that the dominant pole is at the input, the $3db$ bandwidth of this circuit will be given by the following expression as:

$$f3db \approx \frac{A+1}{2\pi Rfb.Cpd} \quad (4)$$

From Equation (4) it can be concluded that the bandwidth of the TIA is greater than that of a simple resistive network by a factor of $A + 1$. To reduce input referred noise current, TIA uses cascode noise from the resistor is directly referred to the input such that the mean squared input referred noise current spectral density is constant and given in equation

$$\frac{d|in|^2}{df} = \frac{d|iRl|^2}{df} = \frac{4KT}{RL} \quad (5)$$

The group delay is defined as the negative of the derivative of the phase of the trans-impedance with respect to frequency [1] common-source topology with feedback resistor. This feedback resistor is implemented to single NMOS transistor in triode operation region because of process variation [6]. The input referred noise current is also dependent on the value of R_L. The

$$Tg(f) = \frac{-1}{2\pi}\left[\frac{d(Zt(f))}{df}\right] \quad (6)$$

A flat group delay means the amplifier has a linear phase response. A flat group delay is important because variations in the group delay with frequency can cause distortions in the output signal.

2. Proposed Design

One of the issues to consider when comparing the two topologies common source and common gate is the effect of the Miller capacitance. **Figure 2** shows a small signal equivalent model of the common source (CS) configuration.

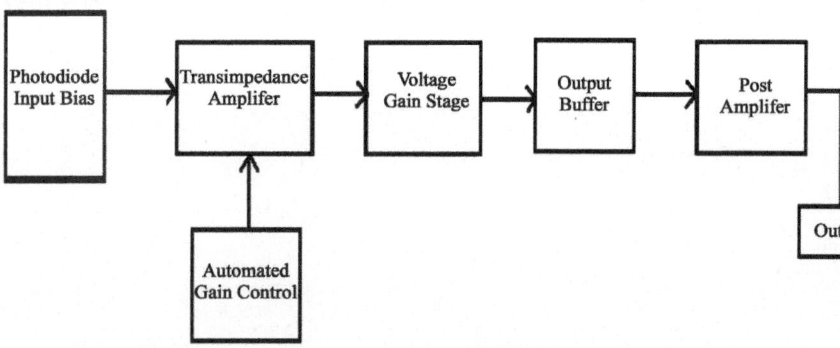

Figure 1. Block diagram of CMOS optical receiver front-end.

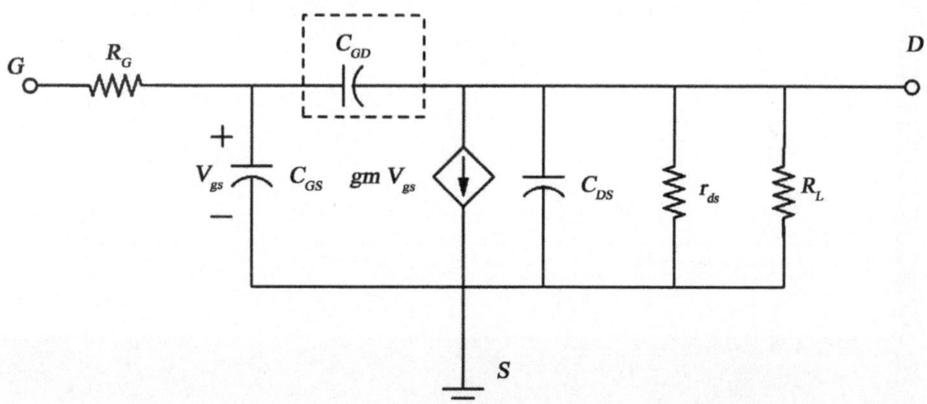

Figure 2. Small signal equivalent model of the CS configuration.

The gate to drain capacitance C_{GD}, known as the Miller capacitance, is connected between the input and output. Miller's theorem allows this capacitance to be replaced with shunt capacitances at the input and output [7]. Miller's theorem can be derived using **Figure 3**. Series admittance is connected between two points with a known voltage gain of K. In order to replace this series admittance with shunt admittances the input and the output, the currents I_1 and I_2 must remain constant during the transformation.

The values of the shunt admittances can now be determined and are shown below.

$$Y_1 = Y(1-K) \quad (7)$$

$$Y_2 = Y(1-1/K) \quad (8)$$

This is an interesting result because as the gain is increased, the input capacitance of the amplifier is increased. This reduces the magnitude of the input pole and reduces the bandwidth of the TIA. This effect can be reduced by using the cascode configuration which minimizes the Miller effect by placing a common gate transistor in series with the common source transistor. A voltage gain stage was added after the trans-impedance stage [8]. This stage uses a simple resistive loaded differential pair. Also, for measurement purposes, an output buffer has been added to the TIA. The output buffer is also a resistive loaded differential pair with output impedance that is matched to 50 Ω. The post amplifier is implemented by a fully-differential two-stage operational trans-conductance amplifier [9,10]. **Figure 4** shows a schematic view of the proposed design.

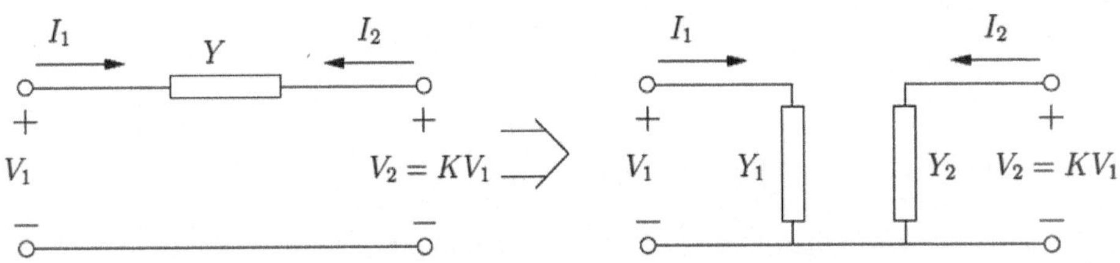

Figure 3. Miller theorem equivalent.

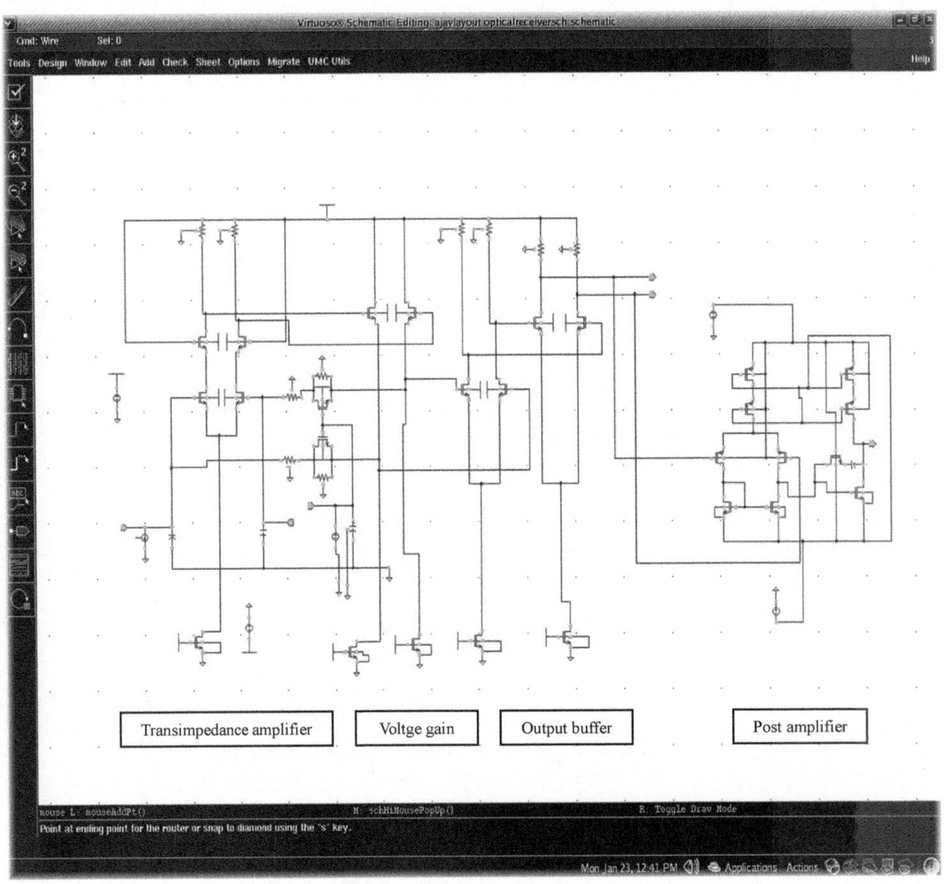

Figure 4. Schematic view of proposed design.

3. Results and Discussions

Figure 5 shows the transient analysis of optical receiver. In this the current at two differential inputs I_6 and C1 is converted into the voltage forms using the trans-impedance amplifier. This converted voltage posses no distortion and expressed as the sinusoidal waves at the differential output which is further shifted its level to 3 V at the output of post amplifier.

3.1. AC Analysis

Figure 6 represents the gain of trans-impedance *i.e.* the ratio of output voltage to input current. At 0 V AGC, the Gain of TIA is 67.26 dBΩ and the corresponding 3 dB bandwidth is 7.03 GHz. This gain is further amplified in this design using post amplifier and amplified gain is 86.70 dBΩ, its graphical representation is given in **Figure 7**.

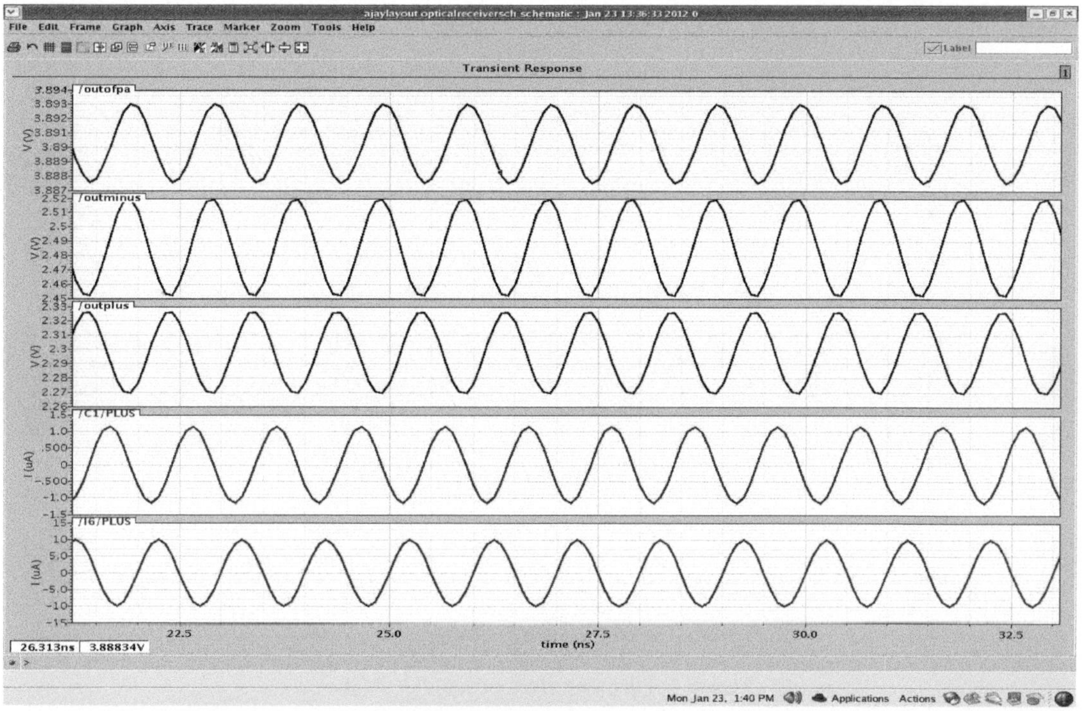

Figure 5. Transient analysis.

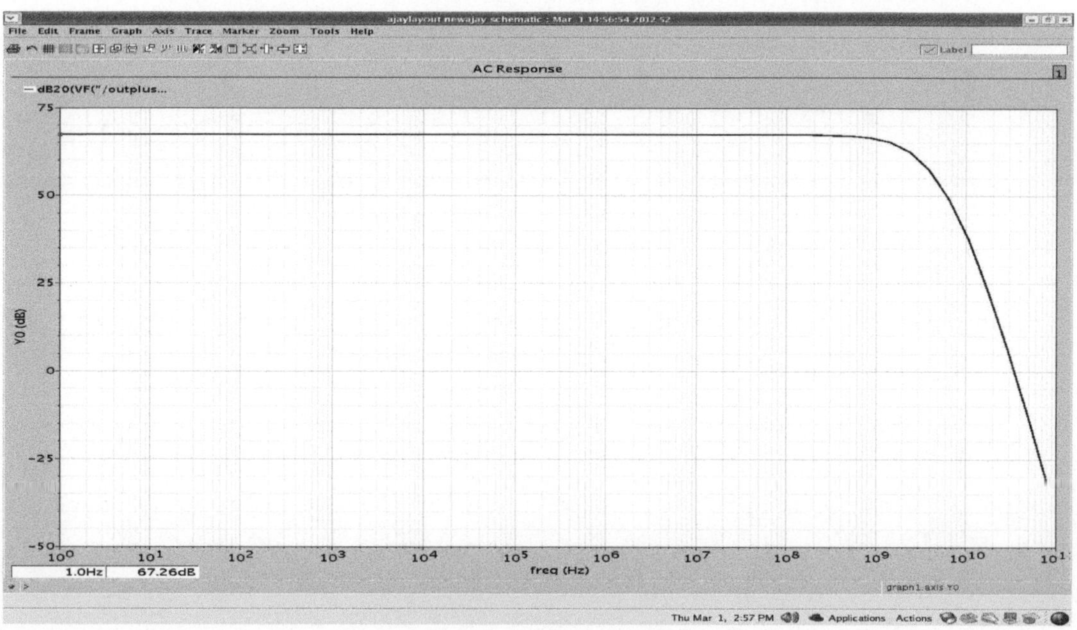

Figure 6. TIA gain at 0 V AGC.

The TIA Gain is 50.65 dBΩ and the corresponding bandwidth is 11.5 GHz at 3 V AGC as shown in **Figure 8**. This gain is increased to 70.09 dBΩ at the output of the post amplifier as represented in **Figure 9** and Phase shift and CMRR of TIA by **Figure 10**.

3.2. Noise Analysis

From **Figure 11**, input referred noise is 9.33 pA/√Hz at 0 V AGC pointed to 1 GHz frequency and it is estimated to its lower value of 43.86 pA/√Hz up to 10 GHz frequency. At 3 V/1 GHz AGC the input referred noise is 31.91 pA/√Hz which exceeds its value to 45.45 pA/√Hz upto 10 GHz *i.e.* the better input noise response as shown in **Figure 12**.

3.3. Group Delay

Figure 13 prescribed that the group delay is 144.48 ps and is maintained constant till 1 GHz frequency and afterwards starts decaying for the higher range of frequencies. **Table 1** presents the comparison of present results with earlier published work.

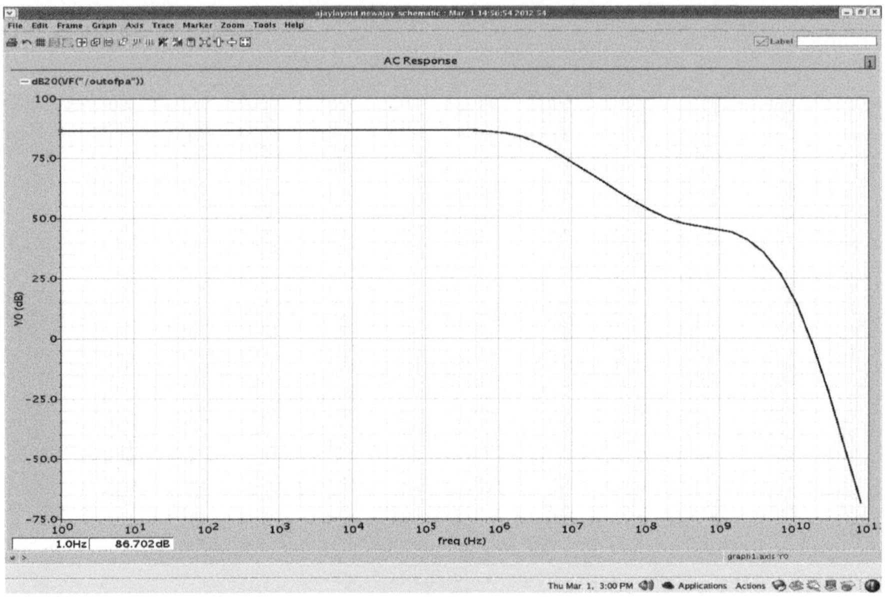

Figure 7. Post amplifier gain at 0 V AGC.

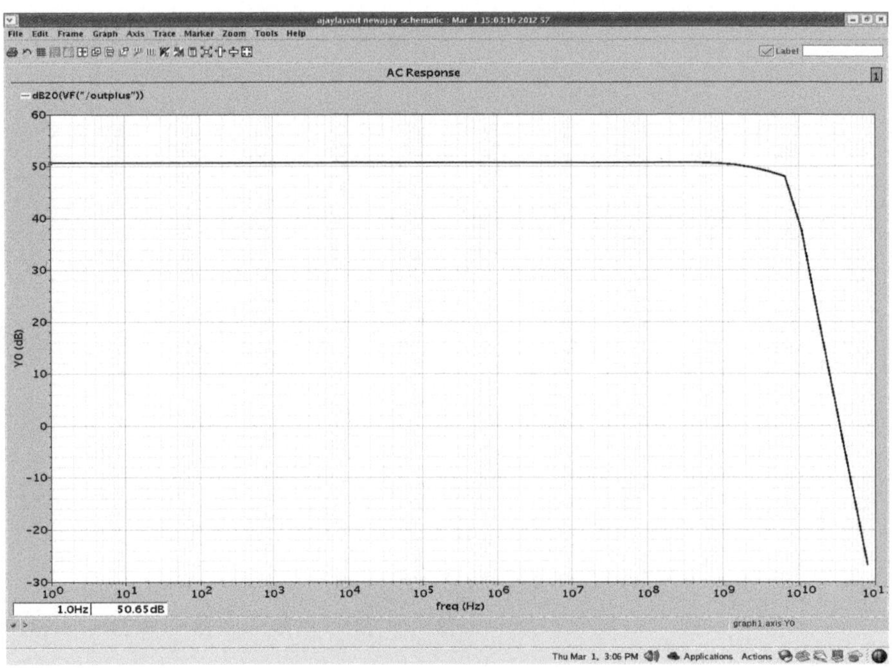

Figure 8. TIA gain at 3 V AGC.

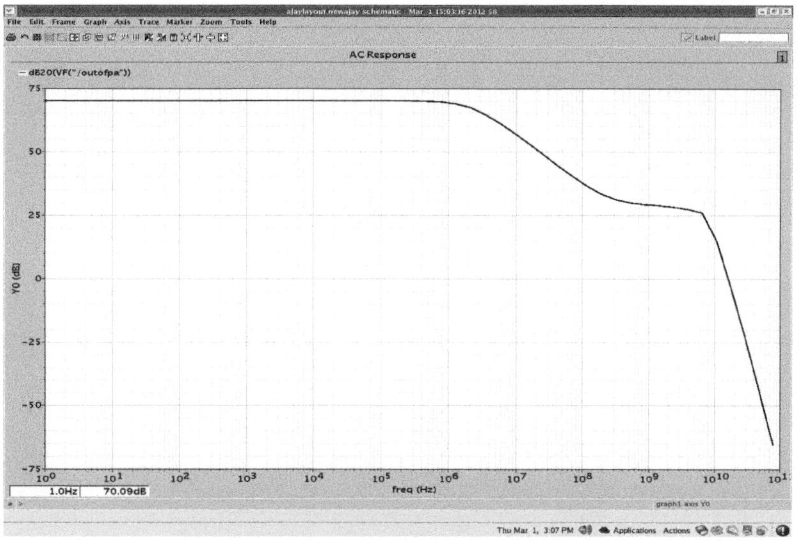

Figure 9. Post amplifier gain at 3 V AGC.

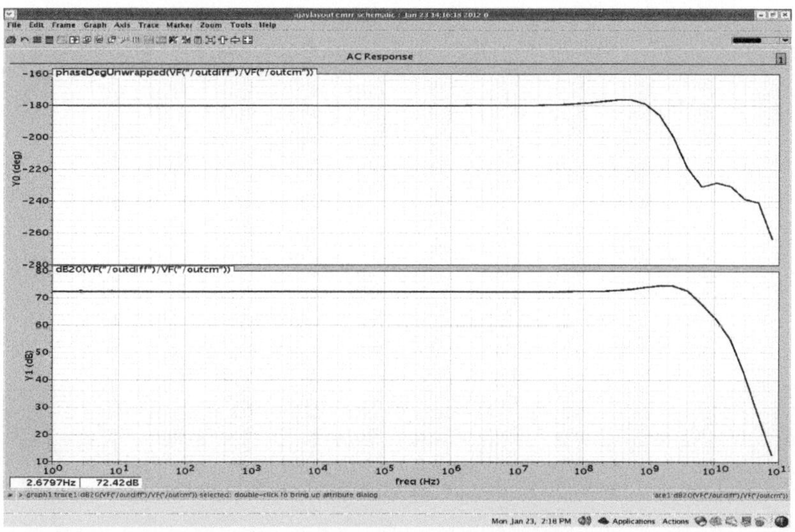

Figure 10. Phase shift and CMRR of TIA.

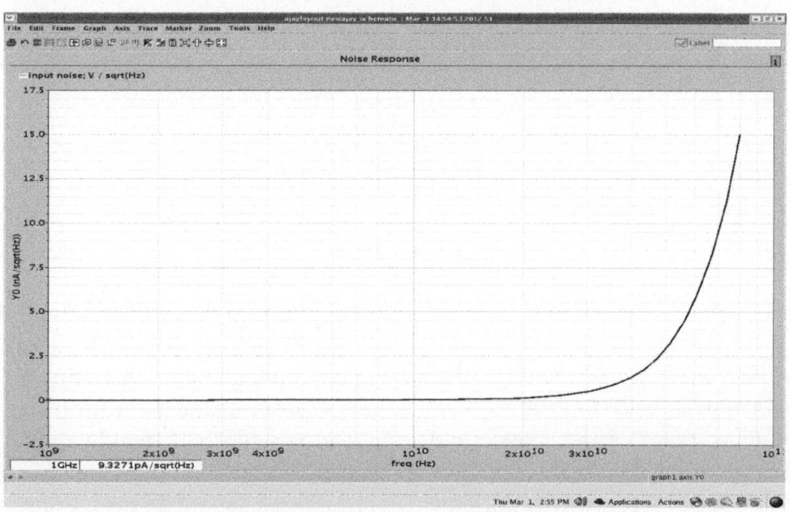

Figure 11. Input referred noise response at 0 V AGC.

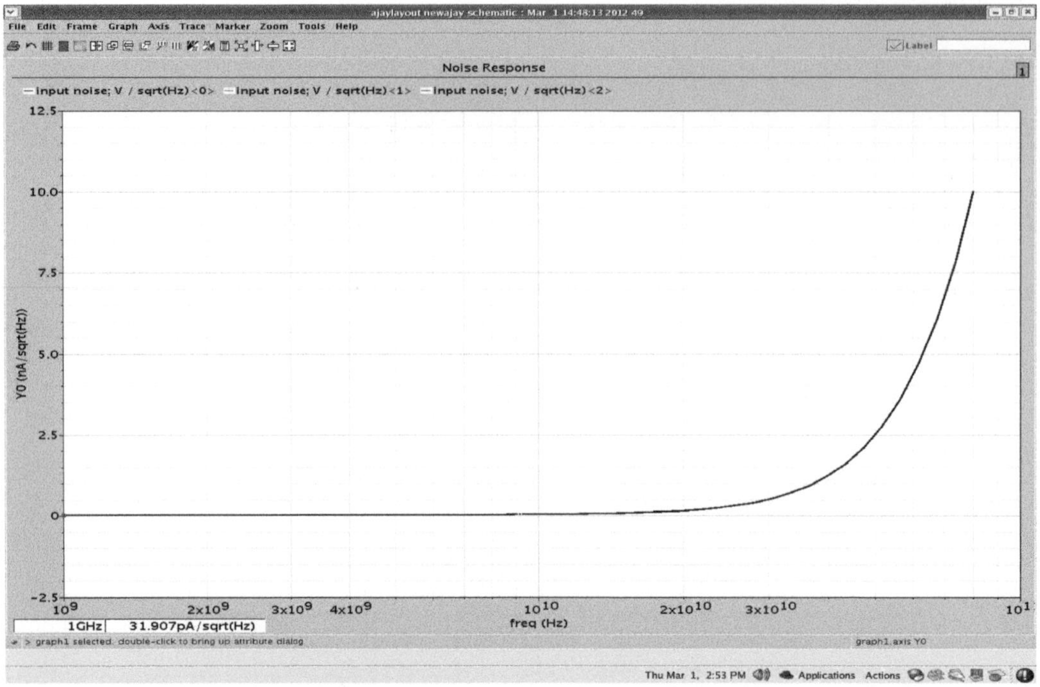

Figure 12. Input referred noise response at 3 V AGC.

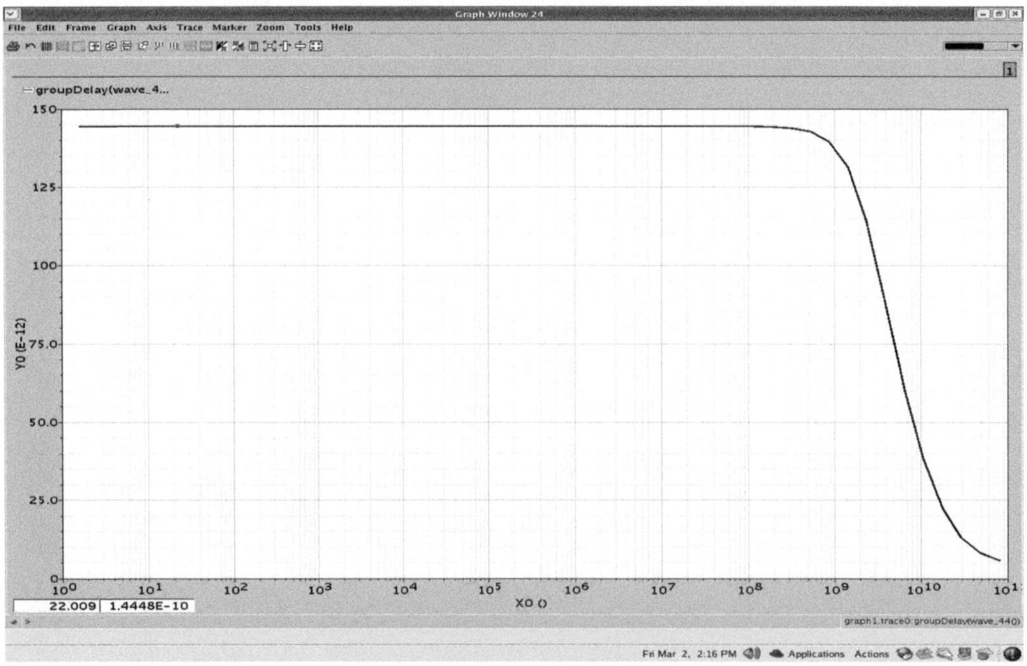

Figure 13. Group delay response.

4. Conclusion

This paper presents a design of an optical receiver that enables to receive the light signals with the help of photodiode. This photodiode thus convert these signals into electric current and current to voltage conversion occurs at TIA gain stage. The TIA is able to generate variable gain ranging from 50.65 dBΩ to 67.26 dBΩ at 0 - 3 V AGC voltage. This gain can further be amplified from 70.09 dBΩ to 86.70 dBΩ using post amplifier which has improved gain value as compare to earlier reported work as given in **Table 1**. From this table it is clear that this work provide high bandwidth of 11.5 GHz and lower input referred noise value of 43.86 pA/√Hz up to 10 GHz frequency. The design has average power dissipation of 76.45 mW with a group delay of 144.48 ps. The

Table 1. Comparison of present results with earlier published work.

Parameters	Ref. [1]		This work	
Gain of TIA at 3 V AGC (automated control voltage)	46 dBΩ		TIA gain	Post amplifier gain
			50.65 dBΩ	70.09 dBΩ
Gain of TIA at 0 V automated control voltage	64 dBΩ		TIA Gain	Post amplifier gain
			67.26 dBΩ	86.70 dBΩ
Bandwidth	Ranges from 5 GHz to 7 GHz		Ranges from 7.03 GHz to 11.5 GHz	
Input referred Noise at 0 V automated control voltage	At 1 GHz	At 10 GHz	At 1 GHz	At 10 GHz
	9 pA/√Hz	68 pA/√Hz	9.33 pA/√Hz	43.86 pA/√Hz
Input referred Noise at 3 V automated control voltage	At 1 GHz	At 10 GHz	At 1 GHz	At 10 GHz
	29 pA/√Hz	68 pA/√Hz	31.91 pA/√Hz	45.45 pA/√Hz
CMRR (common mode rejection ratio)	36 dB		72.42 dB	
Max group delay	170 ps		144.48 ps	
Power dissipation	115 mW		76.45 mW	

simulation results show wide bandwidth, high gain and lower input referred noise, low power dissipation which is suitable in laptop computers, cellular phones, digital cameras, computer peripherals.

5. Acknowledgements

This work has been carried out in SMDP VLSI laboratory of the Electronics and Instrumentation Engineering department of Shri G. S. Institute of Technology and Science, Indore, (M. P.) India. This SMDP VLSI project is funded by Ministry of Information and Communication Technology, Government of India. Authors are thankful to the Ministry for the facilities provided under this project.

REFERENCES

[1] R. D. Bespalko, "Transimpedance Amplifier Design Using 0.18 μm CMOS Technology," Queen's University Kingston, Ontario, 2007.

[2] R. Y. Chen, T. S. Hung and C. Y. Hung, "A CMOS Infrared Wireless Optical Receiver Front-End with a Variable-Gain Fully-Differential Transimpedance Amplifier," *IEEE Transactions on Consumer Electronics*, Vol. 51, No. 2, 2005, pp. 424-429.

[3] S. Hranilovic and D. A. Johns, "A Multilevel Modulation Scheme for High-Speed Wireless Infrared Communications," *Proceedings of IEEE International Symposium on Circuits and Systems* (*ISCAS*), Vol. 6, 1999, pp. 338-341.

[4] D. Guckenberger, J. Schaub and K. Komegay, "A DC Coupled Low-Power Transimpedance Amplifier Architecture for Gb/s Communication System Applications," *Design of Radio Frequency Integrated Circuit* (*RFIC*) *Symposium*, June 2004, pp. 515-518.

[5] B. Razavi, "Design of Integrated Circuits for Optical Communications," McGraw Hill, Boston, 2003.

[6] C. Ciofietal, "A New-Circuit Topology for the Realization of Very Low-Noise Wide-Bandwidth Trans-Impedance Amplifier," *IEEE Transactions on Instrument and Measurement*, Vol. 56, No. 5, 2007, pp. 1626-1631.

[7] A. S. Sedra and K. C. Smith, "Microelectronic Circuits," 4th Edition, Oxford University Press, New York, 1998.

[8] A. Phillip and D. R. Holberg, "CMOS Analog Circuit Design," 2nd Edition, Oxford University Press, New York, 2003.

[9] F. T. Chien and Y. J. Chan, "Bandwidth Enhancement of Transimpedance Amplifier by a Capacitive Peaking Design," *IEEE Journal of Solid State Circuits*, Vol. 34, 1999, pp. 1167-1170.

[10] S. M. Park and H.-J. Yoo, "1.25-Gb/s Regulated Cascade CMOS Transimpedance Amplifier for Gigabit Ethernet Applications," *IEEE Journal of Solid State Circuits*, Vol. 39, No. 1, 2004, pp. 112-121.

16

Parallel Algorithms for Residue Scaling and Error Correction in Residue Arithmetic

Hao-Yung Lo, Ting-Wei Lin

Department of Electrical Engineering, National Tsing Hua University, Hsinchu City, Chinese Taipei.

ABSTRACT

In this paper, we present two new algorithms in residue number systems for scaling and error correction. The first algorithm is the Cyclic Property of Residue-Digit Difference (CPRDD). It is used to speed up the residue multiple error correction due to its parallel processes. The second is called the Target Race Distance (TRD). It is used to speed up residue scaling. Both of these two algorithms are used without the need for Mixed Radix Conversion (MRC) or Chinese Residue Theorem (CRT) techniques, which are time consuming and require hardware complexity. Furthermore, the residue scaling can be performed in parallel for any combination of moduli set members without using lookup tables.

Keywords: Chinese Remainder Theorem (CRT); Error Correction; Error Detection; Parallel Residue Scaling; Residue Number Systems (RNS); Target Race Distance (TRD); Target Residue-Digit Difference

1. Introduction

Because the residue number system (RNS) operations on each residue digit are independent and carry free property of addition between digits, they can be used in high-speed computations such as addition, subtraction and multiplication. To increase the reliability of these operations, a number of redundant moduli were added to the original RNS moduli [RRNS]. This will also allow the RNS system the capability of error detection and correction. The earliest works on error detection and correction were reported by several authors [1-12]. Waston and Hasting [1,2] proposed the single residue digit error correction. Yau and Liu [3] suggested a modification with the table lookups using the method above. Mandelbaum [4-6] proposed correction of the AN code. Ramachandran [7] proposed single residue error correction. Lenkins and Altman [8-10] applied the concept of modulus projection to design an error checker. Etzel and Jenkins [11] used RRNS for error detection and correction in digital filters. In [12-16] an algorithm for scaling and a residue digital error correction based on mixed radix conversion (MRC) was proposed. Recently Katti [17] has presented a residue arithmetic error correction scheme using a moduli set with common factors, *i.e.* the moduli in a RNS need not have a pairwise relative prime.

In this study, we developed two new algorithms without using MRD (Mixed-radix digit) or CRT (Chinese remained Theorem) for speeding-up the scaling processes and simplifying the error detection and correction in RNS. The first algorithm is used for these purposes, through the residue digit difference cyclic property (CPRDD) within the range of $0 \le x \le M_t - 1$, where $M_t = m_1 m_2 \cdots m_n m_{n+1} \cdots m_{n+r}$ with r additional moduli. The moduli $\{m_1, m_2, \cdots, m_n\}$ are called the nonredundant moduli; $\{m_{n+1}, m_{n+2}, \cdots, m_{n+r}\}$ are the redundant moduli. The interval, $[0, M-1]$, is called the legitimate range, where $M = \Pi_{i=1}^{n} m_i$, and the interval, $[M,\ M_t - 1]$, is the illegitimate range, where $M_t = MM_r = M \cdot \Pi_{i=1}^{r} m_{n+i}$, and M_t is the total range. This paper is organized as follows: Section II will describe the scheme the cyclic property of residue digit difference (CPRDD). Section III describes the Target Race Distance (TRD) algorithm and followed by some examples. Section IV discusses residue scaling and error correction using the TRD and CPRDD algorithms. Finally, the conclusion is given in section V.

2. Error Detection and Correction Using Residue Digit Difference Cyclic Property

Any residue digit x_i representation in moduli set $(m_1, m_2, \cdots, m_n)$ has its cyclic length with respect to its module number. For example, if the moduli set is (4, 5, 7,

9), then the cyclic lengths of any residue digits (x_1, x_2, x_3, x_4) are 4, 5, 7 and 9, respectively. Since these cyclic lengths are not equal, they are very difficult to use as tools for error detection and correction. Actually, there exists the property of common (uniform) cyclic length in RNS between residue digital-differences (RDD). Consider three moduli set $(m_1, m_2, m_3) = (2, 3, 5)$. The residue representations and their corresponding digit-differences are shown in **Table 1** and defined as the difference in value between two digits, $d_{ij} = \langle x_i - x_j \rangle_{m_i}$, where d'_{ij} s are all modulo to positive values with respect to m_j if the cycle length of m_j is assigned.

Note that the residue digit-differences $\langle d_{ij} \rangle m_i$ in **Table 1** are obtained from $\langle x_i - x_j \rangle_{m_i}$ if $m_i < m_j$, and from $\langle x_j - x_k \rangle_{m_k}$ if $m_j < m_k$. This difference of $(x_i - x_j)$ or $(x_j - x_k)$ in values may be positive or negative, depending upon $x_i \geq x_j$ or $m_j \geq m_k$ and $x_i < x_j$ or $x_j < x_k$, respectively

All negative values must be modulo to positive values. For example, on starred row 28, as shown in **Table 1**, the digit difference in value for $x_1 = 0$ and $x_3 = 3$ is $d_{13} = 0 - 3 = -3$. It results in $d_{13} = \langle -3 \rangle_2 = 1$

From the cyclic property of residue-digit difference (CPRDD) in RNS, we now have the following theorem.

Theorem 1. For a moduli set $(m_1, m_2, \cdots, m_i, \cdots, m_j, \cdots, m_n)$ and residue representation for $x = (x_1, x_2, \cdots, x_i, x_j, \cdots, x_n)$ in RNS, there exists a cyclic property in differences between two residue digits, $d_{ij} = \langle x_i - x_j \rangle_{m_i}$ or $\langle x_i - x_j \rangle_{m_j}$. The cyclic length can be assigned, either to m_i or m_j, depending upon modulo operation with respect to m_i or m_j.

Proof: Consider the case respective to m_j, the residue-digit difference (RDD) between two digits in $X = (x_1, x_2, \cdots, x_i, x_j, \cdots, x_n)$ can be in general expressed by the equation

$$\langle d_{ij} \rangle = \langle x_i - x_j \rangle_{m_i} = \langle \langle x_i + pm_i \rangle - \langle x_j + qm_j \rangle \rangle_{m_i} \quad (2\text{-}1)$$

where $p = 0, 1, \cdots, (m_j - 1)$
$q = 0, 1, \cdots, (m_i - 1)$
and i, j, p, q are integers.

For simplicity, we only consider the case of $m_i < m_j$ and assume $m_j - m_i = r$, and the case of $m_i > m_j$ can be obtained in a similar way.

The related theorem and algorithm are described as follows.

1) In cycle 0, (the initial cycle), we have
$X = x_j = 0, 1, \cdots, (m_j - 1)$ with $q = 0$,
$\langle d_{ij} \rangle = 0 = \langle x_i - x_j \rangle_{m_i} \neq \langle x_i + p\, m_i - x_j \rangle_{m_i}$ As $x_j = x_i + p$

Table 1. Cyclic property of Residue Digit Difference.

Decimal	$m_1 = 2$ x_1	$m_2 = 3$ x_2	$m_3 = 5$ x_3	$\langle d_{13} \rangle_2$	$\langle d_{23} \rangle_3$	$m_3$cycle
0	0	0	0	0	0	
1	1	1	1	0	0	
2	0	2	2	0	0	0
3	1	0	3	0	0	
4	0	1	4	0	0	
5	1	2	0	1	2	
6	0	0	1	1	2	
7	1	1	2	1	2	1
8	0	2	3	1	2	
9	1	0	4	1	2	
10	0	1	0	0	1	
11	1	2	1	0	1	
12	0	0	2	0	1	2
13	1	1	3	0	1	
14	0	2	4	0	1	
15	1	0	0	1	0	
16	0	1	1	1	0	
17	1	2	2	1	0	3
18	0	0	3	1	0	
19	1	1	4	1	0	
20	0	2	0	0	2	
21	1	0	1	0	2	
22	0	1	2	0	2	4
23	1	2	3	0	2	
24	0	0	4	0	2	
25	1	1	0	1	1	
26	0	2	1	1	1	
27	1	0	2	1	1	5
28*	0	1	3	1	1	
29	1	2	4	1	1	
30	0	0	0	0	0	
31	1	1	1	0	0	Out of Range
32	0	2	2	0	0	

m_i with $p = 0, 1, \cdots, \lfloor m_j / m_i \rfloor$, we have $\langle d_{ij} \rangle_{m_i} = 0$ with m_j's 0s in cycle 0, where $\lfloor x \rfloor$ means the largest integer less than or equal to x.

Thus, the RDD has m_j's "0" in the initial cycle for each modulus, *i.e.*, in cycle 0, $\langle d_{ij}\rangle_{m_i} = (0,0,\cdots,0)$ for all $i \neq j$.

2) Next consider each modulus m_i,

Since $x_i = X - pm_i$ and $x_j = X - qm_j$, then

$\langle d_{ij}\rangle = \langle x_i - x_j\rangle_{m_i} = \langle\langle x_i + pm_i\rangle - \langle x_j + qm_j\rangle\rangle_{m_i}$ where

$x_i = 0,1,2,\cdots,(m_i - 1),0,1,\cdots,r-1$

$x_j = 0,1,2,\cdots,(m_i - 1),m_i,m_i+1,\cdots,m_i+r-1 = (m_j - 1)$

For RDD = 0 (for cycles $0, m_i, 2m_i,\cdots,\lfloor m_j/m_i\rfloor m_i$)

then

$$\langle d_{ij}\rangle_{m_i} = \left((0-0),(1-1),\cdots,[(m_i-1)-(m_i-1)],\langle m_i - 0\rangle_{m_i},\langle (m_i+1)-1\rangle_{m_i},\cdots,\langle m_i+r-1-(r-1)\rangle_{m_i}\right) = (0,0,\cdots,0)$$

with m_j's 0 s.

For RDD = 1 (not necessary in cycle 1),

$$\langle d_{ij}\rangle_{m_i} = \left((1-0),(2-1),\cdots,[m_i-(m_i-1)],\langle (m_i+1)-m_i\rangle_{m_i},\cdots\right) = (1,1,\cdots,1) \text{ with } m_j\text{'s 1 s.}$$

…

For RDD = $m_i - 1$

$$\langle d_{ij}\rangle_{m_i} = \left(((m_i-1)-0),(m_i-1),\langle (m_i+1)-2\rangle_{m_i},\cdots\right) = ((m_i-1),(m_i-1),\cdots,(m_i-1))$$

with m_j's $(m_i - 1)$ s.

Corollary 1. From the above theorem, we can immediately obtain that each cycle in the residue-digit difference of x will start at location 0, and end at location $(m_i m_j m_k - 1) = M_p - 1$.

Corollary 2. It is easily shown that there exists m_i number of cycles with respect to the cyclic length of M_p.

Proof. Since the residue-digit difference of $x = (x_1, x_2,\cdots,x_i,x_j,\cdots,x_n)$ representation is pair-wise, the legitimate range of this pair-wise RDD(x_i,x_j) is $m_i m_j$, (from 0 through $m_i m_j - 1$). From corollary 1, the cyclic length is $m_i m_j$. Thus the number of cycles within this cyclic length for N_i is $N_i = \frac{m_i m_j}{m_i} = m_j$, and for $m_j, N_j = \frac{m_i m_j}{m_j} = m_i$.

Theorem 2. The algorithm of theorem 1 and its corollaries can be extended to two or more pair-wise residue-digit differences.

Proof: consider a three moduli set, we have two pair-wise moduli sets, whose RDD (Residue Digital Difference) is

$$\langle d_{jk}\rangle_{m_k} = \left\langle \langle X\rangle_{m_j} - \langle X\rangle_{m_k}\right\rangle_{m_k} = \left\langle \langle x_j + qm_j\rangle_{m_j} - \langle x_k + sm_k\rangle_{m_k}\right\rangle_{m_k}$$

where m_k is again the referenced module.

Follow the same procedure as step (2) as above.

Assume $m_k = m_j + r_2$, and also pair-wise numbers $x_j = 0,1,2,\cdots,(m_j-2),\ (m_j-1),0,1,\cdots,(r_2-1)$ and

$$x_k = 0,1,2,\cdots,(m_j-2),(m_j-1),m_j,(m_j+1),\cdots,(m_j+r_2-1) = (m_k-1).$$

1) For $q = s = 0, x_j = x_k,\ 0 \le x_j, x_k \le m_k - 1, d_{jk} = 0$ thus

$$\langle d_{jk}\rangle_{m_j} = \underbrace{0,0,\cdots,0}_{m_j\text{'s "0"}},\underbrace{\langle -m_j\rangle_{m_j},\langle -m_j\rangle_{m_j},\cdots,\langle -m_j\rangle_{m_j}}_{r_2\text{'s "}\langle -m_j\rangle_{m_j}\text{"}\ (=0)},$$

This shows that $\langle d_{jk}\rangle_{m_j}$ has also m_k "0"s in cycle 0 of $m_k\ (x_k = 0,\ 1,\ 2,\cdots,m_k - 1)$. The cyclic length is $(m_j)\cdot(m_k) = +r_2$, and the number of cycles for m_j is $m_k\ (= m_j + r_2 = m_j + m_k - m_j \text{ or } = m_j \cdot m_k/m_j)$.

2) For $q \neq s \neq 0$ and $\langle d_{jk}\rangle_{m_j} = h$ (a constant for any RDD), if $x_j \neq x_k$

$$\langle d_{jk}\rangle_{m_j} = \underbrace{\langle (h-0),\ ((h+1)-1),\cdots,<0-(m_j-h)>_{m_j},<1-(m_j-h+1)\rangle_{m_j},\cdots}_{m_k\text{'s"}h\text{"}}$$

This shows that the $\langle d_{jk}\rangle_{m_j} = h$ in any location has also m_k's "h" in cycle i of m_k. The number of cycles for m_j is still $m_k\ (= m_j \cdot m_k/m_j)$. Combining these three moduli (m_i, m_j, m_k) into one set, we have cyclic length $M_p = m_i \cdot m_j \cdot m_k$ (for example, $m_1 \cdot m_2 \cdot m_3 = 2 \cdot 3 \cdot 5 = 30$). The number of cycles for m_1, m_2, m_3 are $N_{m_1} = m_2 \cdot m_3 = 3 \cdot 5 = 15$, $N_{m_2} = m_1 \cdot m_3 = 2 \cdot 5 = 10$, and $N_{m_3} = m_1 \cdot m_2 = 2 \cdot 3 = 6$, respectively. As shown in **Table 1**, the RDD pairs of $(\langle d_{13}\rangle_2, \langle d_{23}\rangle_3)$ are $\underbrace{(0,0),(1,2),(0,1),(1,0),(0,2)}_{\text{All 5 } (=m_3) \text{ pairs in each } m_3 \text{ cycle.}}$, and $(1,1)$

In general, $M_p = m_1 \cdot m_2 \cdots\cdots m_k \cdots\cdots m_n$ and $N_{m_k} = M_p / m_k$ with m_k rows and $(n-1)$RDD in each row.

This completes the proof.

Example 2-1.

Consider a moduli set $(m_1, m_2, m_3) = (4,5,7)$, $X = 9$ and its corresponding residue digits representation set is $(1,4,2)$. The cyclic length is $140 (= 4 \cdot 5 \cdot 7)$ and the number of cycles for m_1, m_2, and m_3 are $N_{m_1} = 35, N_{m_2} = 28$, and $N_{m_3} = 20$, respectively.

Error detection and correction:

Before the CPRDD algorithm used for error detection and correction is described, some basic terms in use must be defined.

Definition 1: Stride distance S_{ij}: It is the incremental or decremental distance between moduli m_i and m_j in absolute value from *i*th cycle to $(i+1)$th cycle.

For example: $S_{23} = |5-7| = 2$.

(1) Error detection

Let the moduli set be $\{m_1, m_2, \cdots, m_k, m_{k+1}, \cdots, m_{k+r}, \cdots\}$ where $m_1, m_2, \cdots, m_k$ are the nonredundant moduli and $m_{k+1}, \cdots, m_{k+r}$ are the redundant moduli. Since the cyclic lengths of CPRDD d_{ij}'s are constant, it is thus easily found that the number of cycles on track L_{ij} from the starting point 0 (or other d_{ij}) to its target position. In turn the distance of RDD's can also be found.

Theorem 3. The number of cycles on track L_{ij} (column d_{ij}) from any starting point (say $\hat{d}_{ij}$) to its target position d_{ij} can be found using the equation below;

$$\left\langle \hat{d}_{ij} + S_{ij} k_{ij} \right\rangle_{m_i} = d_{ij}$$

where S_{ij} = the stride distance between moduli m_i and m_j and k = the number of cycles passing through from starting point $\hat{d}_{ij}$ to the destination, d_{ij} = on track L_{ij} If $\hat{d}_{ij} = 0$, then the number of cycles are equal to the total cycles from the starting point "0" to its target position d_{ij}.

Proof: Since k_{ij} is the number of cycles from 0 to d_{ij} with respect to module m_j, and m_j is the cyclic length, thus $k_{ij} m_j$ is the total distance from the starting point $\hat{d}_{ij} = 0$ to its target position d_{ij}. The remaining distance for d_{ij} on track L_{ij} in the (k_{ij})th cycle must be on the same row of d_{ij} on track L_{ij}. Thus, $\text{RDD}(x_i) = \text{RDD}(x_j) = (k_{ij} m_j + d_{ij})$.

Once the RDD's of $x_1, x_2, \cdots, x_n, x_r$ are found, the error detection and correction for moduli can be found just by comparing the calculated cycles or RDD with the original residue representation, pair-wise so that the error module can be detected.

The procedure for error detection by using CPRDD algorithm is summarized as follows.

1) Choose two most significant (largest) moduli as the referred moduli among the *n* moduli, say m_{n-1} and m_n.

2) Find the skip distance of a cycle $S_{(n-1)(n)} = |m_{n-1} - m_n|$.

3) Find the digit difference $d_{(n-1)(n)} = \langle x_{n-1} - x_n \rangle_{mn}$ from $X = (x_1, x_2, \cdots, x_{n-1}, x_n)$.

4) Create the equation of $\text{RDD}(x_{n-1}, x_n) = \langle d(x_{n-1}, x_n) \rangle_{m_n}$ or

$$\text{RDD}(x_{n-1}, x_n) = \left\langle S_{(n-1)(n)} \cdot k_{(n-1)(n)} \right\rangle_{m_n} = \left\langle d(x_{n-1}, x_n) \right\rangle_{m_n} \tag{2-2}$$

5) Solve for $k_{(n-1)(n)}$ from Equation (2-2) as the $S_{(n-1)(n)}$ and $\langle d(x_{n-1}, x_n) \rangle_{m_n}$? are known. The value of $k_{(n-1)(n)}$ must be less than or equal to $(m_1 \cdot m_2 \cdots\cdots m_k \cdots\cdots m_{n-2})$.

6) Find the corresponding $\text{RDD}(x_{n-1}, x_n)$ distance from the starting point to x_{n-1}.

7) Calculate $(x_1, x_2, \cdots, x_n)$ from RDD$_1$, RDD$_2$, $\cdots$, and check the values of $(x_1, x_2, \cdots, x_n)_1, (x_1, x_2, \cdots, x_n)_2$, and .... If these sets' numbers are equal, then no error occurs; otherwise, error exists.

We take the similar numerical as example 2-1 to verify this algorithm. (CPRDD)

Example 2-2. Assume that a moduli set $(m_1, m_2, m_3, m_4) = (4,5,7,9)$ and number X whose residue representation is $(x_1, x_2, x_3, x_4) = (1,2,6,7) = (97)_{10}$. If an error occurs at $m_2, X = (1,3,6,7)$, the error detection can be described as follows.

Let us begin our procedures from the $\text{RDD}(x_3, x_4) = d(x_3, x_4)$. Since $S_{34}(\text{skip distance of a cycle}) = |m_3 - m_4| = |7-9| = 2$,

$$d_{13} = \langle -5 \rangle_4 = 3, \quad d_{23} = \langle -3 \rangle_5 = 2$$

and $d_{34} = \langle 6-7 \rangle_9 = \langle -1 \rangle_9 = 8$. Then $N(d_{34}) = \langle S_{34} \cdot k_{34} \rangle_9 = \langle 2 \cdot k_{34} \rangle_9 = 8$. Solve for k_{34}, and let $k_{34} < (m_1 m_2) = 20$ within legitimate range $(4 \cdot 5 \cdot 7) = 140$, then $k_{34} = 4, 13$.

The corresponding $\text{RDD}(x_3, x_4)$ primary distances for these two k_{34} are, respectively,

$$\text{RDD}_1(4) = 4 \cdot 7 + 6 = 34$$

$$\text{RDD}_2(13) = 13 \cdot 7 + 6 = 97$$

Thus, the generated results of the residue representation from $\text{RDD}_1(4)$ and $\text{RDD}_2(13)$ are respectively

$$X_1(34) = (x_1, x_2, x_3, x_4) = (2,4,6,7),$$

$$X_2(97) = (x_1, x_2, x_3, x_4) = (1,2,6,7).$$

Since the calculated results of X_1 and X_2 are not identical, there must be errors in one of these moduli. We cannot determine which one is erroneous. To locate the module where the error exists, at least one additional (redundant) module must be used.

The procedure for error correction by using CPRDD algorithm is essential the same as the error detection. However, two additional redundant moduli m_{r_1} and m_{r_2} must be added for one error correction. Note that only one redundant modulus added for error detection.

1) Choose m_{r_1} (or m_{r_2}) as a referred modulus.

2) Find $k_{(1)(r_1)}, k_{(2)(r_2)}, \cdots, k_{(r_1)(r_2)}$ as the same procedures of error detection steps 2-7.

3) Examine the values of $k_{(1)(r_1)}, k_{(2)(r_2)}, \cdots, k_{(r_1)(r_2)}$. If common value exists among, $k_{(1)(r_1)}, k_{(2)(r_2)}, \cdots, k_{(r_1)(r_2)}$, then no error occurs. If there is one and only one, say $k_{(i)(r_1)}$ that has no common value with all other $k_{(j)(r_1)}$, then an error exits in modulus m_i. This completes the error correction procedures.

The following example is illustrated here to verify this algorithm.

Example 2-3. Error correction

As before we can further locate and correct a single error by adding two redundant moduli, m_{r_1} and m_{r_2}. Let us use the same example. The moduli set $(m_1, m_2, m_3, m_4, m_5) = (4,5,7,9,11)$, where m_4 and m_5 are redundant moduli $m_{r_1} = 9$ and $m_{r_2} = 11$, and the residue X representation, $(x_1, x_2, x_3, x_4, x_5) = (1,2,6,7,9) = (97)_{10}$. If a single error occurs at m_3, e.g. $X = (1,2,5,7,9)$, and m_4 is assigned as a reference module, then $\langle d_{14} \rangle_4 = \langle -6 \rangle_4 = 2$, $\langle d_{24} \rangle_5 = \langle -5 \rangle_5 = 0$, $\langle d_{34} \rangle = \langle -2 \rangle_7 = 5$,and $\langle d_{45} \rangle_{11} = \langle -2 \rangle_{11} = 9$. From CPRDD algorithm, we can find the number of cycles for these RDD's.

$$\langle S_{14} k_{14} \rangle_4 = \langle 5 k_{14} \rangle_4 = 2,$$

$$\langle S_{24} k_{24} \rangle_5 = \langle 4 k_{24} \rangle_5 = 0,$$

$$\langle S_{34} k_{34} \rangle_7 = \langle 2 k_{34} \rangle_7 = 5,$$

$$\langle S_{45} k_{45} \rangle_{11} = \langle 2 k_{45} \rangle_{11} = 9$$

Since the cycle length is 9, all above k_{ij} values must be less than $\left\lceil \frac{140}{9} \right\rceil = 16$. Thus we have

$$k_{14} = 2,6,10,14$$

$$k_{24} = 0,5,10,15$$

$$k_{34} = 6,13.$$

$$k_{45} = 10$$

If no errors occur, all k_{ij}'s are equal, *i.e.*, $k_{14} = k_{24} = k_{34} = k_{45}$.

Compared to the above results with pairwise moduli, only $k_{14} = k_{24} = k_{45} = 10$ meets this condition. There exists no such value in k_{34}.

This shows that the module m_3 is faulty, therefore we can correct it as follows: since $k_{14} = k_{24} = k_{45} = 10$, the $\text{RDD} = k_{14} \cdot \text{cycle length} + x_4 = 10 \cdot 9 + 7 = 97$.

Thus $x_1 = \langle 97 \rangle_4 = 1$, $x_2 = \langle 97 \rangle_5 = 2$, $x_3 = \langle 97 \rangle_7 = 6$, $x_4 = \langle 97 \rangle_9 = 7$.

This completes the error correction.

Note that the above CPRDD's for each residue-digit difference, d_{ij}, and k_{ij} can be processed in parallel. In addition, if the referenced module is assigned to the erroneous module by chance, e.g., m_3 this algorithm will fail to locate the error. In this case, there are no k_{ij}'s values that can be found to match this condition. The way to solve the problem is, of course, to assign any other moduli, e.g., m_1 or m_2.

The hardware design for the proposed algorithm in Example 2-3 is shown in **Figure 1**.

3. The Target Race Distance (TRD) Scheme

The conversion or decoding technique from residue representation to X in binary is usually accomplished using the mixed-radix digit (MRD) or Chinese remained theorem (CRT). An optimal matched and parallel converter of this kind can be seen in [13]. The MRD is shown by the following expression with weighted numbers:

$$x = \langle a_1 m_0 + a_2 m_1 + a_3 m_1 m_2 + \cdots + a_n m_1 m_2 \cdots m_{n-1} \rangle_{M_p}$$

$$= \sum_{i=1}^{n} \alpha_i (m_0 m_1 \cdots m_{i-1}) \text{ with } m_0 = 1,$$

where $M_p = m_1 m_2 \cdots m_n = \Pi_{i=1}^{n} m_i$, and $\alpha_i \in [0, m_{i-1}]$ is the mixed-radix conversion (MRC) of x.

Optimization can be obtained using this method, as the accessed table lookup time is exactly equal to the right addition time, after immediate column stage for the tree network of the adders.

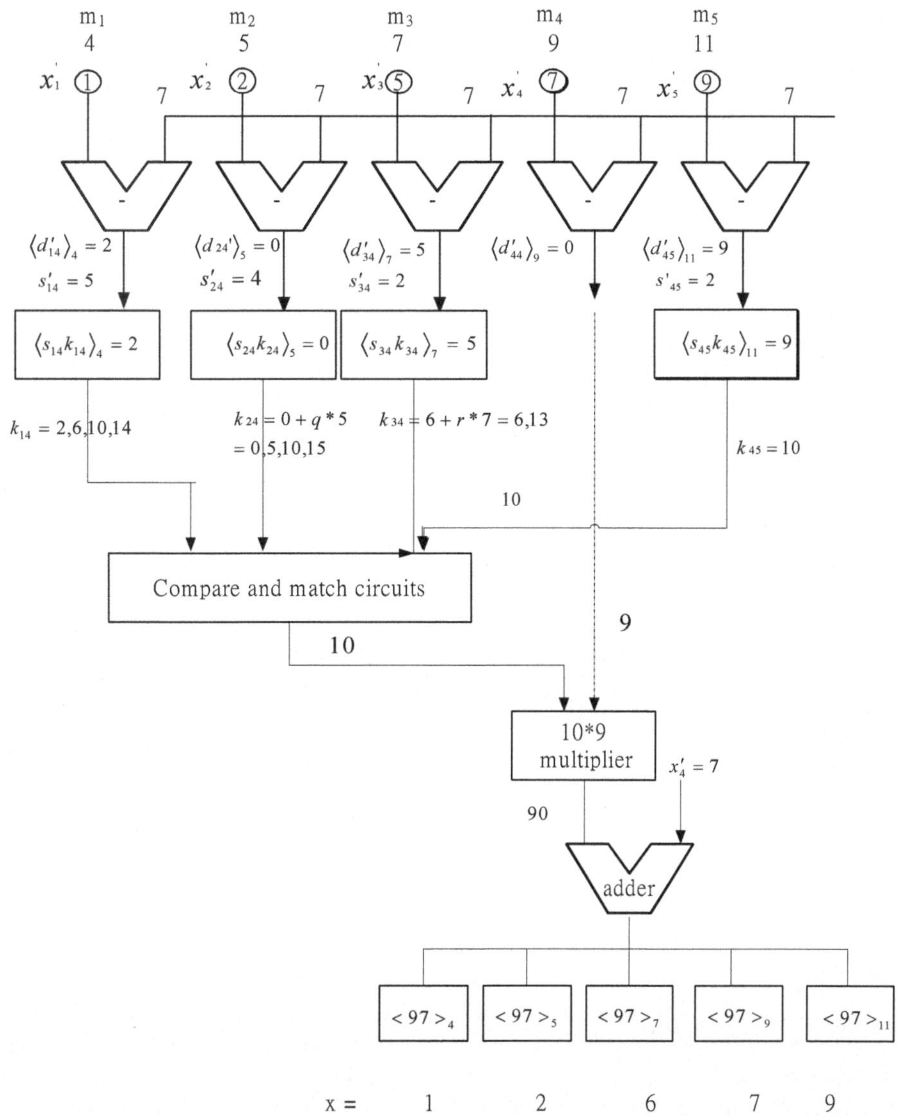

Figure 1. The hardware implementation for the proposed error and correction location algorithm can be accomplished without using lookup tables.

However, time is still consumed reading a large number of lookup tables. Additional hardware complexity is required by the adder-tree networks. An algorithm called the target race distance was with a simpler structure was developed for high-speed conversion.

TRD algorithm

Suppose each residue number in the $\mathrm{RNS}\langle X_i\rangle_{m_i}$ has its own track L_i, and the distance over track L_i from 0 (starting point) to X_i (end point) through k_i cycles can be expressed using

$$D_i = x_i + k_i m_i,\ k_i = 0,1,\cdots,(m_i - 1),.$$

Obviously, the primary (no multiples of m$_i$) distance of x_i is $(D_i)_{\min} = x_i(k_i = 0)$. To obtain the X from its residue representation of $x_1, x_2, \cdots, x_r$, we must find a target such that $x_1, x_2, \cdots, x_r$ traversing the same distances over tracks $l_1, l_2, \cdots, l_r$ respectively, *i.e.* when the TRD distance of each target x_i is reached, then $D_1 = D_2 = \cdots = D_r$. The TRD distance of X can be found from the following theorem:

Theorem 4. Consider the simple case of two moduli sets (m_1, m_2). Its residue representation and targets are x_1 and x_2 respectively. Let $(D_1)_p$ be the primary distance of residue x_1 from 0 to x_1 on the track L_1, and $(D_2)_p$ be the primary distance of x_2 from 0 to x_2 on track L_2. Then the TRD distance for these two residues x_1 and x_2 that have the same TRD distances can be obtained by the following equation.

$$\mathrm{TRD}(x_1, x_2) = (x_1 + k_1 \cdot m_1) = (x_2 + k_2 \cdot m_2) \qquad (3\text{-}1)$$

In addition, k_1 can be calculated from the equation

$$\langle x_1 + k_1 \cdot m_1 \rangle_{m_2} = (D_2)_p = x_2$$

where m_1 is the cyclic length of x_1, and k_1 is number of cycles, all of the integers,

$$k_1 = 0,1,2,\cdots,m_2 - 1.$$

Proof: It is easy to show that the above $\mathrm{TRD}(x_i,\ x_{i+1})$ is the common target distance of x_1 and x_2, Since

$$\langle x_1 + k_1 m_1 \rangle_{m_1} = x_1$$

And $\langle x_2 + k_2 m_2 \rangle_{m_2} = x_2 = \langle x_1 + k_1 m_1 \rangle_{m_2} = X$,

thus $\mathrm{TRD}(x_1, x_2) = \langle x_1 + k_1 m_1 \rangle_{m_2} = (x_1 + k_1 m_1) = X$ is the TRD distances for both of x_1 and x_2.

Corollary: It is evident that the above theorem can be extended to n moduli set $(m_1, m_2, \cdots, m_n)$ and residue number $(x_1, x_2, \cdots, x_n)$. The corresponding TRD of $(x_1, x_2, \cdots, x_n)$ are therefore

$$\begin{aligned}\mathrm{TRD}(x_1, x_2, \cdots, x_n) = {} & (x_1 + k_1 m_1) + (x_2 + k_2 m_1 m_2) \\ & + (x_3 + k_3 m_1 m_2 m_3) \\ & + \cdots (x_{n-1} + k_{n-1} m_1 m_2 \cdots m_{n-1})\end{aligned}$$

In addition, k_i can be solved from the following equations.

$$\langle x_1 + k_1 \cdot m_1 \rangle_{m_2} = x_2$$

$$\cdots$$

$$\langle x_i + k_i \cdot m_1 m_2 \cdots m_i \rangle_{m_{i+1}} = x_{i+1}$$

where $k_i = 0,1,\cdots,(m_{i+1} - 1)$

Note that $x_1, x_2, \cdots, x_n$ are the targets of moduli $m_1, m_2, \cdots, m_n$ respectively and the $\mathrm{TRD}(x_1, x_2, \cdots, x_n)$ is the distance that has equal track lengths, *i.e.* $L_1 = L_2 = \cdots = L_n = L$. That is; $\langle L \rangle_{m_1} = x_1, \langle L \rangle_{m_2} = x_2, \langle L \rangle_{m_3} = x_3, \cdots, \langle L \rangle_{m_n} = x_n$.

Example 3-1 Let the moduli set be $(m_1, m_2, m_3, m_4) = (4,\ 5,\ 7,\ 9)$ and the residue representation be $(x_1, x_2, x_3, x_4) = (3,\ 1,\ 2,\ 5)$. The procedures to find the TRD distance can be described as follows:

1) Find the primary distance (D_1) of residue $x_1 = (D_1)_p = \langle 3 \rangle_{m_2}$ since $m_2 > m_1$ and $\langle 3 + k_1 \cdot 4 \rangle_5 = 1$ is required, thus $k_1 = 2$, and $\mathrm{TRD}(x_1, x_2) = (3 + 2 \cdot 4) = 11$

2) Repeat the procedure 1 to find the number of cycles k_2 and k_3 and the last TRD distances (destinations), $\mathrm{TRD}(x_1, x_2, x_3)$ and $\mathrm{TRD}(x_1, x_2, x_3, x_4)$.

Since $\hat{x}_3 = \langle 11 \rangle_7 = 4$

$\langle \hat{x}_3 + k_2 \cdot 4 \cdot 5 \rangle_7 = 2$

$\langle 4 + k_2 \cdot 4 \cdot 5 \rangle_7 = 2$

$\therefore k_2 = 2$

thus $\mathrm{TRD}(x_3) = 2 \cdot 4 \cdot 5 = 40$

and $\mathrm{TRD}(x_1 x_2 x_3) = 11 + 40 = 51$

$\hat{x}_4 = \langle 51 \rangle_9 = 6$

$\langle 6 + k_3 \cdot 4 \cdot 5 \cdot 7 \rangle_9 = 5$

$\therefore k_3 = 7$

thus $\mathrm{TRD}(x_4) = 7 \cdot 4 \cdot 5 \cdot 7 = 980$

and $\mathrm{TRD}(x_1 x_2 x_3 x_4) = 51 + 140 \cdot 7 = 1031$

The final TRD distance is the common distinction of this system for targets x_1, x_2, x_3 and x_4 *i.e.* $\mathrm{TRD}(x_1 x_2 x_3 x_4) = 1031 = X$. This result can be verified as follows:

$\langle 1031 \rangle_4 = 3, \langle 1031 \rangle_5 = 1, \langle 1031 \rangle_7 = 2$ and $\langle 1031 \rangle_9 = 5$

Figure 2 Shows the TRD's on tracks l_1, l_2, l_3 and l_4 respectively.

Error detection and correction by TRD algorithm

A redundant residue number system with $r = 1$ redundant moduli will allow detection of any single error [4,14]. Consider the moduli set $(m_1, m_2, m_3, m_4) = (4,5,7,9)$ and the correct residue representation $X(x_1, x_2, x_3, x_4) = (1,2,6,7) = 97$. Let us

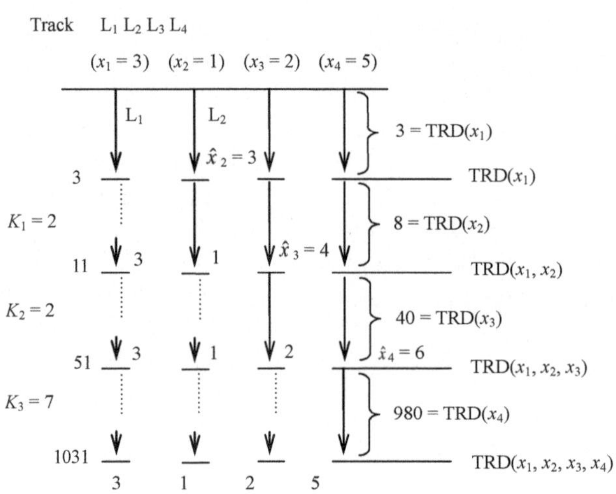

Figure 2. TRD's on track L_1, L_2, L_3 and L_4.

assume that $m_4 = 9$ is the redundant moduli with a single error $X_F(x_1, x_2, x_3, x_4) = (1,3,6,7)$ residue representation. The TRD theorem can be used to detect this error. We find that final TRD for x_1, x_2, x_3 and x_4 does not fall into the legitimate range as follows *i.e.*
$R_F > (4 \cdot 5 \cdot 7) = 140$

$$\mathrm{TRD}_1(x_1 = 1) = 1$$
$$\mathrm{TRD}_2(x_2 = 3) = 12(k_1 = 3)$$
$$\mathrm{TRD}(x_1 x_2) = \mathrm{TRD}_1 + \mathrm{TRD}_2 = 13$$
$$\hat{x}_3 = \langle 13 \rangle_7 = 6 = x_3$$
$$\mathrm{TRD}_3(x_3 = 6) = 0$$
$$\hat{x}_4 = \langle 13 \rangle_9 = 4$$
$$\langle 4 + k_3 \cdot 140 \rangle_9 = 7$$
$$k_3 = 6$$
$$\therefore \mathrm{TRD}_4(x_4 = 7) = 6 \cdot 140 = 840.$$

The final TRD distance
$\mathrm{TRD}(x_1, x_2, x_3, x_4) = 13 + 840 = 853 > 140$. If we need to locate and correct this module error, another redundant module must be added. Let us assume that $m_5 = 11$ for this requirement in the above residue representation.

The current redundant moduli set is
$(m_1, m_2, m_3, m_4, m_5) = (4, 5, 7, 9, 11)$ and the correct residue representation is
$x = (x_1, x_2, x_3, x_4, x_5) = (1, 2, 6, 7, 9) = (97)$. Let us assume that $m_4 = 9$ and $m_5 = 11$ are the redundant moduli. With a single error
$x_F = (x_1, x_2, x_3, x_4, x_5) = (1,3,6,7,9)$. The TRD theorem can again be used to locate and correct this error. We find that final TRD's for $(x_1, x_2, x_4, x_5) = (1,3,7,9)$ dose not fall in the legitimate range, but other final TRD's for $(x_1, x_2, x_3, x_4) = (1,6,7,9)$ do, falls in the legitimate range:

1) TRD for x_1, x_2, x_4 and x_5

$$\mathrm{TRD}_1(x_1 = 1) = 1$$
$$\mathrm{TRD}_2(x_1, x_2) = 13$$
$$\mathrm{TRD}_4(x_1, x_2, x_4) = 133$$
$$\hat{x}_5 = \langle 133 \rangle_{11} = 1$$
$$\langle 1 + k_4 \cdot 180 \rangle_{11} = 9 = x_5, k_4 = 2$$
$$TRD_5(x_1, x_2, x_4, x_5) = 133 + 360$$
$$= 493 > 140(\text{out of legitimate range}).$$

2) TRD for x_1, x_3, x_4 and x_5

$$\mathrm{TRD}_1(x_1 = 1) = 1$$
$$\mathrm{TRD}_3(x_3 = 6) = 12; k_2 = 3$$
$$\hat{x}_4 = \langle 13 \rangle_9 = 4$$
$$\langle 4 + k_3 \cdot 28 \rangle_9 = 7, k_3 = 3$$
$$\mathrm{TRD}_4(x_1, x_3, x_4) = 13 + 84 = 97$$
$$\hat{x}_5 = \langle 97 \rangle_{11} = 9 = x_5$$
$$\therefore TRD_5(x_1, x_3, x_4, x_5)$$
$$= 97 < 140(\text{within legitimate range}).$$

Thus, the error is located at module m_2 and must be corrected to $x_2 = \langle 97 \rangle_5 = 2$.This algorithm can also be used for multiple error corrections. However, at least three redundant moduli are required. The procedures are similar.

4. Scaling with Error Correction

The above proposed algorithm used for error detection and correction has the advantage of not requiring lookup tables. No CRT (Chinese residue theorem) decoding processes are required. However, it is still time consuming and requires extensive hardware complexity for each module having multiple-value inputs to the match unit and selecting a correct one as a output. To improve this drawback, an optimal matching algorithm is proposed here for the error correction. The following two theorems will be used and an example follows.

Theorem 5. Let m_1 and m_2 be two relative prime numbers in RNS for module 1 and module 2 respectively. Then there must exist the relation represented by the equation $\langle m_1 x_1 \rangle_{m_2} - \langle m_2 x_2 \rangle_{m_1} = |k|$, where
$0 \le \langle m_1 x_1 \rangle_{m_2} \le m_2$, $0 \le \langle m_2 x_2 \rangle_{m_1} \le m_1$ so that $0 \le |k| \le m_2$, assuming $m_2 > m_1$. The x_1, x_2 and k are restricted to integers.

Proof: As a first step, let $k = 0$. It is easily seen that $x_1 = m_2$ and $x_2 = m_1$ will be satisfied. Next consider $k \ne 0$. Since there are two different pair combination $\langle m_1 x_1 \rangle_{m_2} \le m_2$ and $\langle m_2 x_2 \rangle_{m_1} < m_1$, thus the difference between $m_1 x_1$ and $m_2 x_2$ of k will always be satisfied for $0 \le |k| \le m_2$, where k is restricted in integers.

Theorem 6. If the values of m_1 and m_2 and k in the equation $\langle m_1 p_1 \rangle_{m_2} - \langle m_2 p_2 \rangle_{m_1} = |k|$ are known, then p_1 and p_2 can always be determined from equation
$\langle (m_2 - m_1) p_1 \rangle_{m_1} = k$ or $\langle (m_2 - m_1) p_2 \rangle_{m_2} = k$, where p_1, p_2 and k are within the range: $0 \le p_{1k} \le (m_1 \text{ or } m_2)$

Proof: Let the difference value of $|m_2 - m_1|$ be equal to d, then d will be the integers within the range between 0 and m_2, *i.e.*, $p_1 = 0,1,2,\cdots,(m_1 - 1)$, or
$p_2 = 0,1,2,\cdots,(m_2 - 1)$. These two expressions show that we can always select an integer value p, within the interval between 0 and $(m_1 - 1)$ or $(m_2 - 1)$ to satisfy the

conditions $\langle dp_1\rangle_{m_1} = k$ or $\langle dp_2\rangle_{m_2} = k$

Example 4-1 Let $m_1 = 5$, and $m_2 = 7$. Find the minimum values of p_1 and p_2 respectively from the following equation :

$$7p_1 - 5p_2 = 3$$

Since $m_1 = 5$ and $m_2 = 7$, we have
$d = m_2 - m_1 = 7 - 5 = 2$,
and

$$\langle 2p_1\rangle_5 = 3 \quad (4\text{-}1),$$

or

$$\langle 2p_2\rangle_7 = 3 \quad (4\text{-}2),$$

from Equation (4-1)

$\langle 2p_1\rangle_5 = 3$ so $p_1 = 4$, $(2\cdot 4) - 5 = 3$,

from Equation (4-2)

$\langle 2p_2\rangle_7 = 3$, so $p_2 = 5$ for $(2\cdot 5) - 7 = 3$.

This result can be verified by substituting $7\cdot 4 - 5\cdot 5 = 3$ into the above equation. Theorem 6 is very useful as shown in the following example.

In Theorem 3 of Section III, the number of cycles on track L_{ij} from the starting point "0" to its target position "d_{ij}" can be expressed by setting $\hat{d}_{ij} = 0$, *i.e.*

$$\left\langle s_{ij}k_{ij}\right\rangle_{m_i} = d_{ij}, \text{ or } \left\langle s_{ij}k_{ij} + p_i m_i\right\rangle_{m_i} = d_{ij} \quad (4\text{-}3),$$

where s_{ij} is the module i stride distance referring to module j. Similarly, the number of cycles on track l_{jk} from the starting point "0" to its target position "x_k" can be expressed by setting $\hat{d}_{ij} = 0$, *i.e.*;

$$\left\langle s_{jk}k_{jk}\right\rangle_{m_2} = d_{jk} \text{ or } \left\langle s_{jk}k_{jk} + p_k m_j\right\rangle_{m_j} = d_{jk} \quad (4\text{-}4)$$

Since, from theorem 3, the cyclic length of the residue digits differences reference to module m_j is constant (uniform), then there must exist a condition,
$c_{ij}\cdot s_{ij}\cdot k_{ij} = c_{jk}\cdot s_{jk}\cdot k_{jk}$ Eliminating the above terms from Equations (4-3) and (4-4),

$$c_{ij}\cdot p_i\cdot m_i - c_{jk}\cdot p_{jk}\cdot m_k = c_{ij}\cdot d_{ij} - c_{jk}\cdot d_{jk} = D'_{ik}$$

$$p'_i m_i - p'_i m_k = D'_{ik}$$

where $p'_i = c_{ij}p_i$, $p'_k = c_{jk}p_k$ and
$D'_{ik} = c_{ij}\cdot d_{ij} - c_{jk}\cdot d_{jk}$

Example 4-2

Let the moduli set $(m_1, m_2, m_3, m_4, m_5) = (4,5,7,9,11)$
$x = (x_1, x_2, x_3, x_4, x_5) = (1,2,6,7,9)$, and the error
$x' = (x'_1, x'_2, x'_3, x'_4, x'_5) = (1,2,5,7,9)$, the error occurs at m_3.

Follow the same procedures of the Example 4-1 to use this algorithm.

$$\langle S_{14}k_{14}\rangle_4 = \langle 5k_{14}\rangle_4 = 2, \text{ or } \langle 5k_{14} + 4p_1\rangle_4 = 2 \quad (4\text{-}5)$$

$$\langle S_{24}k_{24}\rangle_5 = \langle 4k_{24}\rangle_5 = 0, \text{ or } \langle 4k_{24} + 5p_2\rangle_5 = 0 \quad (4\text{-}6)$$

$$\langle S_{34}k_{34}\rangle_7 = \langle 2k_{34}\rangle_7 = 5, \text{ or } \langle 2k_{34} + 7p_3\rangle_7 = 5 \quad (4\text{-}7)$$

$$\langle S_{45}k_{45}\rangle_{11} = \langle 2k_{45}\rangle_{11} = 9, \text{ or } \langle 2k_{45} + 11p_5\rangle_{11} = 9 \quad (4\text{-}8)$$

Eliminating $5k_{14}$ and $4k_{24}$ from Equation's (4-5) and (4-6)

$16p_1 - 25p_2 = 8$,

$p_1 = 13$, and $p_2 = 8$,

solve for k_{14} from (4-5),

$$\langle 5 * k_{14} + 4 * 13\rangle_4 = 2,$$

$$\therefore\ k_{14} = 10,$$

or $4\cdot k_{24} = 5\cdot 8 = 40$,

$\therefore\ k_{24} = 10$, .

Check from Equation (4-5),

$$\langle 2\times 10\rangle_{11} = 9,\ \langle 2\times 10\rangle_7 = 6 \neq 5.$$

This shows that the error occurs at module m_3. From this result, we can immediately obtain $\langle 2\times 10\rangle_7 = 6$. Noting that it may happen that the assigned referenced memory moduli falls coincidentally with error memory module m_3. In this occurrence, we cannot find the correct (integers) values of P_1 and P_2 within the legitimate range. It seems that this algorithm can only detect error. To complete the error correction procedure, we can simply change the referenced module to any other and follow the same procedure as before. This guarantees that the proposed algorithm in Theorem 4 will also work well in this case. The hardware structure for illustrating this algorithm is shown in **Figure 3**.

The proposed TRD (target Race Distance) scheme used for error correction can be used for scaling and assigning numbers in a residue number system. A redundant residue number system (RRNS) is defined as before in an RNS with r additional moduli. The moduli $\{m_1, m_2, \cdots, m_i, \cdots, m_k\}$, are called the nonredundant moduli, while the extra r moduli, $\{m_{k+1}, m_{k+2}, \cdots, m_{k+r}\}$ are the redundant moduli. The interval, $[0, M_k - 1]$, is called the legitimate range where $M_k = \prod_{i=1}^{k} m_i$ and the interval, $[M_k, M_{kr} - 1]$, is the illegitimate range, where $M_{kr} = M_k M_r = M_k \prod_{i=1}^{r} m_{k+i}$ is the total range. In the RRNS, the negative numbers within the dynamic range are represented as states at the upper extreme of the total range, which is part of the illegitimate range. The positive members are mapped to the interval $\left[0, \frac{(M_k - 1)}{2}\right]$, if M_k is odd, or $\left[0, \frac{M_k}{2}\right]$, if M_k is even. The negative numbers are mapped to the interval

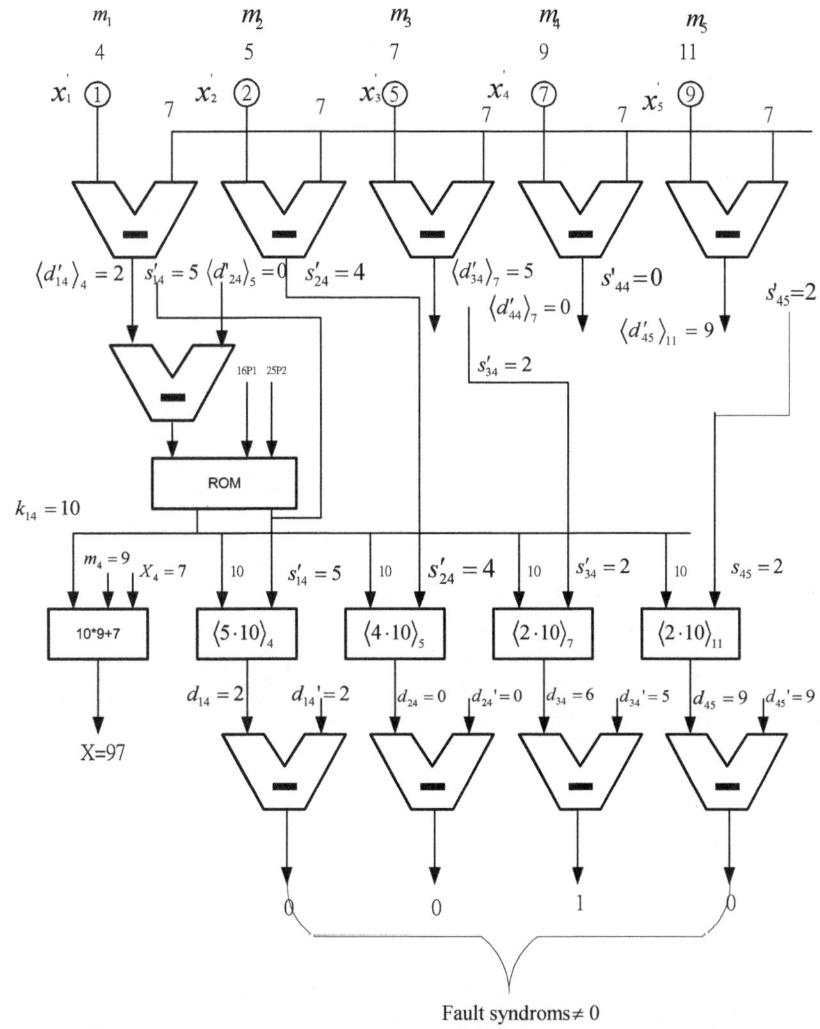

Figure 3. In the block diagram using optimal matching between multiples P_i m and $P_k m_k$, the residue digits are corrected by x_i - x_4 = d_{i4}.

$\left[M_{kr}-\frac{(M_k-1)}{2}, M_{kr}-1\right]$ if M_k is odd or

$\left[M_{kr}-\frac{M_k}{2}, M_{kr}-1\right]$ if M_k is even [14].

The one-to-one correspondence between the integers of the dynamic range and the states of the legitimate range in the RRNS can be established using a polarity shift. [11], The polarity shift is defined as below.

$$X_p = X+\frac{M_k}{2} \text{ for } M_k \text{ even} = X+\frac{M_k-1}{2} \text{ for } M_k \text{ odd.}$$

where X_p denotes the value X after a polarity shift and $X \in \left[-\frac{M_k}{2}, \frac{M_k}{2}\right]$ if M_k is odd, so that $X_p \in [0, M_k]$, a polarity shift needs to be performed prior to correcting or scaling since X_p belongs to the legitimate range. If a single residue digit error $e_j = \{0,0,\cdots,e_j,0,\cdots,0\}$ is introduced and corresponds to modules m_j, then, after a polarity shift.

$$x'_p = \langle X_p + E_j \rangle_{Mkr} = \left\langle X_p + \frac{M_{kr}}{m_j}\langle w_j e_j \rangle_{m_j} \right\rangle M_{kr}$$

$$= \left\langle X_p + \frac{M_{kr}}{m_j} e'_j \right\rangle_{Mkr},$$

where w_j is the multiplicative inverse of $\frac{M_{kr}}{m_j}$ moduli m_j *i.e.* $\left\langle \left(\frac{M_{kr}}{m_j}\right)_{wj} \right\rangle_{m_j} = 1$ and $e'_j = \langle w_j e_j \rangle_{m_j}$ The x'_p denotes a single residue digit error and must fall within the illegitimate range, $m_k \le x'_p < m_{kr}$ [11].

Since $\left[\frac{x'_p}{M_k}\right] < M_r \left(x'_p < m_k \cdot m_r\right)$, and can be represented uniquely by $\{a_{k+1}, a_{k+2}, \cdots, a_{k+r}\}$,where a_{k+r}'s are the coefficient from the Chinese Remainder Theorem (CRT), *i.e*, $x'_p = \sum_{i=1}^{k+r} a_i \prod_{j=i-1}^{k+r-1} m_j$, where $m_0 = 1, 0 \le a_i \le m_i$. Note that the redundant digits $a_{k+1}, a_{k+2}, \cdots, a_{k+r}$ are zeros if no error is introduced, while at least one redundant digit is not equal to zero if a single error is introduced. Therefore, it has the same meaning that $\left[\frac{x'_p}{M_k}\right]_r$ or $\{a_{k+1}, a_{k+2}, \cdots, a_{k+r}\}$ is used to be the entries of the error correction.

1) $M_r > m_i m_{k+s}$, $1 \le i \le k$, $1 \le s \le r$, and ,
2) $M_r > 2m_i m_j - m_i - m_j$, $1 \le i, j \le k$, and $i \ne j$

Although the errors detection and correction described in section II have been simplified the processes due to no need of CRT conversion. It is still hardware complex and time consuming for the residue scaling operation. To improve this drawback, a direct residue-scaling algorithm can be used. It is flexible and direct to detect and prevent the errors. The flexibility means that the scaling factor can be arbitrary chosen any single module such as m_i , *i.e.* not necessarily beginning from $m_1, m_2, \cdots$ to m_k . in order. The direct capability means no requirement for CRT extension processes for decoding or lookup tables. The following theorem (theorem 7) and example are clarified.

Theorem 7. If the scaling factor K is one of the module set $= \{m_1, m_2, \cdots, m_i, \cdots, m_k, \cdots, m_{k+r}\}$ and the residue digits are $\{x_1, x_2, \cdots, x_i, \cdots, x_k, \cdots, x_{k+r}\}$, respectively, then the residue digit x_i scaled by a factor

$m_{j,i\ne j,} = \left(\frac{x_i}{m_i}\right) = y_i$ can be obtained using the equation

$$\langle m_i y_i \rangle_{m_j} = x_i \qquad (4\text{-}9).$$

Proof: It is easy to show that when $m_1 \ne m_2$, and Equation (4-9) is divided by m_i on both side, we have

$$\left\langle \frac{m_i y_i}{m_i} \right\rangle_{m_j} = \frac{x_i}{m_i} = y_i \qquad (4\text{-}10).$$

Example 4-3. For convenient comparison of the proposed TRD algorithm to other schemes such as appeared in [14], we take the same numerical example in [11]. Let the moduli set $\{m_1, m_2, m_3, m_4, m_5, m_6\} = \{2, 5, 7, 9, 11, 13\}$, where $\{m_1, m_2, m_3, m_4\}$ are regular moduli and $\{m_5, m_6\}$ are redundant moduli. Then $M_k = 2\cdot 5\cdot 7\cdot 9 = 630$, $M_r = 11 * 13 = 143$, $M_{kr} = M_k M_r = 630\cdot 143 = 90090$, and $X_p \in \left[-\frac{M_k}{2}, \frac{M_k}{2}\right] = [-315, 315]$. The sufficient conditions for correcting single residue digits errors are

1) $M_r > \{m_i m_{k+s}\}_{\max}$ $i = 1, 2, 3$, or 4, $s = 1$, or 2, $k = 4$, The maximum $m_i m_{k+s} = m_4 \cdot m_6 = 9\cdot 13 = 117 < 143(M_r)$, and
2) $M_r > \{2m_i m_j - m_i - m_j\}_{\max}$ $i, j = 1, 2, 3,$ or 4,

$$\text{The max } \{2m_i m_j - m_i - m_j\} = 2m_3 m_4 - m_3 - m_4$$
$$= 2\cdot 7\cdot 9 - 7 - 9$$
$$= 110 < 143 (= M_r).$$

Thus the moduli set satisfies the necessary and sufficient conditions for correcting single errors digit. Assume $X = -311 = \{1, 4, 4, 4, 8, 1\}$ and a single digit error $e_2 = 4$ is introduced, then $X' = \{1, 3, 4, 4, 8, 1\} = 53743$. After a polarity shift, $X'_p = X' + \frac{M_k}{2} = \{0, 3, 4, 4, 4, 4\}$.

Follow the same procedures as shown in Example 4-2. CPRDD is applied for correction without the need for using a table.

1) Assign the moduli $m_4 = 9$ as the reference moduli, the following residue digit references and its corresponding CPRDD equations: $\langle sk_{ij} \rangle_{m_i} = \langle d_{ij} \rangle_{m_i}$ are obtained

$$\langle d_{14} \rangle_2 = 1, \langle 7k_{14} \rangle_2 = 1;$$
$$\langle d_{24} \rangle_5 = 4, \langle 4k_{24} \rangle_5 = 4;$$
$$\langle d_{34} \rangle_7 = 0, \langle 2k_{34} \rangle_7 = 0;$$
$$\langle d_{45} \rangle_{11} = 7, \langle 2k_{45} \rangle_{11} = 7;$$
$$\langle d_{46} \rangle_{13} = 3, \langle 4k_{46} \rangle_{13} = 3.$$

2) Choose two highest digit difference as one pair for equal target race distance e.g. $\langle 2k_{45} \rangle_{11} = 7$ and $\langle 4k_{46} \rangle_{13} = 3$. Then the true primary RDD equations are

$$\langle 2k_{45} + 11p_1 \rangle_{11} = 7 \qquad (4\text{-}11),$$

$$\text{And } \langle 4k_{46} + 13p_2 \rangle_{13} = 3 \qquad (4\text{-}12),$$

where p_1 and p_2 are selected so that the two RDD are equal distances.

3) Eliminating k terms in Equation's (4-11) and (4-12) by putting $k_{45} = k_{46}$

$(22p_1) - (13p_2) = 11$, $p_1 = \frac{11 + (13p_2)}{22}$ where

$p_2 = 11$, then $p_1 = \frac{11 + 143}{22} = 7$.

4) Substituting p_1 and p_2 into equations (4-9) and (4-10) respectively, we have $2k_{45} = -11\cdot 7 + 7 = -70$, then $k_{45} = -35$, and $4k_{46} + 13\cdot 11 = 3$, also, $k_{46} = \frac{-143 + 3}{4} = -35$.

5) Checking other three RDD's

$$\langle 4k_{24}\rangle_5 = 4 - \rangle\langle 4\times(-35)\rangle\rangle_5 = 4,$$

$$\langle 2k_{34}\rangle_7 = 0 - \rangle\langle 2\times(-35)\rangle\rangle_7 = 0,$$

$$\langle 7k_{14}\rangle_2 = 1 - \rangle\langle 7\times(-35)\rangle\rangle_2 = 1.$$

The only different module residue occurs on module number at $m_2 = 5$, *i.e.*, $x'_{24} - x_{24} = 4 - 0 = 4$. The three target distances, can be from any module residue, say, (except $m_2 = 5$), $m = 7$. $x = -9\cdot 35 + 4 = -311$.

The residue representation of X is therefore, $X = \{1,4,4,4,8,1\}$. If a single digit error $e_2 = 4$ is introduced, then, $X' = 53743 = \{1,3,4,4,8,1\}$. The corresponding error is therefore

$$e_2 = \{(x'_1 - x_1),(x'_2 - x_2),(x'_3 - x_3),\cdots,(x'_6 - x_6)\}.$$

$$\therefore\ X' - X = \{0,-1,0,0,0,0\} = \{0,4,0,0,0,0\}$$

After a polarity shift,

$$x'_p = x' + \frac{M}{2} = 54058 = \{1,3,4,4,8,1\} + \{0,3,4,4,4,4\}$$

and the scaling factor $\frac{1}{K}$ to x'_p is $\left\lceil \frac{x'_p}{K} \right\rceil = 85(M_k = 2\cdot 5\cdot 7\cdot 9 = 630)$. The final step must use a lookup table to obtain the result, $\left\lceil \frac{x_p}{K} \right\rceil$ [13].

For verifying our proposed algorithm, the table of the corresponding $\left\lceil \frac{x'_p}{K} \right\rceil$ is not required as in [13]. The processes for finding and correcting a single error based on our method are described below.

1) Find the residue digit difference to a selected module, say m_4 as before $x' = 53743 = \{1,3,4,4,8,1\}$. For verifying that our proposed algorithm detects and corrects single error without using a table, the same numerical example is used to describe the procedure as follows:

$$d_{14} = \langle -3\rangle_2 = 1,$$

$$d_{24} = \langle -1\rangle_5 = 4,$$

$$d_{34} = \langle 0\rangle_7 = 0,$$

$$d_{45} = \langle -4\rangle_{11} = 7,$$

$$d_{46} = \langle 3\rangle_{13} = 3.$$

Then

$$\langle 7k_{14}\rangle_2 = 1,$$

$$\langle 4k_{24}\rangle_5 = 4,$$

$$\langle 2k_{34}\rangle_7 = 0,$$

$$\langle 2k_{45}\rangle_{11} = 7,$$

$$\langle 4k_{46}\rangle_{13} = 3.$$

2) Choose two highest digit differences as one pair for equal target race distances. e.g.

$\langle 2k_{45}\rangle_{11} = 7$ and $\langle 4k_{46}\rangle_{13} = 3$, the following two equations can be obtained:

$$\langle 2k_{45} + 11p_1\rangle_{11} = 7 \quad \text{(4-13a)}$$

$$\langle 4k_{46} + 13p_2\rangle_{13} = 3 \quad \text{(4-13b).}$$

3) Eliminating k terms in (4-13a) and (4-13b) by putting $k_{45} = k_{46}$

$22p_1 - 13p_2 = 0$ then $p_1 = 13$ and $p_2 = 22$.

4) Substituting p_1 and p_2 into Equation's (4-13a) and (4-13b) respectively, we have $2k_{45} + 11\cdot 13 = 0$, then $k_{45} = \frac{-143}{2}$, and $4k_{46} + 13\cdot 22 = 0$, also, $k_{46} = \frac{-143}{2}$

$X'_p = X' + \frac{M}{2}$	$X_p = X + \frac{M}{2} = -311 + 315 = 4$
$\langle 1+1\rangle_2 = 0$	$\langle 4\rangle_2 = 0$
$\langle 0+3\rangle_5 = 3$	$\langle 4\rangle_5 = 4$
$\langle 0+4\rangle_7 = 4$	$\langle 4\rangle_7 - 4$
$\langle 0+4\rangle_9 = 4$	$\langle 4\rangle_9 = 4$
$\langle 7+8\rangle_{11} = 4$	$\langle 4\rangle_{11} = 4$
$\langle 3+1\rangle_{13} = 4$	$\langle 4\rangle_{13} = 4$

Obviously, the error is located at $m_2 = 5$ thus $e_2 = X'_p - X_p = 3 - 4 = -1$.

Furthermore, the CPRDD algorithm can be used directly and in parallel for residue scaling and error correction. Thus the process is greatly speeded up.

Example 4-4 For convenient comparison, the same numeric example as in [13] is illustrated here. Consider $\{m_1, m_2, m_3, m_4, m_5, m_6\} = \{2,5,7,9,11,13\}$, and scaling factor $K = m_1 \cdot m_2 = 2\cdot 5 = 10$. If an input $X = -205 = \{1,0,5,2,4,3\}$ and a single residue digit error $e_3 = 1$, corresponding to $m_3 = 7$,

Then $X' = 25535 = \{1,0,6,2,4,3\}$.

After a polarity shift,

$$x'_p = x' + \frac{M}{2} = 25850 = \{0,0,6,2,0,6\}$$

1) Dividing by $m_1 = 2$ after subtracting $x'_{p1} = 0$ from $x_2, x_3, \cdots, x_6$

$\langle 2p_2\rangle_5 = 0$, this leads $p_2 = 0$,

$\langle 2p_3 \rangle_7 = 6, p_3 = 3$,

$\langle 2p_4 \rangle_9 = 2, p_4 = 1$,

$\langle 2p_5 \rangle_{11} = 0, p_5 = 0$,

$\langle 2p_6 \rangle_{13} = 6, p_6 = 3$.

2) Dividing by $m_2 = 5$ after subtracting $x'_{P_2} = 0$ from $x_3, x_4, \cdots, x_6$

$\langle 5k_3 \rangle_7 = 3, k_3 = 2, 9, 16, \cdots;$

$\langle 5k_4 \rangle_9 = 1, k_4 = 2, 11, 20, \cdots;$

$\langle 5k_5 \rangle_{11} = 0, k_5 = 0, 11, 22, \cdots;$

$\langle 5k_6 \rangle_{13} = 3, k_6 = 11, 24, 37, \cdots.$

Since from above only k_3 does not match with all other's k_i, *i.e.* $k_3 \cap k_{4,5,6} = 0$ and $k_4 \cap_5 \cap k_6 = 11$. Therefore, there occurs an error at $m_3 = 7$. Once this error is detected, it is easily found and corrected from the above equations, $\langle 5k_3 \rangle_7 = \langle 5 \cdot 11 \rangle_7 = 6$, which in turn $p'_3 = k_3 = 6$ and $X = \langle 2p'_3 \rangle_7 = \langle 2 * 6 \rangle_7 = 5$.

$m_i = (2 \;\; 5 \;\; 7 \;\; 9 \;\; 11 \;\; 13)$

$x = -205, x_i = (1 \;\; 0 \;\; 5 \;\; 2 \;\; 4 \;\; 3)$

$630/2 = 315 = \frac{M}{2} = (1 \;\; 0 \;\; 0 \;\; 0 \;\; 7 \;\; 3)$

$-205 + 315, x + \frac{M}{2} = (0 \;\; 0 \;\; 5 \;\; 2 \;\; 0 \;\; 6) = 110$

that $(p_1 p_2 p_3 p_4 p_5 p_6) = (0 \;\; 0 \;\; 5 \;\; 2 \;\; 0 \;\; 6)$

Divided by "2",

$\langle 2p'_2 \rangle_5 = p_2 = 0, p'_2 = 0;$

$\langle 2p'_3 \rangle_7 = p_3 = 5, p'_3 = 6;$

$\langle 2p'_4 \rangle_9 = p_4 = 2, p'_4 = 1;$

$\langle 2p'_5 \rangle_{11} = p_5 = 0, p'_5 = 0;$

$\langle 2p'_6 \rangle_{13} = p_6 = 6, p'_6 = 3;$

$(p'_2 p'_3 p'_4 p'_5 p'_6) = (0 \;\; 6 \;\; 1 \;\; 0 \;\; 3) = 55 (= 110/2).$

Divided by "5"

$\langle 5p''_3 \rangle_7 = p'_3 = 6, p''_3 = 4;$

$\langle 5p''_4 \rangle_9 = p'_4 = 1, p''_4 = 2;$

$\langle 5p''_5 \rangle_{11} = p'_5 = 0, p''_5 = 0;$

$\langle 5p''_6 \rangle_{13} = p'_6 = 3, p''_6 = 11$

$(p''_3 p''_4 p''_5 p''_6) = (42011) = 11 (= 55/5).$

The hardware structure of this example for the residue scaling is shown in **Figure 4**.

Actually this algorithm can be divided by any arbitrary moduli.

Example 4-5

Divided by any arbitrary moduli, say $m_4 = 9$, it must subtract $x'_{p4} = 2$ from X

$\langle x'_{p1} - x'_{p4} \rangle_2 = \langle 0 - 2 \rangle_2 = 0,$

$\langle x'_{p2} - x'_{p4} \rangle_5 = \langle 0 - 2 \rangle_5 = \langle -2 \rangle_5 = 3,$

$\langle x'_{p3} - x'_{p4} \rangle_7 = \langle 6 - 2 \rangle_7 = 4,$

$\langle x'_{p5} - x'_{p4} \rangle_{11} = \langle 0 - 2 \rangle_{11} = 9,$

$\langle x'_{p6} - x'_{p4} \rangle_{13} = \langle 6 - 2 \rangle_{13} = 4,$

Then

$\langle 9k_1 \rangle_2 = 0, k_1 = 0;$

$\langle 9k_2 \rangle_5 = 3, k_2 = 2;$

$\langle 9k_3 \rangle_7 = 4, k_3 = 2;$

$\langle 9k_5 \rangle_{11} = 9, k_5 = 1;$

$\langle 9k_6 \rangle_{13} = 4, k_6 = 12;$

check $(25850 - 2)/9 = 2872$.

This results

$x'_1 = \langle 2872 \rangle_2 = 0, x'_2 = \langle 2872 \rangle_5 = 2, x'_3 = \langle 2872 \rangle_7 = 2,$

$x'_5 = \langle 2872 \rangle_{11} = 1$, and $x'_6 = \langle 2872 \rangle_{13} = 12$.

It can be seen from above that $x'_1 = k_1, x'_2 = k_2, x'_3 = k_3, x'_4 = k_4 = 0, x'_5 = k_5$, and $x'_6 = k_6$, which are equal each other as expected.

Example 4-6

For processing two residue scalings and error corrections in parallel, we take Example 4-4 as an illustration. Let scaling factor $K = 2 * 5 = 10$, *i.e.*, the first residue scaling factor is 2 and the second one is 5 or verse versa. It is easily shown that the extended CPRDD algorithm is used and can be completed in one cycle. That is

$\langle 10p'_2 \rangle_5 = p_2 = 0, p'_2 = 0, 1, \cdots, 11, 12, \cdots;$

$\langle 10p'_3 \rangle_7 = p_3 = 5, p'_3 = 4, 11, 18, \cdots;$

$\langle 10p'_4 \rangle_9 = p_4 = 2, p'_4 = 2, 11, 20, \cdots;$

$\langle 10p'_5 \rangle_{11} = p_5 = 0, p'_5 = 0, 11, 22, \cdots;$

$\langle 10p'_6 \rangle_{13} = p_6 = 6, p'_6 = 11, 24, 37, \cdots.$

The result is identical

$$X_p = \frac{x + \frac{M}{2}}{10} = \frac{-205 + 315}{10} = \frac{110}{10} = 11, \; i.e.,$$

$\langle 11 \rangle_7 = 4, \langle 11 \rangle_9 = 2, \langle 11 \rangle_{11} = 0$, and $\langle 11 \rangle_{13} = 11$, which are identical results as shown in Example 4-4.

Example 4-7 For error correction

$m_2 \; m_3 \; m_4 \; m_5 \; m_6 \quad (5 \;\; 7 \;\; 9 \;\; 11 \;\; 13)$

$$\frac{x + \frac{M}{2}}{2} = (0 \;\; 3^* \;\; 1 \;\; 0 \;\; 3) = 55, \; k_{ij} = \frac{55}{11}$$

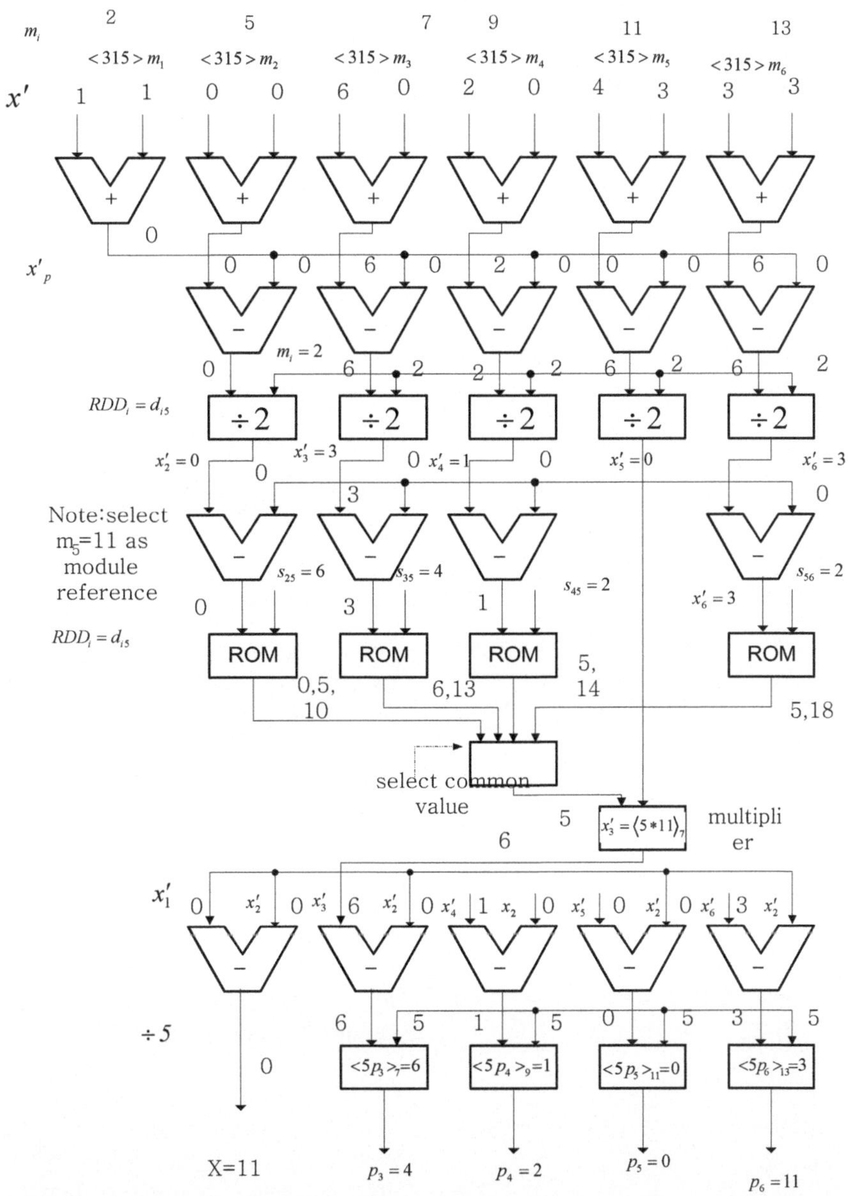

Figure 4. Hardware structure of the residue scaling number for Example 4-4.

$d_{25}=0,|S_{25}|=6,\langle 6\cdot k_{25}\rangle_5=0,k_{25}=0,5;$

$d_{35}=3,|S_{35}|=4,\langle 4\cdot k_{35}\rangle_7=3,k_{35}=6,13;$

$d_{45}=1,|S_{45}|=2,\langle 2\cdot k_{45}\rangle_9=1,k_{45}=5,14;$

$\langle d_{56}\rangle_{13}=\langle -3\rangle_{13}=10,|S_{56}|=2,\langle 2\cdot k_{56}\rangle_{13}=10,k_{56}=5,18;$

the correct $k_{i,5}=5$ $\therefore$ $\mathrm{RDD}=11\cdot 5=55$, and $\langle k'_{35}\rangle_7=\langle 55\rangle_7=6.$

This shows $k_{25}=k_{45}=k_{56}(=5)k_{35}$. Therefore the error correction is made by $d_{35}=\langle 4\cdot 5\rangle_7=6$, and $X_3=d_{35}+X_5=6+0=6$, which corresponds to the value in Example 4-4, in scaling factor $k=10$, (dividing by "5" part).

From above results, this checks that scaling $\frac{x}{k}=11-\frac{315}{10}=-20.5$ which is within the accuracy of the residue scaling factor.

In a general case, $x_i\neq x_j\neq 0$, this time we must modify the subtraction of x_i and x_j from the *X*, before the process of the scaling. If $k=m_i\cdot m_j$ is the scaling factor, then the subtraction must change to $X^{\Delta}=X-\left(x_i+x_j^{\Delta}\right)$, where $x_j^{\Delta}=m_ik_i$ so that $x_j=x_i+\langle m_ik_i\rangle_{m_j}$ or $\langle m_ik_i\rangle_{m_j}=x_j-x_i$. Let us consider the following example:

Example 4-8

$X=\{1,0,2,0,3,5\}=135$ of moduli set

$\{2,5,7,9,11,13\}$. The scaling factor $K = m_1 \cdot m_3 = 2 \cdot 7 = 14$. is assumed. Then, residue $x_j^{\Delta} = m_1 k_i = 2k_i$, and k_i can be found from $\langle 2k_i \rangle_7 = x_1 - x_3 = (2-1) = 1$. $\left(k_i < \frac{M_r}{m_i m_j} \right) k_i = 4$. Thus $x^{\Delta}_3 = 2 \cdot 4 = 8$, and $(x_i + x^{\Delta}_j) = 1 + 8 = 9$.

Alternatively, it could be from other module m_3, $x^{\Delta}_j = \langle 7k_j \rangle_2$, where $\langle 7k_j \rangle_2 = x_1 - x_3 = \langle 1-2 \rangle_2 = \langle -1 \rangle_2 = 1$, and $7k_j = 7 \therefore (x_i + x_j^{\Delta}) = (2+7) = 9$ which has the same number to be subtracted.

$m_1 m_2 m_3 m_4 m_5 m_6$

m_i 2 5 7 9 11 13

x_i 1 0 2 0 35 = 135

$x_i - 9 = \langle -8 \rangle_2 \langle -9 \rangle_5 \langle -7 \rangle_7 \langle -9 \rangle_9 \langle -6 \rangle_{11} \langle -4 \rangle_{13}$

0 1 0 0 5 9 135 − 9 = 126

From CPRDD algorithm, the scaling processes are performed as before, we then have the following results by scaling factor $K = m_1 \times m_3 = 2 \times 7 = 14$;

$\langle 14P_2 \rangle_5 = 1, P_2 = 4, 9, \cdots;$

$\langle 14P_3 \rangle_7 = 0, P_3 = 0, 1, 2, \cdots, 9, \cdots;$

$\langle 14P_4 \rangle_9 = 0, P_4 = 9, 18, \cdots;$

$\langle 14P_5 \rangle_{11} = 5, P_5 = 9, \cdots;$

$\langle 14P_6 \rangle_{13} = 9, P_6 = 9, \cdots$

Thus $X = 9$, which is exactly the value $\frac{126}{14} = 9$ and is the most closed to $\left\lfloor \frac{135}{14} \right\rfloor = 9$.

This result can be checked using sequential steps as follows:

For $x_i - 1, i = 1, 2, 3, \cdots, 6;$

$m_i = (2 \quad 5 \quad 7 \quad 9 \quad 11 \quad 13)$

$x_i = (1 \quad 0 \quad 2 \quad 0 \quad 3 \quad 5)$

$x_i - 1 = (0 \quad -1 \quad 1 \quad -1 \quad 2 \quad 4)$

Divided by 2:

$\langle 2k_2 \rangle_5 = -1, k_2 = -3;$

$\langle 2k_3 \rangle_7 = 1, k_3 = 4;$

$\langle 2k_4 \rangle_9 = -1, k_4 = -5;$

$\langle 2k_5 \rangle_{11} = 2, k_5 = 1;$

$\langle 2k_6 \rangle_{13} = 4, k_6 = 2.$

$m_i (5 \quad 7 \quad 9 \quad 11 \quad 13)$

$(k_2 k_3 k_4 k_5 k_6)$

$k_i = (-3 \quad 4 \quad -5 \quad 1 \quad 2)$

$k_i - 4 = (\langle -7 \rangle_5 \langle 0 \rangle_7 \langle -9 \rangle_9 \langle -3 \rangle_{11} \langle -2 \rangle_{13})$

30 0 811

Divided by 7:

$\langle 7q_2 \rangle_5 = \langle -2 \rangle_5 = 3, q_2 = 9;$

$\langle 7q_3 \rangle_7 = 0, q_3 = 9;$

$\langle 7q_4 \rangle_9 = 0, q_4 = 9;$

$\langle 7q_5 \rangle_{11} = \langle -3 \rangle_{11} = 8, q_5 = 9;$

$\langle 7q_6 \rangle_{13} = 11, q_6 = 9.$

This result of $q = \left\lfloor \frac{135}{14} \right\rfloor = 9$ shows that the CPRDD algorithm has the capability of parallel processing operations in residue scaling and error corrections, *i.e.*, any combination moduli scaling factors for Ks of moduli set $\{m_1, m_2, \cdots, m_k\}$ can be performed simultaneously.

5. Conclusions

The arithmetic operations in the residue number system for addition, subtraction, and multiplication can be speeded up by using its parallel processing properties. However, some difficult operations, such as error detection and correction, must go through conversion or decoding processes from the residue representation to the regional binary number x. This is because the decoding technique is usually accomplished using the mixed-radix digit (MRD) or Chinese Remained Theorem (CRT), which are time consuming processes requiring hardware complexity. We proposed two algorithms for scaling and error correction without the need for lookup tables or increasing the encoding process.

The Cyclic property of the Residue-Digit Difference (CPRDD) algorithm can detect and correct errors from the RNS cyclic property. Any residue moduli set has a specific cycle length, which can be obtained from the individual residue number, difference, each pair, to a reference memory module m_i. Once the cyclic length is known, then the original value x is easily found, and in turn, the errors can be detected and corrected.

The TRD (Target Race Distance) algorithm combined with CPRDD is used for scaling and for error detection and correction. The scaling results and error correction can be directly performed by these two algorithms without using MRD or CRT. Thus, the decoding process is significantly reduced, and the hardware structure is greatly simplified. Several examples are illustrated and verified for these two algorithms.

REFERENCES

[1] R. W. Watson, "Error Detection and Correction and Other Residue-Interacting Operations in a Redundant Residue Number System," University of California, Berkeley,

1965.

[2] R. W. Watson and C. W. Hastings, "Self-Checked Computation Using Residue Arithmetic," *Proceedings of the IEEE*, Vol. 54, No. 12, 1966, pp. 1920-1931.

[3] S. S. S. Yau and Y. C. Liu, "Error Correction in Redundant Residue Number Systems," *IEEE Transactions on Computers*, Vol. C-22, No. 1, 1973, pp. 5-11.

[4] D. Mandelbaum, "Error Correction in Residue Arithmetic," *IEEE Transactions on Computers*, Vol. C-21, No. 6, 1972, pp. 538-545.

[5] F. Barsi and P. Maestrini, "Error Correcting Properties of Redundant Residue Number Systems," *IEEE Transactions on Computers*, Vol. 22, No. 3, 1973, pp. 307-315.

[6] F. Barsi and P. Maestrini, "Error Detection and Correction by Product Codes in Residue Number Systems," *IEEE Transactions on Computers*, Vol. 23, No, 9, 1974, pp. 915-924.

[7] V. Ramachandran, "Single Residue Error Correction in Residue Number Systems," *IEEE Transactions on Computers*, Vol. C-32, No. 5, 1983, pp. 504-507.

[8] W. K. Lenkins and E. J. Altman, "Self-Checking Properties of Residue Number Error Checkers Based on Mixed Radix Conversion," *IEEE Transactions on Circuits and Systems*, Vol. 35, No. 2, 1988, pp. 159-167.

[9] W. K. Lenkins, "Residue Number System Error Checking Using Expanded Projection," *Electronics Letters*, Vol. 18, No. 21, 1982, pp. 927-928.

[10] W. K. Lenkins, "The Design of Error Checkers for Self-Checking Residue Number Arithmetic," *IEEE Transactions on Computers*, Vol. C-32, No. 4, 1983, pp. 388-396.

[11] M. H. Etzel and W. K. Jenkins, "Redundant Residue Number Systems for Error Detection and Correction in Digital Filters," *IEEE Transactions on Acoustics Speech and Signal Processing*, Vol. 28, No. 10, 1980, pp. 588-544.

[12] C. C. Su and H. Y. Lo, "An Algorithm for Scaling and Single Residue Error Correction in Residue Number Systems," *IEEE Transactions on Computers*, Vol. 39, No. 8, 1990, pp.1053-1064.

[13] H. Y. Lo, "An Optimal Matched and Parallel Mixed-Radix Converter," *Journal of Information Science and Engineering*, Vol. 10, 1994, pp. 411-421.

[14] A. P. Shenoy and R. Kumaresan, "Fast Base Extension Using a Redundant Modus in RNS," *IEEE Transactions on Computers*, Vol. 38, No. 2, 1989, pp. 152-161.

[15] E. D. Diclaudis, G. Orlandi and F, Piazza, "A Systolic Redundant Residue Arithmetic Error Correction Circuit," *IEEE Transactions on Computers*, Vol. 42, No. 4, 1993, pp. 427-433.

[16] S. S. Wang and M. Y. Shau, "Single Residue Error correction Based on K-Term M_j-Projection," *IEEE Transactions on Computers*, Vol. 44, No. 1, 1995, pp. 129-131.

[17] R. S. Katti, "A New Residue Arithmetic Error Correction Scheme," *IEEE Transactions on Computers*, Vol. 45, No. 1, 1996, pp. 13-19.

Assisted Channel Estimation with Dynamic Allocation Pilot Subcarriers for OFDM Communications

Boudali Ouarzazi[1], Atika Menhaj-Rivenq[1], Iyad Dayoub[1], Marion Berbineau[2]

[1]Department OAE, Institut d'Electronique de Microelectronique et de Nanotechnologie, Université de Valenciennes et du Hainaut-Cambrésis,Valenciennes, France; [2]Université Lille Nord de France, Villeneuve d'Ascq, France.

ABSTRACT

In this paper, a transmission system based on OFDM (Orthogonal Frequency Division Multiplexing) technique and channel estimation is proposed. The idea is to enhance the assisted channel estimation without data rate loss. A solution based on frequency diversity is proposed. With the same number of pilots, the global system performances are better for higher mobility speed. The main results will be presented in the case of high mobility context.

Keywords: Doppler Effect; Channel Estimation; OFDM; Rayleigh Channel; BEM; Mobility

1. Introduction

The channel estimation is a fundamental step in a transmission system whose performances are directly related to the severity of this step in the reception process. The estimation quality is highly dependent on variations of the channel propagation. In the literature, one can find three main types of channel estimation techniques: supervised or assisted techniques, blind techniques and semi-blind techniques with decision feedback [1]. The blind estimation technique does not require insertion of pilot subcarriers in the frame. It uses some statistical properties of the received signal (*i.e.* redundancy due to the insertion of cyclic prefix, higher order statistics) and directly addresses the problem of loss of spectral efficiency [2-5]. However, this technique often requires significant computation time and isn't appropriate for this case of high mobility. The estimation technique with a semi-blind decision feedback using both known data (*i.e.* pilot subcarriers) and estimated data sent with a forward link. This estimation is very sensitive to the detection of errors on the estimated symbols [1,4], and in particular in the context of the high mobility with rapid channel variations. The assisted channel estimation is based on inserting, in the OFDM frame, symbols which are known by the receiver and called pilot symbols. The associated subcarriers are called the subcarrier pilots. Training sequence corresponds to the set of pilot subcarriers dedicated to the channel estimation. It consists in a pattern robust to the frequency selectivity of the channel. It is suitable for quasi-stationary channels for which the channel coefficients vary little from one OFDM symbol to another. The insertion of pilot "comb type" consists in reserving some pilot subcarriers only to channel estimation for all transmission duration. This pattern is robust to the channel temporal variations from an OFDM symbol to another [6]. The main disadvantage of the channel assisted estimating technique is the spectral efficiency loss due to the insertion of pilots between the transmitted data. However, we chose this technique despite the loss of spectral efficiency. In the broadband challenge for high mobility, the hostile characteristics nature of mobile propagation channel requires the use of pilots. The rapid variations of the channel require a minimum of channel knowledge by pilots in order to get a correct detection datum.

In this paper, we will focus on two aspects which influence strongly the channel estimation: pilot's geometry used and the algorithms applied to estimate the transfer function gains associated to data subcarriers. In the following section, we rely on the work of [7,8], using a model of BEM (Basis Expansion Modeling) channel development. Then, we propose a pilot subcarriers configuration to improve the estimation for rapid changes in time channel. Our objective is to enhance the spectral efficiency with the same number of pilots in order to increase the data rate.

The paper is organized as follows. Section II formulates the data model for OFDM system. Section III presents the chosen propagation channel model and its Basis Expansion Modeling (BEM) approach. In Section IV, this approach is applied to channel estimation with a dynamic allocation of pilot subcarriers. Simulation results are shown in Section V. Finally, we conclude the presented study in section VI.

Notation: x is a real scalar, X is a vector, $\boldsymbol{X}$ is a matrice. $\boldsymbol{X}^H$, $\boldsymbol{X}^+$, $\boldsymbol{X}^*$, $\boldsymbol{X}^T$ represent respectively hermetian, opposite, conjugate and transpose of the matrice $\boldsymbol{X}$. $X(k)$ is the kth element of vector X and $\boldsymbol{X}(i,k)$ are the ith and k^{th} entries of matrice $\boldsymbol{X}$. $\otimes$ is the circular convolution operator.

2. System Model

The multipath effect introduced by the propagation channel causes interferences between symbols when using a transmission type series. A good solution to overcome the effects of the channel is to use a parallel or multicarrier transmission. The OFDM technique consists to convert the serial coded data stream on parallel blocks and to transmit simultaneously these blocks over several orthogonal carriers using the Inverse Fast Fourier Transform (IFFT). In the transmitter side, the data stream is grouped into blocks of N data symbols, called OFDM symbol, and are represented by the vectors $X_i(k)$ (k,i: are frequency and OFDM symbol index). Then, an IFFT is performed on each data symbols block:

$$S_i(t)=\frac{1}{N}\sum_{k=1}^{N}X_i(k)\cdot e^{j2\pi\cdot f_k\cdot t}, \tag{1}$$

where $X_i(k)$ represents the complex data (M-PSK or M-QAM) which modulates the frequency f_k, and $f_k=k/T_b$. T_b is the symbol OFDM duration. By sampling the continuous signal $S_i(t)$ at time sampling T_e ($t=n\cdot T_e$ and $T_b=N\cdot T_e$), we obtain the following equation:

$$S_i(n)=\frac{1}{N}\sum_{k=1}^{N}X_i(k)\cdot e^{j2\pi\cdot\frac{n\cdot k}{N}}. \tag{2}$$

The received signal $R_i(n)$ is generally the sum of a convolution of the channel impulse response $h(n)$ with the signal $S_i(n)$ and an additive white Gaussian noise $V_i(n)$. It is implicitly assumed that the transmitter and the receiver are perfectly synchronized:

$$R_i(n)=\sum_{l=1}^{L}h_l(n)\otimes S_i(n-l)+V_i(n), \tag{3}$$

where $\otimes$ is the symbol of the circular convolution operator and $V_i(n)$ are samples of additive white gaussian noise (AWGN). At the receiver side, the cyclic prefix is removed and the data symbols $\boldsymbol{Y}$ are obtained by performing the FFT operation:

$$\boldsymbol{Y}=\boldsymbol{H}\cdot\boldsymbol{X}+\boldsymbol{V}=\boldsymbol{F}\boldsymbol{H}^t\boldsymbol{F}^H\cdot\boldsymbol{X}+\boldsymbol{V}, \tag{4}$$

where $\boldsymbol{Y}=\boldsymbol{Y}(i,k)$, $\boldsymbol{X}=\boldsymbol{X}(i,k)$, $\boldsymbol{V}=\boldsymbol{V}(i,k)$, $\boldsymbol{H}^t$ and $\boldsymbol{H}$ represent the channel matrices respectively in time and frequency domain for each OFDM symbol. The channel matrix $\boldsymbol{H}$ is not diagonal because of the channel temporal variations during an OFDM symbol (*i.e.* $\boldsymbol{H}(k,z)$, k transmitted and z received carriers). It is expressed using the following equation [9,10]:

$$\boldsymbol{H}(k,z)=\frac{1}{\sqrt{N}}\sum_{n=1}^{N}\boldsymbol{H}^t(k,n)\cdot e^{-j2\pi\frac{(k-z)n}{N}}, \tag{5}$$

where $\boldsymbol{H}^t(k,n)$ represents the Fourier transform of the impulse response $h_l(n)$:

$$\boldsymbol{H}^t(k,n)=\frac{1}{\sqrt{N}}\sum_{l=1}^{L}h_l(n)\cdot e^{-j2\pi\frac{kl}{N}}, \tag{6}$$

with L is the total number of paths in the multipath fading channel.

3. Propagation Channel Model

Since many years, different approaches for mobile radio channels modeling and simulation have been developed in the literature. But, no specific channel model is proposed for the mobile communications taking into account very high velocities such those encountered for very high speed train. The model of Clarke [11], which is the mathematical model of reference, and the simplified version of Jakes [12] are the most widely used to model variations in the signal channels called Rayleigh fading.

Simulation of a fading channel with Doppler calls for the generation of a Gaussian random process in order to mimic the sampled version of the channel waveform. These simulations are widely used in the process of designing efficient and robust next generation wireless communication and broadcasting systems. One important aspect to be considered is the accuracy of such simulation methods. Statistically, the provided discrete channel response samples should meet certain requirements related to the spectrum and time-domain variations. Hence, some quality measures have been introduced such as the probability density function (PDF), the cumulative distribution function (CDF), the autocovariance function (ACF), the level crossing rate (LCR) and the average duration of fades (ADF). Generated samples should meet the requirements in terms of quality measures during the predefined simulation period. The task of designing a channel simulator should tackle these issues while keeping a reasonable computation complexity and memory requirements.

State-of-the-art Rayleigh fading simulators can be divided into three main categories: the sum of sinusoids (SOS) method, filtering white Gaussian variables method and the inverse fast Fourier transform (IFFT) method. In [13], *Pop and Beaulieu* showed that the Jakes simulator is not wide-sense stationary in some cases and proposed other modifications. However, although the number of sinusoids is considered close to infinity, the higher order statistics of the simulator do not follow the theoretical statistical model of Clarke. This was also shown in [14,15]. In [16], Xiao and Zheng have proposed a new model for a sum of sinusoids Rayleigh fading channel that generates a non-deterministic signal with statistical properties converging to those of Jakes model even with a small number of sinusoids N_s.

The following equations give the fading process of this simulator:

$$Z(t)=Z_c(t)+jZ_s(t) \tag{7}$$

$$Z_c(t)=\frac{2}{\sqrt{N_s}}\sum_{n=1}^{N_s}\cos(\omega_d\cdot t\cdot\cos\alpha_n+\psi_n) \tag{8}$$

$$Z_s(t)=\frac{2}{\sqrt{N_s}}\sum_{n=1}^{N_s}\cos(\omega_d\cdot t\cdot\sin\alpha_n+\psi_n) \tag{9}$$

$$\alpha_n=\frac{2\pi n-\pi+\theta}{4N_s}, \tag{10}$$

where $\omega_d=2\pi F_d$; α_n, ψ_n and θ are random variables statistically independent and uniformly distributed (iid) on $[-\pi,\pi]$ for any number of sinusoids N_s. α_n, ψ_n represent respectively the arrival angle in the direction of movement of the receiver and the initial phase.

This model is selected for the proposed system simulations presented in this paper. In the next section, we describe the approximation of the complex channel gains by a development in basis functions. The goal is to model the temporal channel variations during an OFDM symbol and to determine their evolutions during to the transmission duration.

Basis Expansion Modeling

The idea of Basis Expansion Modeling (BEM) is described in [8,17,18-22]. The use of such a development is particularly useful to reduce the number of channel parameters to be estimated on a symbol time. For a block transmission during N symbols, the channel is characterized by $(N\times L)$ samples, L is the total number of broadband channel paths according to the representation given by the relation (5) in which each path has a different realization at each time. The BEM model represents the total samples of the channel complex gains $(L\times N)$ with a minimum of basis function coefficients BEM $(l\times Q)$. In this model, each path of the channel $h_l(n)$ is represented by:

$$h_l(n)=\boldsymbol{B}\cdot Cb_l(n)+\varepsilon_l(n), \tag{11}$$

where $\boldsymbol{B}$ is the matrix approximation of the basis functions and Q is the order the BEM development.

$Cb_l(n)=Cb_l=\left[Cb_{l,0},Cb_{l,1},\cdots,Cb_{l,Q-1}\right]^{\mathrm{T}}$ contains the Q coefficients corresponding to the basic N samples $h_l(n)$ and ε_l is the modeling error that's often neglected and dependent on the number Q. For the basis function matrix $\boldsymbol{B}$, there are several approaches: the Discrete Spheroidal Sequences (DSS-BEM), the Karhunen-Loeve (KL-BEM), the Complex Exponential (CE-BEM) or the Polynomial (P-BEM) [8,18,23-25]. Subsequently, we will focus on the P-BEM.

Considering the modeling error ε_l equal to zero (*i.e.* $h_l(n)=\boldsymbol{B}\cdot Cb_l(n)$), the correlation matrix of the basis coefficients $Cb_l(n)=\boldsymbol{B}^+\cdot h_l(n)$ is defined by:

$$\boldsymbol{R}_{Cb_l}=\boldsymbol{B}^+\boldsymbol{R}_{h_l}\boldsymbol{B}^{+H}. \tag{12}$$

Regarding the channel estimation that is modeled by a BEM development, we only need to estimate the basis coefficients. We chose to model the channel path by a P-BEM [26]. In this case, $\boldsymbol{B}$ has the following expression $\boldsymbol{B}_{n,q}=n^q$ where $n=0,\cdots,N_{-1}$ and $q=0,1,\cdots,Q-1$.

$$\boldsymbol{B}_{n,q}=\begin{bmatrix}1 & 1 & \cdots & 1\\ 0^1 & 1^1 & \cdots & (N-1)^1\\ \vdots & \vdots & \ddots & \vdots\\ 0^{Q-1} & 1^{Q-1} & \cdots & (N-1)^{Q-1}\end{bmatrix}^{\mathrm{T}}. \tag{13}$$

4. Proposed Channel Estimation Technique Channel Estimation with Dynamic Allocation of Pilots

In this section we consider the estimation of a time-varying channel with pilot subcarriers whose positions change from one OFDM symbol to another. We perform time channel estimation instead of the frequency estimation. Indeed, the last estimation based on conventional methods such as ML (Maximum Likelihood, MMSE (Minimum Mean Square Error) or LS (Least Square), gives poor performances because the estimated channel matrix is no longer diagonal in this context [8,9,26].

We propose to vary the positions of the pilot carriers in the OFDM symbol, in order to scan the whole spectrum of the OFDM symbol (see **Figure 1**). The aim is, firstly, to avoid the estimation of the channel on the same pilots positions, which could be lost by fading, and secondly, to obtain an adaptive estimation of the channel. This is the proposed technique called Dynamic Allocation of Pilots (DAP).

Figure 2 illustrates the operating mechanism of the DAP. We note here that before applying the IFFT signal $X_i(k)$, it consists of complex payload $C_i(k)$ (QPSK or 16QAM) and pilots $P_i(k)$. The whole is bounded by guard bands $0_i(k)$ (Zero Padding).

In this context, we consider an expansion in the P-

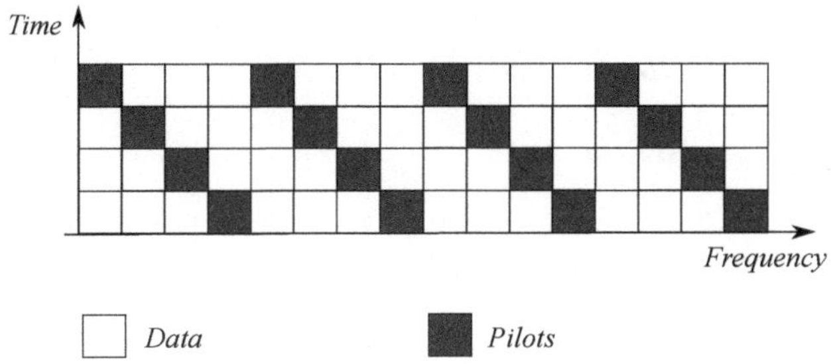

Figure 1. OFDM frame structure with pilot carrier offset.

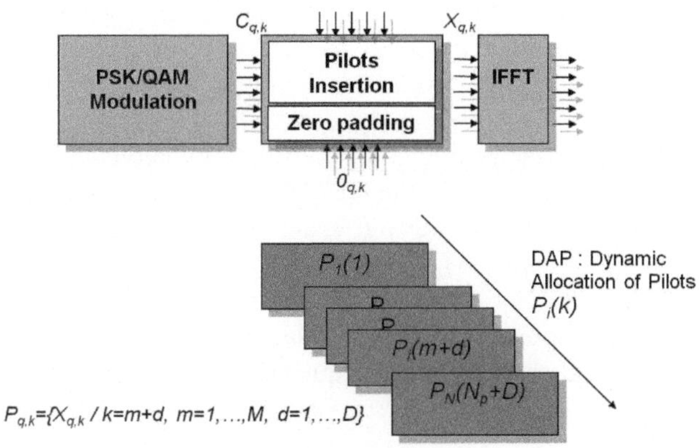

Figure 2. Block diagram of the DAP.

BEM basic functions for the channel model in order to reduce system complexity while maintaining an acceptable level of performance. Equation (14) can be rewritten as $\boldsymbol{Y} = \boldsymbol{W}\cdot h + V$ [27,28]. So the ith symbol can be written as:

$$Y_i(n) = \frac{1}{N}\sum_{n=0}^{N-1} h(n)\cdot X_i(k)\cdot \mathrm{e}^{-j\frac{2\pi n(i-k)}{N}} + V_i(n) = \frac{1}{N}\sum_{n=0}^{N-1} \boldsymbol{W}_i(n)\cdot h(n) + V_i(n), \tag{14}$$

or $W_i(n) = \sum_{k=0}^{N-1}\Omega_i(k,n)\cdot X_i(k)$ and $\Omega_i(k,n) = \mathrm{e}^{-j\frac{2\pi n(i-k)}{N}}$.

Assuming that N_p is the number of pilot subcarriers placed at positions $\Psi = \{X(1),\cdots,X(m),\cdots,X(N_p)\}$, and we know $X(k=m)$, where $m \in \{k/k = 1:D:N\}$, $m \in \{1,\cdots,N_p\}$ and D is the space between two successive pilot subcarriers. It is then possible to reformulate $\boldsymbol{W}(i,n)$ as the sum of two matrices such as:

$$\boldsymbol{W}_i(n) = \sum_{k=m}^{N-1}\Omega_i(k,n)\cdot X_i(k) + \sum_{k\neq m}^{N-1}\Omega_i(k,n)\cdot X_i(k) = \boldsymbol{W}_i^{k\in\Psi}(n) + \boldsymbol{W}_i^{k\notin\Psi}(n) = \boldsymbol{W}_i^{P}(n) + \boldsymbol{W}_i^{data}(n), \tag{15}$$

where $\boldsymbol{W}_i^P(n)$ corresponds to the pilots and $\boldsymbol{W}_i^{data}(n)$ is the useful data.

By replacing $\boldsymbol{W}_i(n)$ in Equation (14), we obtain:

$$Y_i(n) = \boldsymbol{W}_i^P(n)\cdot h + \boldsymbol{W}_i^{data}(n)\cdot h + V_i(n) \quad \text{or} \quad Y_i = \boldsymbol{W}_i^P\cdot h + \boldsymbol{W}_i^{data}\cdot h + V_i. \tag{16}$$

Now, we will use the pilot positions to estimate the channel impulse response h. Let

$\tilde{Y} = \left[Y_i(1),\cdots,Y_i(m),\cdots,Y_i(N_p)\right]^{\mathrm{T}}$ and

$\tilde{V} = \left[V_i(1),\cdots,V_i(m),\cdots,V_i(N_p)\right]^{\mathrm{T}}$ corresponding to the positions of the pilots. We can build N_p linear equations using the following equation:

$$\begin{pmatrix} Y_i(1) \\ \vdots \\ Y_i(N_p) \end{pmatrix} = \begin{pmatrix} W_i^P(1,0) & \cdots & W_i^P(1,N\cdot L-1) \\ \vdots & \ddots & \vdots \\ W_i^P(N_p,0) & \cdots & W_i^P(N_p,N\cdot L-1) \end{pmatrix}\cdot h + \begin{pmatrix} V_i(1) \\ \vdots \\ V_i(N_p) \end{pmatrix} \tag{17}$$

or

$$\tilde{Y} = \tilde{\boldsymbol{W}}\cdot h + \tilde{V}. \tag{18}$$

To find the impulse response h as a least squares solution of the system, we'll replace it with its basis coefficients Cb. Omitting the indices, the Equation (18) becomes [27,28]:

$$\tilde{Y} = \tilde{\boldsymbol{W}}\cdot B\cdot Cb + \tilde{V}, \tag{19}$$

hence the expression of the estimated channel h becomes:

$$\hat{h} = \boldsymbol{B}\cdot Cb = \boldsymbol{B}\left(\tilde{\boldsymbol{W}},\boldsymbol{B}\right)^{+}\tilde{Y}. \tag{20}$$

5. Simulation Results

In the following, we consider an OFDM system. The total number of subcarriers is 128 in which subcarriers are deployed for data transmission and subcarriers are used like pilots. We set to zero the middle DC carrier and the lateral subcarriers placed on left and on right of the OFDM symbol spectrum. A Cyclic Prefix (CP) with samples is used and the OFDM symbol duration is 72 µs. For the channel model simulation, we have chosen the Zheng model with a Jakes spectrum for mobile radio channel with respect to the motion between the transmitter and the receiver. Assuming that the synchronization between the transmitter and the receiver is perfect at receiver, the Bit Error Rate (BER) is evaluated for three normalized Doppler frequencies F_1 = 0.0003, F_2 = 0.0006 and F_3 = 0.0008 corresponding to a mobile speed v of 140, 280 and 350 *kmph* at a carrier frequency f_0 = 2.5 GHz.

Figures 3 and **4** illustrate the evolution of BER with increasing the value of the normalized Doppler frequency. Concerning the QPSK modulation, in a channel where the temporal variations are quantified by F_1 and F_2 and energy per bit $\frac{E_b}{N_o} \geq 25dB$, The DAP improvement is about 1 or 2 dB. The DAP proposed method tracks channel variations better than the conventional case with fixed pilots. On the other hand, for the value F_3 (350 *kmpl*), the DAP offers a lower gain of 0.5 dB. This is explained by the rapid variations in the channel and changing the pilot positions doesn't achieve the same performance as the values for F_1 and F_2.

Concerning the 16 QAM and for three values of the normalized Doppler frequency, the effect of mobility on performance is significant. The DAP is limited by a gain which does not exceed 0.25 dB. This can be due to the size of the constellation of 16 QAM.

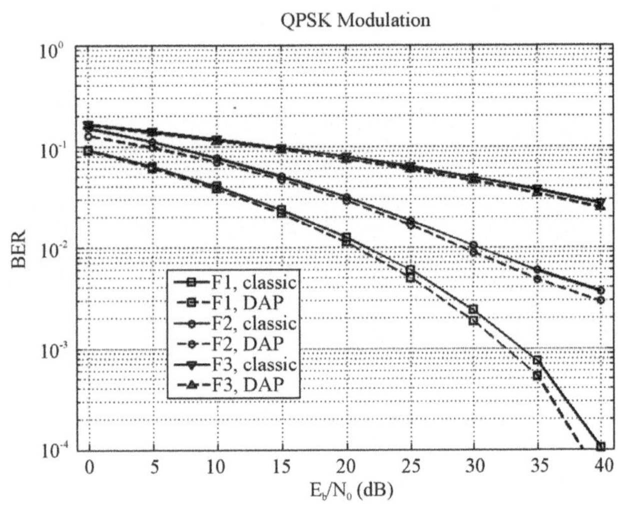

Figure 3. Comparison of classical and DAP channel estimations for QPSK modulation.

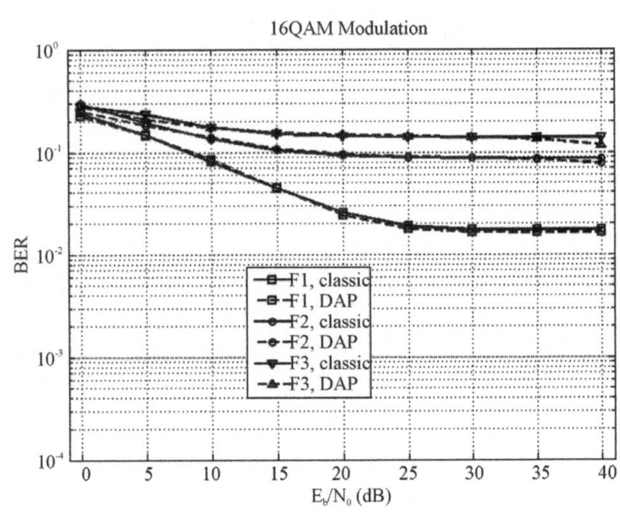

Figure 4. Comparison of classical and DAP channel estimations for 16 QAM modulation.

6. Conclusion

In this paper we proposed a solution to minimize the Doppler Effect damage and consequence on the performances of OFDM communication systems. OFDM technology provides efficient estimation of time-variant channels. In order to estimate the channel, pilot subcarriers are inserted into the OFDM symbol. The channel is estimated in the time domain and the temporal variations of each path are modeled by BEM. A pilot structure based on varying progressively the positions of pilot subcarriers from a symbol to another, is proposed. The pilot subcarrier positions are allocated dynamically during the transmission of the OFDM symbols with the DAP. The estimated time-varying channel is better and therefore the system performance will be improved in presence of the Doppler Effect. Using the DAP, the results show that the time-varying channel estimation is improved and the severe consequences of the Doppler Effect are reduced.

7. Acknowledgements

We thank the International Campus on Safety and Intermodal Transportation program (CISIT), French National Research Agency ANR and Transports Terrestrial Promotion Northern France (I-Trans).

REFERENCES

[1] L. Hanzo, M. Munster, B. Choi and T. Keller, "OFDM and MCCDMA for Broadband Multi-User Communications, WLANs and Broadcasting," John Wiley and Sons, Hoboken, 2003.

[2] I. Bradaric and A. P. Petropulu, "Performance of Training-Based OFDM Systems in the Presence of Time Varying Frequency Selective Channels," *IEEE International Conference on Acoustics, Speech, and Signal Processing*, Vol. 4, No. 29, 2004, pp. 737-740.

[3] X. Cai and A. N. Akansu, "A Subspace Method for Blind Channel Identification in OFDM Systems," *IEEE International Conference on Communications*, New Orleans, 18-22 June 2000, pp. 929-933.

[4] M. de Courville, "Utilisation de Bases Orthogonales pour l'Algorithmique Adaptative et l'Égalisation des Systèmes Multiporteuses," Ph.D. Thesis, École Nationale Supérieure de Télé-communications, Rennes, 1996.

[5] R. W. Heath and G. B. Giannakis, "Blind Channel Identi-Fication for Multirate Precoding and OFDM Systems," *Proceedings of* 13*th International Conference on Digital Signal Processing*, Santorini, 2-4 July 1997, pp. 383-386.

[6] T. Yamamura and H. Harada, "High Mobility OFDM

Transmission System by a New Channel Estimation and ISI Cancellation Scheme Using Characteristics of Pilot Symbol Inserted OFDM Signal," *Proceedings of IEEE 50th Vehicular Technology Conference*, Amsterdam, 19-22 September 1999, pp. 319-323.

[7] Z. Tang, G. Leus, R. C. Cannizzaro and P. Banelli, "Pilot Assisted Time-Varying OFDM Channel Estimation," *IEEE Transactions on Signal Processing*, Vol. 55, No. 5, 2006, pp. 2226-2238.

[8] T. Tian, D. Gang and J. Jun, "Estimation of Time Varying Channels for Pilot Assisted OFDM Systems," *Journal of China Universities of Posts and Telecommunications*, Vol. 14, No. 2, 2007, pp. 94-99.

[9] R. C. Cannizzaro, P. Banelli and G. Leus, "Adaptive Channel Estimation for OFDM Systems with Doppler spread," *Proceedings of IEEE Conference on Signal Processing Advances in Wireless Communications*, Cannes, 2-5 July 2006, pp. 1-5.

[10] S. Sezginer and Y. El Hajj Shehadeh, "An Iterative Channel Estimator for Fast-Varying Channels Using Successive OFDM Symbols," *IEEE 20th International Symposium on Personal, Indoor and Mobile Radio Communications*, Tokyo, 13-16 September 2009, pp. 2404-2408.

[11] R. H. Clarke, "A Statistical Theory of Mobile Radio Reception," *Bell System Technical Journal*, Vol. 47, No. 6, 1968, pp. 957-1000.

[12] W. C. Jakes, "Microwave Mobile Communications," 2nd Edition, Wiley-IEEE Press, 1994.

[13] M. F. Pop and N. C. Beaulieu, "Limitations of Sum-of Sinusoids Fading Channel Simulators," *IEEE Transactions on Communications*, Vol. 49, No. 4, 2001, pp. 699-708.

[14] P. Dent, G. E. Bottomley and T. Croft, "Jakes Fading Model Revisited," *IEEE Electronics Letters*, Vol. 29, No. 13, 1993, pp. 1162-1163.

[15] S. A. Fechtel, "A Novel Approach to Modeling and Efficient Simulation of Frequency-Selective Fading Radio Channels," *IEEE Journal on Selected Areas in Communications*, Vol. 11, No. 3, 1993, pp. 422-431.

[16] C. Xiao, Y. R. Zheng and N. C. Beaulieu, "Novel Sum-of Sinusoids Simulation Models for Rayleigh and Rician Fading Channels," *IEEE Transactions on Wireless Communications*, Vol. 5, No. 12, 2006, pp. 3667-3679.

[17] P. Banelli, R. Cannizzaro and L. Rugini, "Data Aided Kalman Tracking for Channel Estimation in Doppler Affected OFDM Systems," *Proceedings of IEEE International Conference on Acoustics, Speech and Signal Processing*, Honolulu, 15-20 April 2007, pp. 133-136.

[18] H. Senol, H. Cirpan and E. Panayirci, "A Low Complexity KL Expansion Based Channel Estimator for OFDM Systems," *EURASIP Journal on Wireless Communications and Networking*, Vol. 2005, No. 2, 2005, pp. 136-174.

[19] Z. Tang, G. Leus and P. Banelli, "Pilot-Assisted Time-Varying OFDM Channel Estimation Based on Multiple OFDM Symbols," *Proceedings of IEEE 7th Workshop on Signal Processing Advances in Wireless Communications*, Cannes, 2-5 July 2006, pp. 1-5.

[20] S. Tomasin, A. Gorokhov, H. Yang and J. P. Linnartz, "Iterative Interference Cancellation and Channel Estimation for Mobile OFDM," *IEEE Transactions on Wireless Communications*, Vol. 4, No. 1, 2005, pp. 238-245.

[21] B. Yang, K. Letaief, R. Cheng and Z. Cao, "Channel Estimation for OFDM Transmission in Multipath Fading Channels Based on Parametric Channel Modeling," *IEEE Transactions on Communications*, Vol. 49, No. 3, 2001, pp. 467-479.

[22] G. B. Giannakis and C. Tepedelenlioglu, "Basis Expansion Models and Diversity Techniques for Blind Identification and Equalization of Time-Varying Channels," *Proceedings of the IEEE*, Vol. 86, No. 10. 1998, pp. 1969-1986.

[23] T. Whitworth, M. Ghogho and D. McLernon, "Optimized Training and Basis Expansion Model Parameters for Doubly-Selective Channel Estimation," *IEEE Transactions on Wireless Communications*, Vol. 8, No. 3, 2009, pp. 1490-1498.

[24] T. Zemen and C. F. Mecklenbrauker, "Time-Variant Channel Estimation Using Discrete Prolate Spheroidal Sequences," *IEEE Transactions on Signal Processing*, Vol. 53, No. 9, 2005, pp. 3597-3607.

[25] H. Hijazi and L. Ros, "OFDM High Speed Channel Complexgains Estimation Using Kalman Filter and QR-Detector," *IIEEE International Symposium on Wireless Communication Systems*, Reykjavik, 21-24 October 2008, pp. 26-30.

[26] H. Hijazi and L. Ros, "Joint Data QR-Detection and Kalman Estimation for OFDM Time-Varying Rayleigh Channel Complex Gains," *IEEE Transactions on Communications*, Vol. 58, No. 1, 2010, pp. 170-178.

[27] W.-G. Song and J.-T. Lim, "Pilot-Symbol Aided Channel Estimation for OFDM with Fast Fading Channels," *IEEE Transactions on Broadcasting*, Vol. 49, No. 4, 2003, pp. 398-402.

[28] W.-G. Song and J.-T. Lim, "Channel Estimation and Signal Detection for MIMO-OFDM with Time Varying Channels," *IEEE Communications Letters*, Vol. 10, No. 7, 2006, pp. 540-542.

18

UWB System Based on the Modified Gegenbauer Function in MISO Channel

A. Okassa M'foubat, C. Tatkeu, F. Elbahhar

Université Lille Nord de France, Lille, France.

ABSTRACT

In this paper, we propose a new multi-user Rake receiver, based on the interference mutualization with a matrix representation for Multiple Input Single Output MISO channel. The proposed system used the Modified Gegenbauer functions in order to generate the signal and to ensure the multi users transmission system. The new proposed receiver allows, using the temporal and special diversity, to avoid the interferences between symbols and to improve the system performances in terms of Bit Error Rate BER and interferences between users with a low algorithm complexity. The proposed solution is based on the classical Rake receiver associated with the equalizer receiver. In order to adapt the Rake approach, in single detection case and in multi users Ultra Wide Band environment, we propose a multi-user Rake receiver using the matrix form. Our proposed system is evaluated in terms of channel effects and multi users' interferences.

Keywords: UWB; MISO Channel; Rake Receiver; LMMSE; Modified Gegenbauer Function

1. Introduction

In recent years a considerable interest arose in ultra-wideband (UWB) communication systems, due to their appealing features and release of the spectral mask from the Federal Communications Commission (FCC) [1]. This technology has many potential advantages such as high data rate, low probability of interception and detection, low complexity, low cost, reduced average power consumption, weak sensitivity to the near-far problem and immunity to interferences [2-4]. The UWB system communicates using short-duration pulses lower than nanosecond. Several waveforms can be used, like Gaussian pulses, monocycle pulses or waveforms based on orthogonal polynomials like Hermite and Gegenbauer functions [5-7]. The orthogonal functions allow, without orthognals code used in classical multi users systems, to share the channel propagation between the users. The previous works have shown that the Modified Gegenbauer functions or MGF give good results in terms of BER in the case of Additive White Gaussian Noise (AWGN) channel [8]. In order to introduce the multi channel effects, the modified Saleh-Valenzuela (S-V) model was adopted in 2003 as UWB channel reference model by the IEEE 802.15.3a [9]. The modelling of UWB channels is based on the measurement of indoor propagation environment, as the main commercial applications will be indoor communications. The main distinguishing features of UWB channel propagation are its extremely multipath-rich profile and non-Rayleigh fading amplitude characteristics. The UWB channel is characterised by the impluse response with high path number and delays between 50 to 150 ns [10].

For high data rate, the superposition of symbols in the receiver destroys the signal and generates intersymbol interferences. However, the Rake receivers, using the spatial diversity (antenna) and temporal diversity improve the receiver performance and maximise the signal to noise ratio at the receiver output.

Several type of Rake receivers have been proposed such as the Arake receiver that combines all the signal paths [11-13]. This receiver is not easy to implement because UWB channel is characterised by a large number of multipath.

However a feasible implementation can be obtained using a selective Rake (Srake), which combines the multipath components with a higher power. Maximal Ratio Combining MRC receiver Rake uses the path with higher signal noise ratio SNR [14]. This receiver is optimal in the AWGN channel, without multi users interference and without interference symbols. For multi users interferences and symbol interferences cases, others receivers are proposed to improve the BER values but with higher algorithm complexity [15,16].

In our approach, in order to have a low cost solution, we propose a multi-user Rake receiver in Multiple input Single Output MISO channel based on the matrix representation. The new receiver has a computational complexity equal to the product of the user numbers M and the symbol numbers Q transmitted in a packet of data frame. In this case, the complexity is lower compared to the others receivers cases.

The results obtained and compared to different receivers show that our approach improves the performances in terms of BER, gives better performances with a lower complexity algorithm.

This paper is organized as follows. In the first section, we describe the UWB system and structure of the MISO channel. The second section highlights the main drawbacks of using single detection receivers based on Rake receivers MRC in a multi user. The third section presents our approach proposed and in the fourth section we give simulation results and discussions. At the end, a conclusion is drawn with prospects.

2. The Proposed UWB System

In order to exploit the diversity benefits, the receiver must be able to combine different transmitted signals. The presented methods assume that different signals to be combined are received through different branches [17]. In case, the generated signals are the packets Q, with equiprobability bits modulated using the Binary Phase Shift Keying modulation (BPSK). If we consider Q packet of data frame, the transmitted signal for user k is given by the following Equation:

$$x_k(t)=\sum_{q=0}^{Q-1}\sqrt{E_b}b_q^{(k)}G_k(t-qT_s-\tau_k) \quad (1)$$

where E_b is the bit energy and T_s is the symbol period. The zero mean i.i.d data symbol $\{b_q^{(k)}\}$ are passed through a unit energy pulse shaping $G_k(t)$ includes the effects of transmit antenna. In this equation, each Gegenbauer order is assigned to each active user. The used pulses $G(t)$ are defined by the recurrence relation:

$$G_{k,\beta}(x)=2\left(1+\frac{k+\beta-1}{k}\right)xG_{k-1,\beta}(x)-\left(1+\frac{k+2\beta-2}{k}\right)G_{k-2,\beta}(x) \quad (2)$$

where $x\in[-1,1]$ is the time in nanosecond, $k>1$ the order of the Gegenbauer function and β is the parameter defining the Gegenbauer polynomials family. The results in [17] show that $\beta=1$ gives the best performance of UWB system. The orthogonality condition is satisfied for all $m\neq n$

$$\int_{-1}^{1}G_{n,\beta}(x)\cdot G_{m,\beta}(x)\cdot w(x)\mathrm{d}x=0 \quad (3)$$

The Gegenbauer polynomials may be used in UWB systems to construct MGF pulses with narrow widths. For this purpose, they are multiplied by the square root the weight function $w(x,\beta)$. These functions, that satisfy the Equation (3), are given by the following formula:

$$G_u(x)=\frac{G_n(x)\sqrt{w(x)}}{\int_a^b w(x)G_n(x)^2\mathrm{d}x} \quad (4)$$

We consider a MISO channel with each transmitted (antenna) is composed of a MGF pulse as shown in **Figure 1**, the signal $x_k(t)$ through the channel propagation. In the output receiver, the signal received by the station is the sum of signal of all users can be written:

$$r(t)=\sum_{k=0}^{M-1}\left(x_k(t)*h_k(t)\right)+\eta(t) \quad (5)$$

where $\eta(t)$ is the Additive White Gaussian Noise AWGN with zero mean and variance σ_n^2; $h_k(t)$ is the impulse response associated with the kth user.

The stochastic channel models, used to evaluate the physical layer of UWB, are adopted by the committee 802.15.3a especially for the intra-building environment, short-range (up to 10 m) for high date speed communications (>100 Mbit/s). These models are defined by an impulse response and given in the following equation:

$$h_k(t)=X_k\sum_{l=0}^{L}\sum_{i=0}^{P}\alpha_{i,l}^{(k)}\delta\left(t-T_l^{(k)}-\tau_{i,l}^{(k)}\right) \quad (6)$$

where $\alpha_{i,l}^{(k)}$ is the gain coefficient of the ith ray within the lth cluster. $T_l^{(k)}$ is the delay of lth cluster for desired user. $\tau_{i,l}^{(k)}$ is the delay of the ray relative to the cluster arrival time $T_l^{(k)}$ of the desired user's. X_k represents the log-normal shadowing; L and P indicate respectively the number of resolvable path and the number of rays of each cluster. To simplify the analysis, we can write the impulse response in another form as:

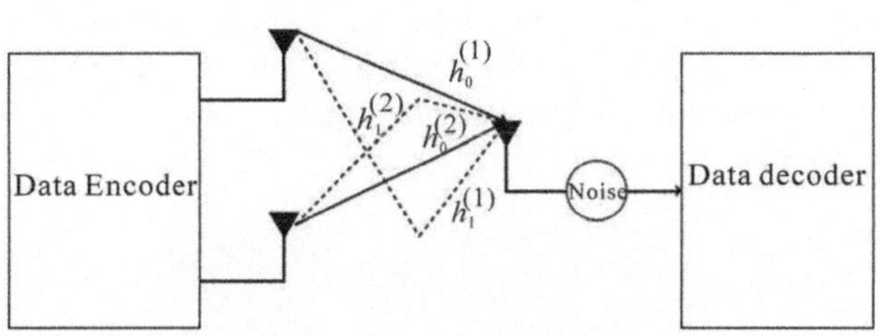

Figure 1. Illustration of model.

$$h_k(t)=\sum_{l=0}^{L_p-1} h_l^{(k)}\delta\left(t-\tau_l^{(k)}\right) \quad (7)$$

where L_p is the total number of rays, $h_l^{(k)}$ and $\tau_l^{(k)}$ are the gain and delay introduced by the lth ray. As the number of signal sample $x_k(t)$ is $QN = Q\times N$ is the sample of pulse $G_k(t)$; then the convolution operation between the signal $x_k(t)$ and the impulse response $h_k(t)$, therefore the number of signal $r(t)$, signal sample is $F = QN + L_p - 1$: assuming that j is the desired user and $b_0^{(j)}$ is the symbol of the j desired user. The received signal given by 5 can be written by the following equation:

$$\begin{aligned} r(t) &= b_0^{(j)}\tilde{h}_j\left(t-t_j\right)+\sum_{q\neq 0}^{Q-1} b_q^{(j)}\tilde{h}_j\left(t-qT_s-t_j\right) \\ &+\sum_{k\neq j}^{M-1} b_0^{(k)}\tilde{h}_j\left(t-t_k\right)+\sum_{k\neq j}^{M-1}\sum_{q=0}^{Q-1} b_q^{(j)}\tilde{h}_k\left(t-qT_s-t_k\right)+w(t) \end{aligned} \quad (8)$$

with

$$\tilde{h}_k(t)=\sum_{l=0}^{L_p-1}\sqrt{E_b}\,h_l^{(k)}G_k\left(t-\tau_l^{(k)}\right) \quad (9)$$

In this equation there are three different terms. The first term corresponds to the useful signal of the desired user, the second term is the symbol interference. This second term interact with the useful signal when the total duration of the response channel denoted T_m is higher than the symbol duration T_s. The third term corresponds to the multi-user interference.

3. Single Detector Receivers: Case of Rake Receivers

Due to the fact that impulse responses have a large number of multiple paths, using Rake receiver [18] method used in several studies to exploit the diversity of impulse and to maximize the energy available for the receiver. By identifying z_i the correlation output of branch i, the output of the receiver, after single-user combination in the case of Rake, is given by the following formula:

$$Y^{(k)}=\sum_{i=1}^{L_r}\alpha_i^{(k)}y_i^{(k)}$$

where L_r is the number of receiver branch and α_i is the weight assigned to the branch i, regardless the combination method. For the branch i, with the delay $\tau_i^{(k)}$, the output of the correlator or conventional receiver is given by :

$$y_i^{(k)}=\int_{\tau_i}^{\tau_i+T_S} r(t)G_k\left(t-\tau_i^{(k)}\right)\mathrm{d}t \quad (10)$$

In the multi user case, where each user transmits in a channel different from its neighbor as expressed by Equation (8), the choice of weights assigned α_i for the branch i, as proposed in several studies, cannot be efficient for estimating the symbols, even though no actual interference between symbols occurs. If we adopt the all Rake (Arake) with MRC and when the channel is known, the gains combinations $\alpha_k=\left[\alpha_0^{(k)} \; \alpha_1^{(k)} \; \cdots \; \alpha_{L_r-1}^{(k)}\right]$ are identical to the channel gain with $\alpha_k=\left[h_0^{(k)} \; h_1^{(k)} \; \cdots \; h_{QN-1}^{(k)}\right]$. In this case $\alpha_i = h_i^{(k)} = 0$ for $i > L_p - 1$ and $L_r = QN$.

4. Novel Approach Proposed

In this section, to improve the reception performances, we propose a multi-user Rake receiver matrix representation using MGF functions. The idea of this approach is to mutualise different interferences as seen in Equation (8), to constitute only a single type of interference. In this case, interference cancellation can be combined easily. So, the signal received from the user k illustrated by **Figure 1** can be formulated as following:

$$r=\begin{bmatrix} r_0 \\ \vdots \\ r_{F-1}\end{bmatrix}=\boldsymbol{H}_k x_{QN}^{(k)} \quad (11)$$

where $x_{QN}^{(k)}=\left[x_0^{(k)} \; x_1^{(k)} \; \cdots \; x_{QN-1}^{(k)}\right]^{\mathrm{T}}$ is the transmitted waveform; $\boldsymbol{H}_k$ is a channel matrix of dimension $F\times QN$; $x_{QN}^{(k)}$ represents spread data and therefore, can be expressed as:

$$x_{QN}^{(k)}=vec\left(\boldsymbol{G}_k\left(\boldsymbol{b}_Q^{(k)}\right)\right)^{\mathrm{T}}=\left(\boldsymbol{1}_Q\otimes\boldsymbol{G}_k\right)\boldsymbol{b}_Q^{(k)} \quad (12)$$

Here $\boldsymbol{b}_Q^{(k)}=\left[b_1^{(k)} \; b_2^{(k)} \; \cdots \; b_Q^{(k)}\right]^{\mathrm{T}}$ is a column vector which represents the symbols transmitted by the user k with the dimension $Q\times 1$; $\boldsymbol{G}_k$ denotes a column vector of the modified Gegenbauer function sampled by N factor; $\boldsymbol{1}_Q$ is the identity matrix of size $Q\times Q$; the operand means the Kronecker product and the operand vec is a column vector which allows the concatenation of the column vectors of a matrix. The channel matrix $\boldsymbol{H}_k$ is expressed by:

$$\boldsymbol{H}_k=\begin{bmatrix} h_0^k & 0 & \cdots & 0 \\ h_1^k & h_0^k & \cdots & 0 \\ \vdots & \vdots & \ddots & \vdots \\ h_{F-1}^k & h_{F-2}^k & \cdots & h_{F-QN}^k \end{bmatrix} \quad (13)$$

We can note that, in $\boldsymbol{H}_k$ element $h_n^k = 0$ for $n > L_p - 1$. The matrix channel $\boldsymbol{H}_k$ is lower triangular due to the causal nature of $h^{(k)}(t)$. For the purposes of Arake receiver, the channel gain vector must be

$\boldsymbol{\alpha}_k = \left[h_0^{(k)} \quad h_1^{(k)} \quad \cdots \quad h_{QN-1}^{(k)} \right]$. Assuming M active users, the received signal is given by this formula:

$$r = \sum_{k=1}^{M} H_k x_Q^{(k)} + \boldsymbol{\eta} \tag{14}$$

where $\boldsymbol{\eta} = \left[\eta_0 \quad \eta_1 \quad \cdots \quad \eta_{F-1} \right]^{\mathrm{T}}$ is a column vector corresponding to the Gaussian noise with zero mean and variance σ_η^2. Similarly to the Rake receiver single detection, here we choose gains of each branch as vector $\boldsymbol{\alpha}_k = \left[h_0^{(k)} \quad h_1^{(k)} \quad \cdots \quad h_{QN-1}^{(k)} \right]$. Thus the channel matrix Arake $\boldsymbol{\alpha}_k$ can be decomposed into a sum of the various paths as following:

$$\boldsymbol{\alpha}_k = h_0^k 1_{QN} + h_1^k z + h_2^k z^2 + \cdots + h_{QN-1}^k z^{(QN-1)} \tag{15}$$

where z is a shift matrix expressed by

$$z_{QN} = \begin{bmatrix} 0 & \cdots & 0 \\ 1 & 0 & \\ & \ddots & 0 \\ 0 & 1 & 0 \end{bmatrix}_{QN \times QN} \tag{16}$$

To extract the symbols, observed variable r^{T} is despread by applying a matrix correlation $G = \left[1_Q \otimes G_1 \quad \cdots \quad 1_Q \otimes G_M \right]$. That gives

$$Y^{\mathrm{T}} = r^{\mathrm{T}} \times G = \sum_{k=1}^{M} \left[x_Q^{(t)} \right]^{\mathrm{T}} \alpha_k^{\mathrm{T}} G + w^{\mathrm{T}} G \tag{17}$$

If we set

$$l_k = \left[1_Q \otimes G_k^{\mathrm{T}} \right] \alpha_k^{\mathrm{T}} \tag{18}$$

It combines the different l_k in a matrix constituted of elements $l = \left[l_1 \quad l_2 \quad \cdots \quad l_M \right]^{\mathrm{T}}$ of dimension $MQ \otimes QN$. So Equation (17) becomes:

$$Y = \left(\left[\left(b_q^{(1)} \right)^{\mathrm{T}} \quad \cdots \quad \left(b_q^{(M)} \right)^{\mathrm{T}} \right] \begin{bmatrix} l_1 \\ \vdots \\ l_M \end{bmatrix} + w^{\mathrm{T}} \right) \times \left[1_Q \otimes G_1 \quad \cdots \quad 1_Q \otimes G_M \right] \tag{19}$$

Let us consider

$$\boldsymbol{\Lambda} = l \times \left[1_Q \otimes G_1 \quad \cdots \quad 1_Q \otimes G_M \right] \tag{20}$$

where $\boldsymbol{\Lambda}$ is mutualisation of different interferences matrix. By substituting the channel matrix Arake $\boldsymbol{H}_k$ by the expression of Equation (15), development of the square matrix $\boldsymbol{\Lambda}$ whose dimension is $QM \otimes QM$ could be written as following:

$$\boldsymbol{\Lambda} = \left[\Psi_1 \quad \cdots \quad \Psi_s \quad \cdots \quad \Psi_{(QN-1)} \right] \begin{bmatrix} M_c^{(0)} \\ \vdots \\ M_c^{(s)} \\ \vdots \\ M_c^{(QN-1)} \end{bmatrix} \tag{21}$$

where $\boldsymbol{\Psi}_s$ and $s \in L_r$ represent respectively the coefficient matrix of the multipath channel and the index of relative delay multipath; the matrix $\boldsymbol{M}_c^{(s)}$ and $\boldsymbol{G}^{(s)}$ give respectively the correlation matrix and matrix of code shifted by a number of lines s upwards. The matrix $\boldsymbol{M}_c^{(s)}$ and $\boldsymbol{\Psi}_s$ are defined as following:

$$\boldsymbol{M}_c^{(s)} = \left[G^T \times G^{(s)} \right] \tag{22}$$

$$\boldsymbol{\Psi}_s = \begin{bmatrix} \left[h_s^1 1_Q \right] & \cdots & \mathbf{0}_{Q,Q} \\ \vdots & \ddots & \vdots \\ \mathbf{0}_{Q,Q} & \cdots & \left[h_s^M 1_Q \right] \end{bmatrix} \tag{23}$$

here $\mathbf{0}_{Q,Q}$ is the zero matrix whose size is $(Q \times Q)$. In this configuration, the equation of the variable decision is encapsulated:

$$Y^{\mathrm{T}} = \begin{bmatrix} y_1 \\ \vdots \\ y_M \end{bmatrix}^{\mathrm{T}} = \sqrt{\xi_b} \boldsymbol{b}_Q^{\mathrm{T}} \Lambda + w^{\mathrm{T}} \times G \tag{24}$$

where $\boldsymbol{b}_Q^{\mathrm{T}} = \left[\left(b_Q^{(1)} \right)^{\mathrm{T}} \quad \cdots \quad \left(b_Q^{(M)} \right)^{\mathrm{T}} \right]$ are random vector of parameters whose realization is to be estimated and has mean zero and covariance matrix is $\boldsymbol{C}_b$ of the dimension symbols $1 \times QM$. Thus, the new variance of noise denoted equal to $\|C\|^2 \sigma_\eta^2$. To restore the transmitted symbols, the optimal data estimation $\hat{b}_Q$ should resolve the approach to the problem of least squares assuming $\hat{b}_Q^{\mathrm{T}} \Lambda \approx Y$ such as

$$\min_{\hat{b}_Q^{\mathrm{T}}} \left\| \hat{b}_Q^{\mathrm{T}} \Lambda - Y \right\| \tag{25}$$

By solving Equation (8), data estimation is obtained by

$$\hat{b}_Q = sign\left(Y \times \boldsymbol{\Lambda}_{LS} \right) \tag{26}$$

With $\boldsymbol{\Lambda}_{LS} = \Lambda^{\mathrm{T}} \left(\Lambda \Lambda^{\mathrm{T}} \right)^{-1}$ the pseudo inverse matrix. This algorithm is referred to using the acronym Rake-LS (Rake least squares). Unfortunately the optimal solution in the least-squares approach is obtained without the Gaussian noise [19-22]. Forcing the interference zeros pooled significantly amplify the noise. The LMMSE approach allows taking into account the noise and the correlation factor in the variable decision, is obtained by minimizing the following equation:

$$\min_{\hat{b}_Q^{\mathrm{T}}} E\left\{ \left\| b_Q^{\mathrm{T}} - \hat{b}_Q^{\mathrm{T}} \right\|^2 \right\} \tag{27}$$

where $E\{\cdot\}$ means the mathematical Esperance. The solution of Equation (27) using [19] gives:

$$b_Q^{\mathrm{T}} = sign\left(Y \times \left(\|G\|^2 \sigma_n^2 C_b^{-1} I_{QM} + \Lambda^{\mathrm{T}} \Lambda \right)^{-1} \Lambda^{\mathrm{T}} \right) \tag{28}$$

This algorithm is referred to using the acronym Rake-

LMMSE [23,24]. In the case of Single Input Single Output SISO channel where all users transmit on a single channel $(H_1 = H_2 = H_M)$ only the matrix Ψ_s is modified and then, becomes:

$$\Psi_s = \left[h_s^1 1_Q\right] \qquad (29)$$

In the presence of multi users interference and lack of interference intersymbol, the proposed approach reduces the correlation matrix to the value $\Lambda = (G^T G)$ and consequently, the complexity of design loads is reduced and thus the matrix Λ has the dimension $(M \times M)$.

5. Simulation Results and Discussions

In this section, we discuss simulation results of ultra wideband system using stochastic channel models adopted by the committee IEEE 802.15.3a. To analyze the results, we studied the case of using the four first orders of modified Gegenbauer functions, where each waveform with duration of 2 ns is assigned to each user. The signal waveforms are sampled at the period $T_e = T_s = 8$. The simulations are performed on Matlab using the Monte Carlo method. Two antenna configurations are analyzed, the case of transmission in SISO channel and MISO channel. Four types of channels IEEE 802.13a noted CM1 to CM4 are used in our simulation. Here we took a frame of data closed by four symbols, *i.e.* the transmission rate at 8 ns.

In the simulations, the following **Figure 2** shows that, according to a user associated with each order of the Gegenbauer polynomials, performances may vary depending on the desired user. By choosing the first 4 orders (1 2 3 4), where order 1 and order 2 is named respectively user N#1 and user N#2. The user N#2 gives good performances compared to the user N#1, and this, whatever the type of receiver used. Also the configuration of the antennas has an impact on the receivers' performances. In this **Figure 2**, we can see that the proposed approach gives better performances compared than conventional and Arake receivers. Also the performances of Rake LMMSE are significantly better than those obtained with the Rake-LS method. **Figure 3** shows the proposed approach performances, while increasing duration of a symbol. We note that the performances degrade as a function of the reduction of the symbol duration.

The **Figure 4** shows the impact of channel types on the algorithm performances. It is observed that the proposed approach is more resistant to channels degradation, compared to the conventional and Arake receivers. Finally, in the absence of the transmission channel, the proposed receiver performances are close to the optimal solution, compared to the conventional receiver. Despite the presence of interferences and lack of interference between symbols, the proposed approach gives performances close to optimal results, *i.e.* in the case of a threshold in a user environment, without interference of intersymbols. This is illustrated on **Figure 5**.

6. Conclusion

In this paper, we proposed a receiver that combines all interferences in a MISO channel using the modified Gegenbauer polynomials. A novel proposed approach based on the matrix representation is given. The simula-

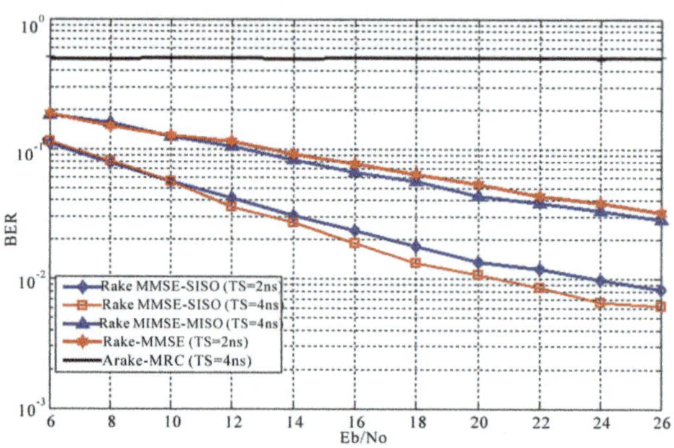

Figure 3. BER versus SNR for 4 symbols in case of CM1 environment with respectively waveforms orders (1, 2, 3, 4).

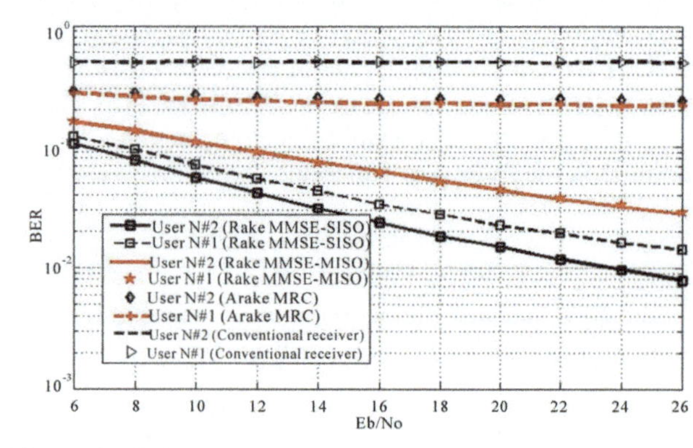

Figure 2. BER versus Eb/No in case of CM1 environment with respectively waveforms functions orders (1, 2, 3, 4).

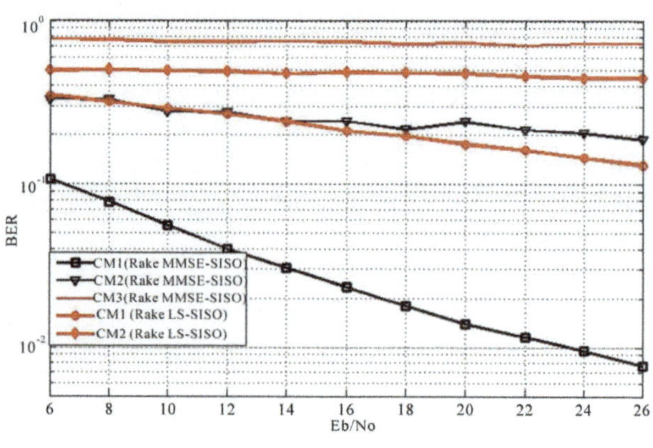

Figure 4. Receiver performances according to type of environment (CM1 to CM3) under user N#2.

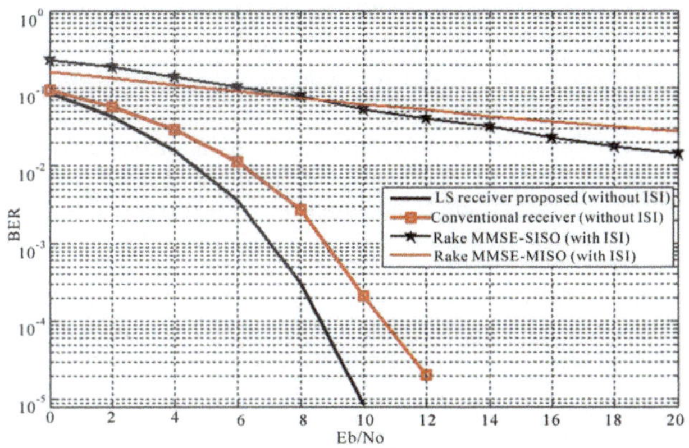

Figure 5. Performances in the presence and absence of the channel under CM1 environment.

tions carried out, show that our approach gives high performances compared to a conventional or ARAKE receivers. Using this new approach, we can achieve a trade-off between performance in terms of bit error rate and computational complexity. Our approach offers a high performance system in terms of data rate and bit error rate with a low cost. In the future work, we will apply these studies on others channel type such as the IEEE 802.15.4a models. We will validate these theoretical and simulation results by tests in real environments.

REFERENCES

[1] FCC Notice of Proposed Rule Making, "Revision of Part 15 of the Commission's Rules Regarding Ultra Wideband Transmission System," ET-Docket 98-153.

[2] M.-G. Benedetto and B. R. Vojcic, "Ultra Wide Band (UWB) Wireless Communications: A Tutorial," *Journal of Communication and Networks, Special Issue on Ultra-Wideband Communications*, Vol. 5, No. 4, 2003, pp. 290-302.

[3] J. Balakrishnan, A. Batra and A. Dabak, "A Multi-Band OFDM System for UWB Communication," 2003 *IEEE Conference on Ultra Wideband Systems and Technologies*, Dallas, 16-19 November 2003, pp. 354-358.

[4] Yang and G. B. Gianniakis, "Ultra Wideband Communications: An Idea Whose Time Has Come," *IEEE Signal Processing Magazine*, Vol. 21, No. 6, 2004, pp. 26-24.

[5] G. T. F. de Abreu, C. J. Mitchell and R. Kohno, "On the Design of Orthogonal Pulse-Shape Modulation for UWB Systems Using Hermite Pulses," *Journal of Communications and Networks*, Vol. 5, No. 4, 2003, pp. 328-343.

[6] B. M. Lachlan, M. Ghavami and R. Kohno, "Multiple Pulse Generator for Ultra-Wideband Communication Using Hermite Polynomial Based Orthogonal Pulses," *IEEE Conference on Ultra Wideband Systems and Technology*, 2002, pp. 47-51.

[7] C.-Y. Tsai and S.-K. Jeng, "Design of a Legendre-Polynomial-Based Orthogonal Pulse Generator for Ultra-Wideband Communications," *Antennas and Propagation Society International Symposium*, Vol. 2B, 2005, pp. 680-683.

[8] F. Elbahhar, A. Rivenq-Menhaj, J. M. Rouvaen, M. Heddebaut and T. Boukour, "Comparison between DS-CDMA and Modified Gegenbauer Functions for a Multi-User Communication Ultra Wide Band System," *IEEE Proceedings Communications*, Vol. 152, 2005, pp. 1021-1028.

[9] A. A. Saleh and R. A. Valenzuela, "A Statistical Model for Indoor Multipath Propagation," *IEEE Journal on Selected Areas in Communications*, Vol. SAC-5, No. 2, 1987, pp. 128-137.

[10] P. Pagani, P. Pajusco and S. Voinot, "A Study of the Ultra-Wideband Indoor Channel: Propagation Experiment and Measurement Results," COST273 TD(030)060, Barcelona, January 2003.

[11] J. Zhang, R. A. Kennedy and T. D. Abhayapala, "Conditions and Performance of Ideal RAKE Reception for Ultra-Wideband Signals in Log-Normal Fading Channels," *International Journal of Wireless Information Networks*, Vol. 10, No. 4, 2003, pp. 193-200.

[12] M. G. Khan, J. Nordberg, A. Mohammed and I. Claesson, "Performance Evaluation of Rake Receiver for UWB Systems Using Measured Channel Industrial Environments," *International Conference on Wireless Broadband on Ultra Wideband Systems*, March 2006.

[13] Z. Xiao, L. Su, D. Jin and L. Zeng, "Performance Comparison of RAKE Receivers in SC-UWB Systems and DS-UWB Systems," *The Institute of Electronics, Information and Communication Engineers (IEICE) Transactions on Communications*, Vol. E93-B, No. 4, 2010, pp. 1041-1044.

[14] A.-J. Liu, W. H. Xing and P. Ma, W.-B. Deng and B. Cao, "Performance of a Low Sampling Rate RAKE UWB Receiver in Multipath Environments," *International Workshop on Intelligent Systems and Applications*, 2009.

[15] S. Verdu, "Multiuser Detection," Cambridge University Press, Cambridge, 1998.

[16] Y. Li, A. F. Molisch and J. Zhang, "Channel Estimation and Signal Detection for UWB," *International Symposium on Wireless Personal Multimedia Communications (WPMC)*, October 2003.

[17] F. Elbahhar, A. Rivenq-Menhaj and J. M Rouvaen, "Multi-User Ultra Wide Band Communication System Based on Modified Gegenbauer and Hermite Functions," *Wireless Personal Communications (Kluwer)*, Vol. 34, No. 3, 2005, pp. 255-277.

[18] A. G. Kleini, D. R. Brown II, D. L. Goeckels and C. R. Johnson, "RAKE Reception for UWB Communication Systems with Intersymbol Interference," *4th IEEE Workshop on Signal Processing Advances in Wireless Communications*, 15-18 June 2003.

[19] J. G. Proakis, "Digital Communication," 3rd Edition, MC-Graw Hill International Editions, 1996.

[20] S. M. Kay, "Fundamentals of Statistical Signal Processing: Estimation Theory," Prentice Hall, Upper Saddle River, 1996.

[21] J. G. Andrews, "Interference Cancellation for Cellular Sys-

tems: A Contemporary Overview," *IEEE Wireless Communications*, Vol. 12, No. 2, 2005, pp. 19-29.

[22] A. Elabed, F. Elbahhar, Y. Elhillali, A. Rivenq and R. Elassali, "UWB Communication System Based on Bipolar PPM with Orthogonal Waveforms," *Wireless Engineering and Technology*, Vol. 3, No. 3, 2012, pp. 181-188.

[23] S. K. Sharma and S. N. Ahmadber, "Performance of Space-Time Coded MMSE DFE for Wideband Code Division Multiple Access (WCDMA)," *International Journal of Communications, Network and System Sciences*, Vol. 2, No. 4, 2009, pp. 276-282.

[24] A. P. Doukeli, A. S. Lioumpas, G. K. Karagiannidis and A. V. Frangos, "Increasing the Efficient of Rake Receivers for Ultra Wide Band Applications," *Wireless Personal Communications*, Vol. 62, No. 3, 2012, pp. 715-728.

19

Reconstruction of UWB Impulse Train by Parallel Sampling of Cascaded Identical RC Filters

Hannu Olkkonen[1], Simo Ahtiainen[1], Kari Jarvinen[1], Juuso T. Olkkonen[2]

[1]Department of Applied Physics, University of Eastern Finland, Kuopio, Finland; [2]VTT Technical Research Centre of Finland, Espoo, Finland.

ABSTRACT

In this framework we present a new method for measurement of the UWB impulse train based on the parallel sampling of the cascaded identical RC filters. We show that the amplitudes and time locations of p sequential impulses can be reconstructed from simultaneous measurement of the outputs from $2p$ cascaded identical RC filters. The parallel sampling scheme has a wide range of applications including the detection of the ultra wideband (UWB) impulses. Due to identical analog RC filters and buffer amplifiers, the parallel sampling scheme is flexible to implement in VLSI applications.

Keywords: Impulse Train; Pulse Sequence; UWB; SVD

1. Introduction

The conventional sampling methods are based on the sequential measurement of signals at equidistant intervals. If the signal contains abrupt changes, such as edges, impulses and other discontinuities, the signal bandwidth is not limited below the Nyquist frequency and the Shannon's sampling theorem is not warranted. Recently the problem to measure and reproduce non-band limited signals has been widely studied in signal processing society. Excellent articles [1-8] concern on the prefiltering and reconstruction of signals under nonideal sampling constraints. The sampling scheme based on the finite rate of innovation (FRI) recovers a variety of transient signal families, which are not band limited, such as impulses, piecewise linear and amplitude modulated edges.

The measurement and recovery of short term transient signals (Diracs) plays a main role in wireless ultra wideband (UWB) technology, where the information is coded to impulse sequences. The Diracs are fed to an analog circuit, which has a specific impulse response. The purpose of the sampling filter is to lengthen the UWB impulses for equidistant sampling by an analog-to-digital converter. Various methods have been developed for reconstruction of the original signal from discrete samples.

Our research group has introduced a sampling scheme, where the signal is fed to the parallel RC filters, whose outputs are sampled simultaneously [9,10]. A variant of the parallel sampling scheme is tailored for detection of edges [11,12]. The parallel sampling scheme can be applied to reproduce transient signals without ad hoc knowledge of the signal waveform [13]. Recently a multichannel sampling (MCS) arrangement was introduced, where the input signal is modulated by a set of sinusoidal waveforms, followed by a bank of integrators [14]. The MCS yields Fourier series coefficients, which enable the reconstruction of the input signal.

In this framework we introduce a new method for measurement and reconstruction of the impulse trains based on the parallel sampling of the identical cascaded RC filters. We show that the amplitudes and locations of p Diracs can be reproduced from simultaneous measurements of $2p$ outputs of cascaded identical RC filters. We compare the reconstruction performance of the present method with the MCS arrangement using an analog electronic circuit simulator.

2. Theoretical Considerations

2.1. Sampling of the Impulse Train

We consider the impulse train consisting of p sequential Dirac distributions

$$I(t) = \sum_{i=1}^{p} A_i \delta(t - t_i) \quad (1)$$

where A_i are amplitudes and t_i the time locations. The impulse train is fed to the N identical cascaded RC filters (**Figure 1**), each having the impulse response

$$h(t) = \alpha \mathrm{e}^{-\alpha t} u(t) \tag{2}$$

$y_k(t)(k = 0,1,\cdots,N-1)$ of the cascaded RC filter network are obtained as where the unit step function $u(t) = 1$ for $t \geq 0$ and $u(t) = 0$ for $t < 0$ and $\alpha = 1/RC$. The output signals

$$y_k(t) = I(t) * \underbrace{h(t) * \cdots * h(t)}_{(k+1)\text{ times}} \tag{3}$$

In Laplace transform domain we have

$$Y_k(s) = I(s)\underbrace{H(s)\cdots H(s)}_{(k+1)\text{ times}} = I(s)\left(\frac{\alpha}{s+\alpha}\right)^{k+1} \tag{4}$$

The inverse Laplace transform then gives

$$\begin{aligned} y_k(t) &= I(t) * \frac{\alpha^{k+1}}{k!} t^k \mathrm{e}^{-\alpha t} u(t) \\ &= I(t) * C_k t^k \mathrm{e}^{-\alpha t} u(t) \end{aligned} \tag{5}$$

where $C_k = \alpha^{k+1}/k!$. Now we can write the convolution integral

$$\begin{aligned} y_k(t) &= \int_0^t \sum_{i=1}^{p} A_i \delta(\tau - t_i) C_k (t-\tau)^k \mathrm{e}^{-\alpha(t-\tau)} u(t-\tau) \mathrm{d}\tau \\ &= C_k \sum_{i=1}^{p} A_i (t - t_i)^k \mathrm{e}^{-\alpha(t-t_i)} u(t - t_i) \end{aligned} \tag{6}$$

The prominent idea in this work is to measure the RC filter outputs $y_k(t)$ simultaneously at a time $t_0 > t_i$ $(i = 1,2,\cdots,p)$. By denoting the time difference $\Delta_i = t_0 - t_i$ and rearranging we obtain

$$u_k = \sum_{i=1}^{p} a_i \Delta_i^k \tag{7}$$

where the notations $u_k = y_k(t_0)/C_k$ and $a_i = A_i \mathrm{e}^{-\alpha\Delta_i}$ are used.

2.2. Reconstruction Algorithm

In the following we develop a reconstruction algorithm for the amplitudes and time locations of the Diracs based on formulation (7). The z transform of the u_k sequence with respect to k yields

$$\begin{aligned} Z\{u_k\} &= \sum_{k=0}^{\infty}\sum_{i=1}^{p} a_i \Delta_i^k z^{-k} = \sum_{i=1}^{p} a_i \sum_{k=0}^{\infty} \left(\Delta_i z^{-1}\right)^k \\ &= \sum_{i=1}^{p} \frac{a_i}{1 - \Delta_i z^{-1}} \end{aligned} \tag{8}$$

Let us define the pole cancellation filter as

$$H_{pc}(z) = 1 + \sum_{n=1}^{p} h_n z^{-n} = \prod_{j=1}^{p}\left(1 - \Delta_j z^{-1}\right) \tag{9}$$

and the $P(z)$ polynomial as

$$\begin{aligned} P(z) &= Z\{u_k\} H_{pc}(z) \\ &= \sum_{i=1}^{p} \frac{a_i}{1 - \Delta_i z^{-1}} \prod_{j=1}^{p}\left(1 - \Delta_j z^{-1}\right) \\ &= \sum_{i=1}^{p} a_i \prod_{\substack{j=1 \\ j \neq ?}}^{p}\left(1 - \Delta_j z^{-1}\right) \end{aligned} \tag{10}$$

Clearly the roots of the pole cancellation filter correspond to the roots of the $P(z)$ polynomial. The impulse response $p_n(n = 0,1,\cdots)$ of the $P(z)$ can be computed from convolution

$$p_n = \sum_{k=0}^{p} u_{n-k} h_k \tag{11}$$

The roots of the $P(z)$ are obtained by setting $p_n = 0$ for $n \geq 0$. This yields the matrix/vector equation

$$\begin{bmatrix} u_{2p-1} \\ u_{2p-2} \\ \vdots \\ u_p \end{bmatrix} + \begin{bmatrix} u_{2p-2} & u_{2p-3} & \cdots & u_{p-1} \\ u_{2p-3} & u_{2p-4} & \cdots & u_{p-2} \\ \vdots & \vdots & \ddots & \vdots \\ u_{p-1} & u_{p-2} & \cdots & u_0 \end{bmatrix} \begin{bmatrix} h_1 \\ h_2 \\ \vdots \\ h_p \end{bmatrix} = \mathbf{0} \tag{12}$$

$$\Leftrightarrow \boldsymbol{u} + \boldsymbol{U}\boldsymbol{h} = \mathbf{0} \Rightarrow \boldsymbol{h} = -\boldsymbol{U}^{-1}\boldsymbol{u}$$

We may observe that the solution of the coefficients of the pole zero cancellation filter requires the knowledge of the $2p$ values of the u_k sequence. The polynomial $\boldsymbol{h} = \begin{bmatrix} 1 & h_1 & h_2 & \cdots & h_p \end{bmatrix}^{\mathrm{T}}$ has roots Δ_i $(i = 1,2,\cdots,p)$. The time locations of the Diracs are then obtained as $t_i = t_0 - \Delta_i$. For solution of the amplitudes we may write (7) in matrix/vector form

$$\begin{bmatrix} u_0 \\ u_1 \\ \vdots \\ u_{2p-1} \end{bmatrix} = \begin{bmatrix} 1 & 1 & \cdots & 1 \\ \Delta_1 & \Delta_2 & \cdots & \Delta_p \\ \vdots & \vdots & \ddots & \vdots \\ \Delta_1^{2p-1} & \Delta_2^{2p-1} & \cdots & \Delta_p^{2p-1} \end{bmatrix} \begin{bmatrix} a_1 \\ a_2 \\ \vdots \\ a_p \end{bmatrix} \tag{13}$$

$$\Leftrightarrow \boldsymbol{u} = \boldsymbol{D}\boldsymbol{a} \Rightarrow \boldsymbol{a} = \boldsymbol{D}^{\#}\boldsymbol{u}$$

where # denotes the pseudoinverse matrix. Now the amplitudes of the impulses are obtained as $A_i = a_i \mathrm{e}^{\alpha\Delta_i}$. To

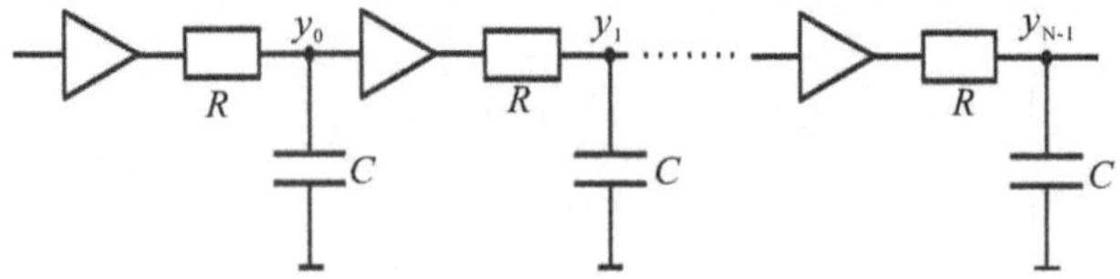

Figure 1. The arrangement for measurement of the impulse train based on the parallel sampling of N cascaded identical RC filters, which are separated by the unity gain buffer amplifiers.

summarize, the reconstruction of p impulses requires the knowledge of the $u_k\left(k=0,1,\cdots,2p-1\right)$ sequence. This needs the simultaneous measurement of at least $2p$ samples from the outputs of the cascaded RC filters. An alternative singular value decomposition (SVD) based null space method is presented in Appendix for the solution of the roots of the pole zero cancellation filter (9).

2.3. Noise Cancellation

The above formulation (8)-(13) is valid only in noise free situation. The noise was rejected by the singular value decomposition (SVD) based subspace method. Let us construct a Hankel matrix

$$\boldsymbol{H}=\begin{bmatrix} u_0 & u_1 & \cdots & u_p \\ u_1 & u_2 & \cdots & u_{p+1} \\ \vdots & \vdots & \ddots & \vdots \\ u_{p-1} & u_p & \cdots & u_{2p-1} \end{bmatrix} \tag{14}$$

where the antidiagonal elements are identical. The singular value decomposition of the matrix $\boldsymbol{H}$ is $\boldsymbol{H}=\boldsymbol{U\Sigma V}^{\mathrm{T}}$, where $\boldsymbol{U}$ and $\boldsymbol{V}$ are unitary matrices. $\boldsymbol{\Sigma}$ is a diagonal matrix consisting of the singular values in descending order. We may decompose $\boldsymbol{H}$ as

$$\begin{aligned}\boldsymbol{H}&=\left[\boldsymbol{U}_s \quad \boldsymbol{U}_n\right]\begin{bmatrix}\boldsymbol{\Sigma}_s & \boldsymbol{0}\\ \boldsymbol{0} & \boldsymbol{\Sigma}_n\end{bmatrix}\left[\boldsymbol{V}_s \quad \boldsymbol{V}_n\right]^{\mathrm{T}} \\ &=\boldsymbol{U}_s\boldsymbol{\Sigma}_s\boldsymbol{V}_s^{\mathrm{T}}+\boldsymbol{U}_n\boldsymbol{\Sigma}_n\boldsymbol{V}_n^{\mathrm{T}}=\boldsymbol{H}_s+\boldsymbol{H}_n\end{aligned} \tag{15}$$

$\boldsymbol{\Sigma}_n$ contains the smallest singular values and hence the matrix $\boldsymbol{H}_n$ can be considered to belong to the noise subspace. The matrix $\boldsymbol{H}_s$ is then related to the noise free signal subspace. The dimension of the signal subspace can be evaluated by several methods [15]. In this work we plotted the logarithm of the singular values in descending order. The logarithmic singular values form clearly separable linear groups. In all cases the dimension of the signal subspace appeared to be p, *i.e.* the number of impulses. The signal matrix $\boldsymbol{H}_s$ is not precisely a Hankel matrix, as some variation occurs in the antidiagonal elements. We reconstructed the noise free Hankel matrix by replacing the antidiagonal elements by their mean values.

2.4. High-Speed Sampling Method

In the case the impulse rate is limited so that only one impulse arrives at the cascaded RC network, the reconstruction algorithm simplifies notably and the sampling of the outputs of the first two RC filters is enough to recover the amplitude and time location of the impulse. Based on (6) we obtain

$$\begin{matrix} y_0=C_0A_1\mathrm{e}^{-\alpha\Delta_1} \\ y_0=C_0A_1\mathrm{e}^{-\alpha\Delta_1}\end{matrix} \Rightarrow \begin{matrix}\Delta_1=C_0y_1/\left(C_1y_0\right) \\ A_1=y_0\mathrm{e}^{\alpha\Delta_1}/C_0\end{matrix} \tag{16}$$

In practice the outputs of at least three RC filters is advisable to measure, since we may then construct the Hankel matrix $\boldsymbol{H}=\begin{bmatrix} y_0 & y_1 \\ y_1 & y_2\end{bmatrix}$ for SVD based noise cancellation.

2.5. Elimination of the Pulse Pile-Up in UWB Transmitter

The practical UWB transmission protocol consists of wireless RF pulses each containing three impulses (**Figure 2**). The current reconstruction algorithm (12, 13) does not take into account the pulse pile-up effect due to the previous pulses in cascaded RC network. Let us consider the situation in which the UWB pulses are transmitted at constant time intervals $nT, n=0,1,\cdots$ and the parallel sampling of the outputs of the RC network also occurs at nT intervals. By denoting the sampled and normalized outputs by $u_k\left[n\right], n=0,1,2,\cdots$ and applying (7), we have

$$\begin{aligned}\hat{u}_k\left[n+1\right]&=u_k\left[n+1\right]-\mathrm{e}^{-\alpha T}\sum_{i=1}^{p}a_i\left(\Delta_i+T\right)^k \\ &\approx u_k\left[n+1\right]-\mathrm{e}^{-\alpha T}u_k\left[n\right]\end{aligned} \tag{17}$$

contains the output due to the impulses occurring at time interval $\left[nT,\left(n+1\right)T\right]$ reduced by the pile-up signal due to the previous UWB pulses at time interval $\left[\left(n-1\right)T,nT\right]$. The approximation in (17) is exact for $k=0$. However, in cases for $k=1,2,\cdots$ the exponential $\mathrm{e}^{-\alpha t}$ is mainly responsible for the descending tail and the approximation error is negligible.

3. Experimental

The reconstruction algorithm was tested by an extensive simulation study. The cascaded network comprised eight identical RC filters which were constructed using an ana-

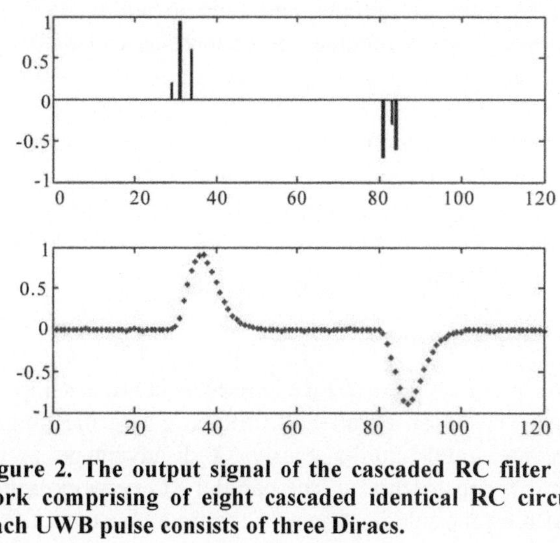

Figure 2. The output signal of the cascaded RC filter network comprising of eight cascaded identical RC circuits. Each UWB pulse consists of three Diracs.

log electronic circuit simulator (Spice). The α parameter varied in the range 0.1 - 0.8. In the absence of noise the reconstruction algorithm (12, 13) recovered the amplitudes and time locations of the impulse train consisting of 1 - 4 Diracs with machine precision. When the random noise was added to the input of the network, the algorithm could reconstruct 1 - 3 Diracs with good accuracy, when the SVD based subspace method was applied to noise cancellation. The error in amplitudes varied between 0.8 - 1.2 percent and in time locations 0.3 - 0.7 percent. In the case of four sequential Diracs the reconstruction algorithm yielded inaccurate results or totally failed.

We made a preliminary comparison with the multichannel sampling (MCS) arrangement [14] using the Spice simulation toolbox. In the case of 1 - 3 sequential impulses the performance of the MCS was clearly poorer than the results obtained by the present RC filter network in the presence of noise below SNR < 45 dB, but significantly higher at SNR > 45 dB. In the case of four sequential impulses the performance of the MCS was significantly higher compared with the present reconstruction method in all noise levels.

The cascaded RC network was implemented using metallic foil resistors and low-noise capacitors. Eight RC filters were separated by unity gain buffer amplifiers (**Figure 1**). The outputs of the RC filters were sampled simultaneously with an eight channel differential 12-bit analog-to-digital converter (ADC) equipped with a sample and hold (S/H) circuit in every channel. For 1 - 3 Diracs the reconstruction error in amplitudes was between 1.1 - 1.4 percent and in time locations 0.6 - 0.9 percent. For four Diracs the reconstruction error was considerably higher, for amplitudes 2.2 - 3.4 percent and for time locations 0.9 - 1.7 percent. In all experimental measurements the SVD based subspace method was used for noise rejection. The high-speed sampling method (16) recovered the amplitudes and time locations of single Diracs with good accuracy using three cascaded RC filters. The amplitude error was between 0.6 - 1.3 percent and the error in time locations 0.1 - 1.1 percent.

The reconstruction of Diracs using the SVD based null space method (Appendix) gave almost identical results, which are not repeated here. The slight difference is probably due to that the matrices $\boldsymbol{H}$ in (14) and $\boldsymbol{U}$ in (17) are different.

4. Conclusions

In our previous work [9] we introduced a novel sampling scheme based on parallel RC filters, where the outputs were measured simultaneously. A disadvantage in the practical construction of the parallel RC network comes from the precise adjustment of the time constants of the RC circuits. For *N* parallel RC circuits the α parameter must obey the condition $\alpha_k = \alpha_0 + k\alpha, k = 0,1,\cdots,N-1$. On the contrary, in present cascaded network all RC filters and buffer amplifiers are identical and much simpler to implement in integrated VLSI circuits. Compared with the previous FRI methods, which are based on the sequential sampling, the outputs of the parallel RC filters are sampled only once per UWB pulse, which contains a restricted number of impulses (**Figure 2**). This eliminates the need for high frequency analog-to-digital converters (ADCs), since the UWB pulse repetition rate can be adjusted to match the ADC's performance.

The formulation of the reconstruction algorithm was warranted by extensive simulation study using Spice toolbox. In the presence of noise the SVD based noise cancellation method had to be applied. The computational complexity of the SVD algorithm is $O(N^3)$, which restricts the real-time applications of the present method to a relatively low transmission rate. However, since the information is coded to both the amplitude and the time locations of the Diracs, the number of transmitted pulses can be significantly lower compared with the conventional UWB methods. As an advantage lower number of transmitted pulses reduces the RF radiation load.

We showed in Spice simulation study that in noise free conditions the *p* Diracs can be reconstructed from sampling the outputs of the 2*p* cascaded identical RC filters. In the presence of noise this theoretical "2*p* rule" is not valid and it seems evident that preferable "2*p* + 1 rule" should be applied.

The high-speed method (16) suits best for reconstruction of single Diracs, whose amplitudes are quantized to only a few levels. For example the impulses with 15 discrete levels can be produced by a 4-bit digital-to-analog converter, which the method reproduces perfectly.

In MCS arrangement the signal is modulated by a set of sinusoidal waveforms, followed by a bank of integrators [14]. The MCS produces Fourier series coefficients, which enable to reconstruct the input signal. However, the MCS is much more complex compared to the circuit consisting of the cascaded identical RC filters. The Spice simulation study revealed that the MCS outperforms the present approach at high noise level, SNR over 45 dB, but inferiors at lower noise environment. To the best of our knowledge the hardware for the implementation of the MCS is not yet available. Hence, in this moment this figure is based only on a simulation study.

In most of the UWB devices information is transmitted by monocycle Gaussian pulses. The FCC restricted the UWB frequency band between 3.1 - 10.6 GHz in year 2002 [16]. The Gaussian pulse train does not meet this constraint and other pulse shapes have been introduced to meet the FCC criteria, e.g. the family of the orthogonal

UWB pulse waveforms [17,18]. The fast digital-to-analog converters (DACs) can produce UWB pulses, where the information is coded via orthogonal base vectors and reproduced by matrix/vector computation.

In this work we have concentrated on the UWB communication device, which transmits pulses consisting of most 1 - 3 impulses. The information is coded to the amplitudes and time locations of the Diracs. Such pulse generators are relatively easy to implement in VLSI [19]. Following [18] the impulse train can be designed so that its spectral density spectrum coincides with the FCC criteria.

5. Acknowledgements

This work was supported by the National Technology Agency of Finland (TEKES).

REFERENCES

[1] P. Marziliano, "Sampling Innovations," Ph.D. Dissertation, Swiss Federal Institute of Technology, Lausanne, 2001.

[2] M. Vetterli, P. Marziliano and T. Blu, "Sampling Signals with Finite Rate of Innovation," *IEEE Transactions on Signal Processing*, Vol. 50, No. 6, 2002, pp. 1417-1428.

[3] I. Maravic and M. Vetterli, "Sampling and Reconstruction of Signals with Finite Rate of Innovation in the Presence of Noise," *IEEE Transactions on Signal Processing*, Vol. 53, No. 8, 2005, pp. 2788-2805.

[4] P. Marziliano, M. Vetterli and T. Blu, "Sampling and Exact Reconstruction of Band Limited Signals with Additive Shot Noise," *IEEE Transactions on Information Theory*, Vol. 52, No. 5, 2006, pp. 2230-2233.

[5] Jovanovic and B. Beferull-Lozano, "Oversampled A/D Conversion and Error-Rate Dependence of Nonband Limited Signals with Finite Rate of Innovation," *IEEE Transactions on Signal Processing*, Vol. 54, No. 6, 2006, pp. 2140-2154.

[6] J. Haupt and R. Nowak, "Signal Reconstruction from Noisy Random Projections," *IEEE Transactions on Information Theory*, Vol. 52, No. 9, 2006, pp. 4036-4048.

[7] Y. C. Eldar and M. Unser, "Nonideal Sampling and Interpolation from Noisy Observations in Shift-Invariant Spaces," *IEEE Transactions on Signal Processing*, Vol. 54, No. 7, 2006, pp. 2636-2651.

[8] P. L. Dragotti, M. Vetterli and T. Blu, "Sampling Moments and Reconstructing Signals of Finite Rate of Innovation: Shannon Meets Strang-Fix," *IEEE Transactions on Signal Processing*, Vol. 55, No. 5, 2007, pp. 1741-1757.

[9] H. Olkkonen and J. T. Olkkonen, "Measurement and Reconstruction of Impulse Train by Parallel Exponential Filters," *IEEE Signal Processing Letters*, Vol. 15, 2008, pp. 241-244.

[10] J. T. Olkkonen and H. Olkkonen, "Reconstruction of Wireless UWB Pulses by Exponential Sampling Filter," *Wireless Sensor Network*, Vol. 2, No. 6, 2010, pp. 462-466.

[11] J. T. Olkkonen and H. Olkkonen, "Sampling and Reconstruction of Piecewise Constant Signals by Parallel CR-Network," *IEEE Transactions on Circuits and Systems II*, Vol. 55, No. 11, 2008, pp. 1159-1162.

[12] J. T. Olkkonen and H. Olkkonen, "Sampling and Reconstruction of Zero-Order Hold Signals by Parallel RC Filters," *Wireless Engineering and Technology*, Vol. 2, No. 3, 2011, pp. 153-156.

[13] H. Olkkonen and J. T. Olkkonen, "Measurement and Reconstruction of Transient Signals by Parallel Exponential Filters," *IEEE Transactions on Circuits and Systems II*, Vol. 57, No. 6, 2010, pp. 426-429.

[14] K. Gedalyahu, R. Tur and Y. C. Eldar, "Multichannel Sampling of Pulse Streams at the Rate of Innovation," *IEEE Transactions on Signal Processing*, Vol. 59, No. 4, 2011, pp. 1491-1504.

[15] E. Biglieri and K. Yao, "Some Properties of Singular Value Decomposition and Their Applications to Digital Signal Processing," *Signal Processing*, Vol. 18, 1989, pp. 277-289.

[16] Y. P. Nakache and A. F. Molisch, "Spectral Shaping of UWB Signals for Time-Hopping Impulse Radio," *IEEE Journal of Selected Areas of Communication*, Vol. 24, No. 4, 2006, pp. 738-744.

[17] B. Parr, B. Cho, K. Wallace and Z. Ding, "A Novel Ultra-Wideband Pulse Design Algorithm," *IEEE Communications Letters*, Vol. 7, No. 5, 2003, pp. 219-221.

[18] H. Olkkonen and J. T. Olkkonen, "Design of Orthogonal UWB Pulse Waveform for Wireless Multi-Sensor Applications," *Wireless Sensor Network*, Vol. 2, No. 11, 2010, pp. 850-853.

[19] M. Miao and C. Nguyen, "On the Development of an Integrated CMOS-Based UWB Tunable-Pulse Transmit Module," *IEEE Transactions on Microwave Theory and Techniques*, Vol. 54, No. 10, 2006, pp. 3681-3687.

Appendix

SVD Based Null Space Solution

Let us write the convolution (11) in the matrix/vector form

$$\begin{bmatrix} u_{2p-1} & u_{2p-2} & \cdots & u_{p-1} \\ u_{2p-2} & u_{2p-3} & \cdots & u_{p-2} \\ \vdots & \vdots & \ddots & \vdots \\ u_p & u_{p-1} & \cdots & u_0 \end{bmatrix} \begin{bmatrix} h_0 \\ h_1 \\ \vdots \\ h_p \end{bmatrix} = \mathbf{0} \tag{18}$$

$\Leftrightarrow \boldsymbol{U}\boldsymbol{h} = \mathbf{0}$

In the following we describe the singular value decomposition (SVD) based null space solution of the unknown vector $\boldsymbol{h}$ in (18). The SDV of the $\boldsymbol{U}$ matrix in (18) yields

$$\boldsymbol{U} = \boldsymbol{L}\boldsymbol{\Sigma}\boldsymbol{R}^{\mathrm{T}} \tag{19}$$

where matrices $\boldsymbol{L} = \begin{bmatrix} \boldsymbol{l}_1 & \boldsymbol{l}_2 & \cdots & \boldsymbol{l}_{p+1} \end{bmatrix}$ and $\boldsymbol{R} = \begin{bmatrix} \boldsymbol{r}_1 & \boldsymbol{r}_2 & \cdots & \boldsymbol{r}_{p+1} \end{bmatrix}$ contain the left and right singular vectors (column vectors) and matrix $\boldsymbol{\Sigma} = \operatorname{diag}\{\sigma_1 \quad \sigma_2 \quad \cdots \quad \sigma_{p+1}\}$ the singular values in descending order. Matrices $\boldsymbol{L}$ and $\boldsymbol{R}$ are unitary, *i.e.* $\boldsymbol{L}^{\mathrm{T}}\boldsymbol{L} = \boldsymbol{I}$ and $\boldsymbol{R}^{\mathrm{T}}\boldsymbol{R} = \boldsymbol{I}$, where $\boldsymbol{I}$ is the identity matrix. Applied to (19) we obtain $\boldsymbol{U}\boldsymbol{R} = \boldsymbol{L}\boldsymbol{\Sigma}$, which yields

$$\boldsymbol{U}\begin{bmatrix} \boldsymbol{r}_1\, \boldsymbol{r}_2 \cdots \boldsymbol{r}_{p+1} \end{bmatrix} = \begin{bmatrix} \sigma_1\boldsymbol{l}_1\, \sigma_2\boldsymbol{l}_2 \cdots \sigma_{p+1}\boldsymbol{l}_{p+1} \end{bmatrix} \tag{20}$$

Finally we may write

$$\boldsymbol{U}\boldsymbol{r}_k = \sigma_k\boldsymbol{l}_k \tag{21}$$

for $k = 1, 2, \cdots, p+1$. Equation (21) forms the basis for the SVD based null space method. By searching very small singular value $\sigma_k \geq 0$, the right singular vector $\boldsymbol{r}_k$ equals vector $\boldsymbol{h}$ in (18) yielding the solution for the roots of the pole cancellation filter (9). In the presence of noise the dimensions of the $\boldsymbol{L}$ and $\boldsymbol{R}$ matrices should be selected so that there appears only one tiny singular value. This can be also achieved by zeroing the rest of the tiny singular values. It should be pointed out that the SVD based null space method does not yield the coefficients of the pole zero cancellation filtering (9) in normalized form, *i.e.* $h_0 \neq 1$. However, we may apply post normalization as $h_k \leftarrow h_k / h_0$ for $k = 0, 1, \cdots, p$.

Embedded System Implementation of Airborne Communication Terminals

Lei Wang[1], Zhizhong Ding[1], Rui Huang[1], Qian Li[1], Hesheng Cheng[2]

[1]Department of Communication Engineering, Hefei University of Technology, Hefei, China; [2]Department of Computer Science and Technology, Hefei Normal University, Hefei, China.

ABSTRACT

Airborne communication terminal is a key unit in Ad hoc network of aircrafts. This paper mainly focuses on its implementation by embedded system, which is based on Samsung S3C2410 chip. System architecture, Linux tailoring and touch-screen driver design are discussed in detail. Considering the requirements of stability and efficiency of the operating system, dynamic driver-loading method was employed firstly and only the necessary library files were transplanted to assist and test. The drivers finally were directly put into kernel configuration and then an integrated kernel was transplanted. Regarding to the problem of positioning issues on touch-screen, which is implemented in this system, an accurate positioning method is also presented.

Keywords: Airborne Communication Terminal; Ad Hoc Network; Embedded System; Touch-Screen

1. Introduction

Air Ad hoc network is a popular trend of the new wireless transmission technology due to that it provides data transport for IP with low latency transport, self network building up without a base station node. It also has the characteristics such as high speed and reliable data transmission link in the case of high-speed mobility. Ad hoc technology has wide application fields, such as disaster emergency, mobile office, personal area network, military wireless communication (e.g. TTNT network of the United States Air Force) etc. [1]. Currently, the latest research covers various topics like: designing a new routing protocol to follow the dynamic changes in the network topology, media access control (MAC) protocol to solve hidden and exposed terminal problems, Ad Hoc networks interconnected cellular network, Ad Hoc network multicast/multicast protocol, control of the power (energy saving), security issues and experiments and application network etc. In this paper, it designs the peer-to-peer mobile communication terminal with a friendly visualization and man-machine interface by use of the Ad hoc network technology.

Airborne communication terminal is mainly used in the data transmission between two aircraft in flight. Considering the factors such as pressure, temperature, atmospheric density, the terminal transmission needs to be real-time, minimum error and stable. The proposed design meets all the requirements.

With the rapid growth of information and electronic products, touch-screen as a new man-machine interface has been widely applied. The widely used interface for terminal control is the keyboard system with imperfect features and unstable program. Based on Samsung's S3C 2410X 16/32-bit RISC microprocessor, this paper introduces the embedded touch-screen technology into this area. Compared with the traditional button and mouse, touch-screen device is more intelligent, visual, and convenient.

This article firstly introduces the hardware design of the embedded system, and then explains the working principle of four-wire resistive touch-screen and equivalent circuit. Furthermore, the airborne communication terminal runs in Auto X/Y Position Conversion Mode and Waiting for Interrupt Mode. Embedded Linux system transplantation is introduced by dynamic or static loading driver method. The fourth part shows the development of touch-screen driver, while the realization of the coordinate positioning is summarized in the last section.

2. Hardware System Design

In this paper, adopting Samsung's S3C2410X (16/32-bit

RISC architecture, main frequency of 203 MHZ) as the microprocessor, the system of airborne communication terminal is comprised of touch-screen, host, control system, RF processing system and antenna. Here the control system is actually a Linux operating system which guarantees the normal work of the host, touch-screen is used as a control display, and the data is finally processed by the RF system will be sent through antenna. The overall system block diagram is shown as **Figure 1**.

This type of processor supports touch-screen interface, four external transistors, and an external voltage source. Besides, the ten-bit CMOS ADC of the S3C2410X is a recycling type of equipment with eight-channel analog inputs. In the process of sampling, the software will reset the special register, and the touch-screen controller will automatically control the touch-screen interface to open, or each MOS to close. At the end, it orderly completes the X coordinate and the Y coordinate collection. **Figure 2** shows the functional block diagram of the touch-screen module circuit.

AIN-7 is used for X coordinate input, and AIN-5 is used for Y coordinate input. Four control signals are connected with four external transistors to master touch-screen pads (XP, XM, YP, YM), that just decide four kinds of touch-screen interface modes: Normal Conversion Mode, Separate X/Y Position Conversion Mode, Auto X/Y Position Conversion Mode, Waiting for Interrupt Mode [2]. In this experiment, the last two of them are used, and the specific conditions are as shown in **Table 1**.

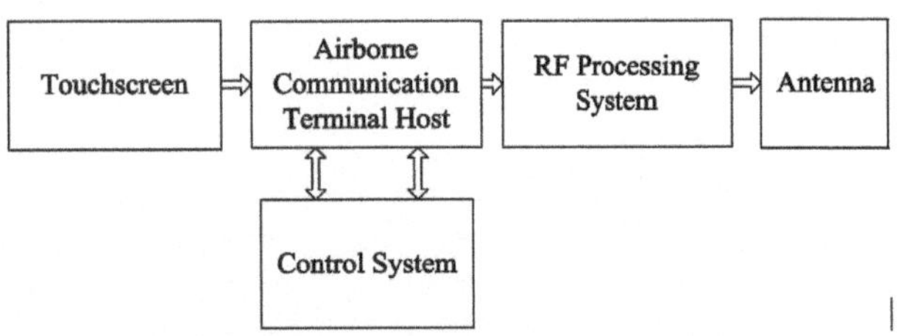

Figure 1. The overall system design.

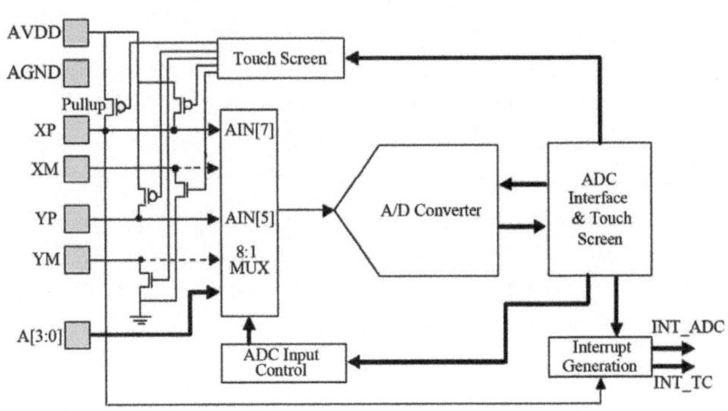

Figure 2. ADC and touch-screen interface functional block diagram.

Table 1. Working modes of touch-screen.

MODE		XP	XM	YP	YM
Auto X/Y Position Conversion Mode	X Position Conversion	External Voltage	GND	AIN [5]	Hi-Z
	Y Position Conversion	AIN [7]	Hi-Z	External Voltage	GND
Waiting for Interrupt Mode		Pull-Up	Hi-Z	AIN [5]	GND

3. Processing of Linux Kernel

Linux device driver is part of the kernel. A driver module can be compiled and loaded into the kernel in two ways, and they are dynamic loading of modules and direct kernel configuration options. Loading driver module dynamically is relatively simple and firstly introduced: the source file of driver must be compiled into object file through cross-compiler. Then, with the command "insmod, lsmod, and rmmod", the driver module can be loaded into the kernel flexibly [3].

What we need is just a complete kernel. The second way uses direct kernel configuration options, putting driver into kernel statically. Similar to this approach and the specific steps are as follows.

3.1. Kernel Configuration

Generally speaking, the second way can control the size of the kernel and has an advantage of convenient debugging, what we need is a stable, complete kernel. Meantime, the kernel of 2.6.32 version has no configuration options of S3C2410X touch-screen [4]. Thus, this article introduces the method of operation in detail:

1) Choose a place for driver code, put the file S3C2410-ts.c into the directory drivers/touchscreen.
cd drivers
mkdir touchscreen
2) Add the file Kconfig in/drivers/touchscreen, and the editorial content is as follows:
menuconfig INPUT_TOUCHSCREEN
bool "Touchscreens"
if INPUT_TOUCHSCREEN
config TOUCHSCREEN_S3C2410
tristate "Samsung S3C2410 touchscreen input driver"
---help---
If you want TOUCHSCREEN support, you should say Y here. If unsure, say N.
To compile this driver as a module, choose M here.
endif
3) Add the file drivers/touchscreen/Makefile, the main content is as follows:
obj-$ (CONFIG_TOUCHSCREEN_S3C2410) += S3C2410-ts.o
4) Amend the file drivers/Makefile, add:

obj-$ (CONFIG_TOUCHSCREEN_S3C2410) + = touchscreen/

5) Amend the file drivers/Kconfig, add:
   source "drivers/touchscreen/Kconfig"
6) At last, add the following content in the file arch/arm/Kconfig:
   source "drivers/touchscreen/Kconfig"

By the researches mentioned before, the option of "Samsung S3C2410 touchscreen input driver" will appear in the kernel configuration (**Figure 3**).

The next work is to compile the kernel:

#make clean, clean some obsolete data to guarantee a pure kernel.

#make dep, establish file dependencies.

#make zImage, this step will generate a kernel named "zImage" in arch/arm/boot directory.

At this moment, the object file "zImage" can be transplanted into the embedded system. In terms of practicality, vivi is the suitable transfer tool which is often used in this experiment.

3.2. Transplantation & Downloading

In order to make the Linux kernel run on the host properly, the cross compiling environment shall be firstly established. Before starting the system, the focus of the work is to put the pre-compiled kernel into a suitable position. Downloading various parts of Linux to the target board, it must take the following sequence: Boot Loader, Kernel, and Root File System [5].

1) Boot Loader: As mentioned earlier, vivi as a kind of Boot Loader is chosen for the suitable transplanted tool. Boot Loader is the first piece of code executed by the embedded system which undertakes the tasks of loading the operating system image into the memory, initializing the hardware devices, and building the map, etc.
2) Kernel: Kernel works in the bottom of operating system which controls all the resources of the host. Currently, there are two types of kernels in market, Linux 2.4 and Linux 2.6. In this paper, Linux 2.6.32 is enough, and of course, it must be configured in advance.
3) Root File System: Because of using the Nand Flash in the target board, the file root_init.cramfs is selected for File System that replaces the file ramdisk. image. gz. Root File System mainly stores some system files and application files to complete lots of necessary functions, and it also requires cutting to support the second extended file system.

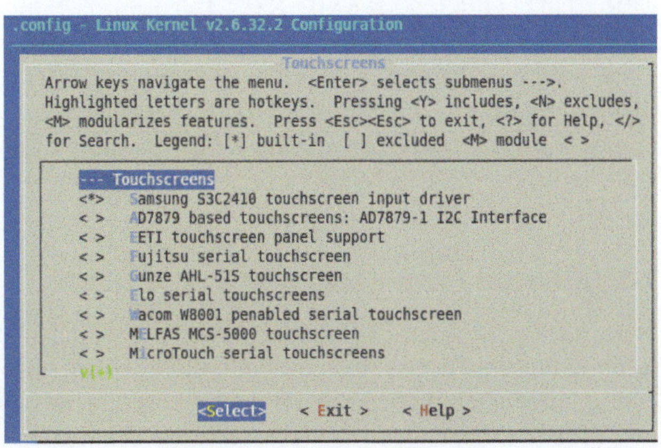

Figure 3. Kernel configuration.

In the process of download, the instructions we used in vivi environment are as follows:

bon part 0 192K 1M/* memory partition*/

load flash vivi x

load flash kernel x

load flash root x/* load three parts into board in xmodem*/

After the system powers or resets, CPU will read the files in accordance with the pre-set order as shown in **Figure 4**.

4. Touch-Screen Driver Implementation

At present, the most popular touch-screen is four-wire resistive touch-screen. It is comprised of several layers, two of which are separated by plenty of small dielectric particles are uniform, transparent, conductive and face-to-face ITO film. People called them X electrode and Y electrode, in other words, the upper line and the lower line. When user touches the surface of the screen, the two layers will meet together, and be connected at that point, finally generate a closed circuit. Meantime, through the lower line, the device controller applies a voltage to the X axis. Then the probe in the upper line detects the voltage of the X position, and it can be used to calculate the X coordinate of this contact point [6]. Similarly, to get Y coordinate, controller changes the direction of applied voltage. By now, the position of contact point is clear. The equivalent circuit diagram is as shown in **Figure 5**.

However, when there is no pressure on the surface of touch-screen, the whole circuit is open.

The touch-screen driver is actually a middle layer between the hardware and application. Linux system supports three types of hardware devices driver: character, block, and network, and what we research is character device-touch-screen. The character device driver is the most basic, also the most common driver of embedded

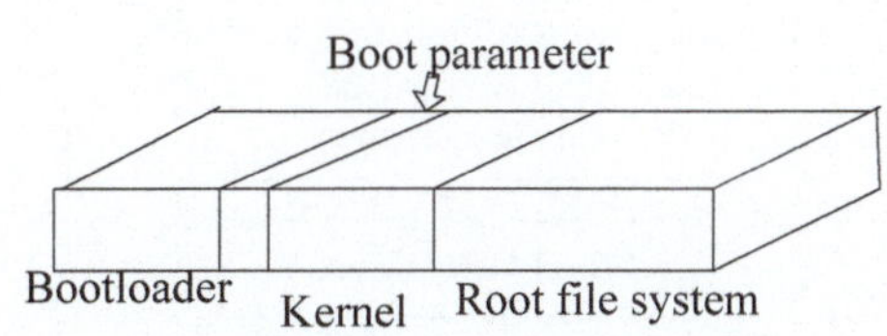

Figure 4. System file location.

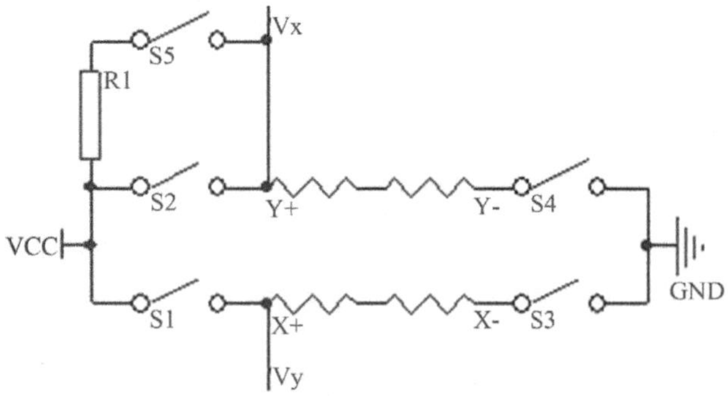

Figure 5. The equivalent-circuit diagram of touch-screen.

Linux. It is very powerful, and can describe almost all the equipment except for the device which can mount file system. In addition, the character device does not pass through the system cache, which can access through read (), write (), open (), close () and so on. It reads and writes in byte, and only executes sequential access.

Figure 6 is the flowchart of driver interface, and it describes the overall framework of the device drivers [7].

Next, specific analysis of the touch-screen driver of airborne communication terminal comes from four aspects: data structure of the driver, hardware initialization, interrupt processing and data acquisition process.

4.1. Analysis of the Data Structure

Members of the structure in the touch-screen device contain a buffer, spinlock, wait queue and the pointer fasync_struct.

```
typedef struct
{
unsigned int penStatus; /*3 states: up, down, sample*/
TS_RET buf[MAX_TS_BUF]; /*Avoid overrun*/
unsigned int head, tail; /* Point to buffer */
wait_queue_head_t wq; /*Achieve the asynchronous event notification mechanism in kernel*/
spinlock_t lock; /*An exclusive equipment to replace the process switching*/
```

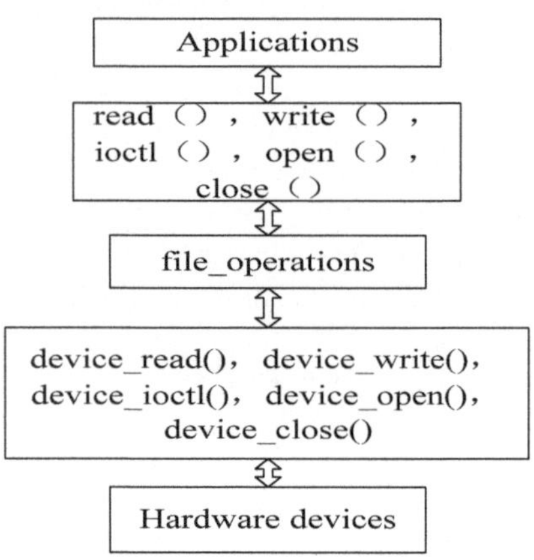

Figure 6. The flowchart of driver interface.

```
#ifdef USE_ASYNC
struct fasync_struct *aq; /*Asynchronous operation mode */
#endif
struct cdev *cdev;
}TS_DEV;
```

Besides, in the system's core, the access of I/O device is performed through a specific entry point. This special set of the entry point is just provided by the structure file_operations of the device driver. The structure file_operations as an important data structure actually in charge of the connection between the syscall and driver. The system reads the corresponding function pointer in the structure file_operations, and gives the control to the device operation function to complete the open, release, read, write and control functions of the device driver [8].

```
static struct file_operations S3C2410_fops =
{
owner: THIS_MODULE,
open: S3C2410_ts_open,
read: S3C2410_ts_read,
release: S3C2410_ts_release,
# ifdef USE_ASYNC
fasync: S3C2410_ts_fasync,
#endif
poll: S3C2410_ts_poll,
}
```

4.2. Hardware Initialization

Hardware initialization function static int _init S3C2410 _ts_init(void) mainly completes the application of device number, add cdev, request interrupt, set up the touch-screen control pins(YPON, YMON, XPON, XMON), etc. Besides, variable ret can't be ignored, which is defined as follows:

ret = request_irq(IRQ_ADC_DONE, S3C2410_isr_adc, SA_INTERRUPT, DEVICE_NAME, S3C2410_isr_adc); /* register ADC interrupt */

ret = request_irq(IRQ_TC, S3C2410_isr_tc, SA_INTER-RUPT, DEVICE_NAME, S3C2410_isr_tc); /* register touch-screen interrupt */

ret = register_chrdev(0, DEVICE_NAME, &S3C2410_fops); /* register the character device*/

In a word, the variable ret plays a key role in the program register. With these settings, the device driver file has been initialized.

4.3. Interrupt Processing

When the touch-screen is pressed down, the kernel will respond to the interrupt and enter into the interrupt processing mode. Interrupt is effective for the setting of falling edge. To make sure the touch-screen is continuously pressed and the system won't be triggered constantly, the coming interrupt will be masked.

There are two types of interrupts in touch-screen driver: contact-point (INT-TC), position conversion interrupt (INT-ADC). When the first interrupt has happened, if the previous state is PEN_UP, it will start the X/Y Position Conversion. What' more, lift interrupt which is put into INT-TC interrupt processing will call tsEvent() function to complete waiting line and releasing signal. When the interrupt of X/Y position conversion occurs, the program firstly gets the spinlock of the touch-screen device, then, reads the X/Y coordinate and puts them into the buffer, at last, releases that lock. The process is as shown in **Figure 7**.

4.4. Data Acquisition Process

The data acquisition depends on two functions: S3C2410 _get_XY(), start_ts_adc(). The calls of these two functions come from position conversion interrupt handler and timer interrupt handler in turn.

After the interrupt of X/Y position conversion occurs, the system calls the function S3C2410_get_XY() to read the coordinates of X, Y, and puts them into the buffer. The contact-point values are obtained through two conversion data registers: ADCDAT0 and ADCDAT1, as follows:

$$\begin{aligned} X &= (ADCDAT1\&0x3ff) \\ Y &= (ADCDAT0\&0x3ff) \end{aligned} \tag{1}$$

In the process of acquiring data, there is a vital function tsEvent_raw(). When the touch-screen's state is PEN_DOWN, the system calls tsEvent_raw() to complete these works: filling the buffer, awaking the wait queue, and releasing the asynchronous notification signal, otherwise, clearing the head of the buffer. Besides, timer interrupt handler calls the start_ts_adc() to convert the position of X/Y. That's the whole process of the data acquisition.

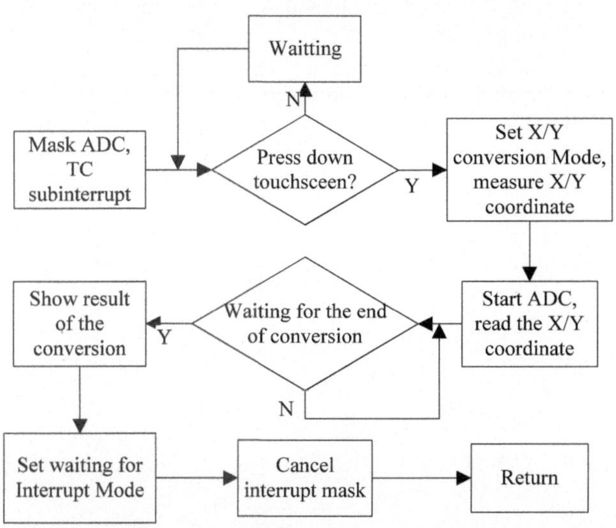

Figure 7. The process of interrupt.

5. Realization of Coordinate Positioning

The coordinates of the touch-screen can apply a variety of different calculation methods, such as multiple-point mean value method, secondary square of processing method and so on. This article uses the first method for the good performance of the touch-screen selected. There are two steps that users must convert the touch-screen data into touch coordinates: calibration, conversion. The terminal is designed to provide 3.5 inches Samsung TFT screen. At the beginning, it needs to obtain two maxima and two minima from the four vertices of the touch-screen, then determine the X, Y directions, and assume that the value range of x and y coordinates are [Xmin, Xmax], [Ymin, Ymax]. According to the **Figure 8**, the X, Y directions can be fixed by the following **Table 2**.

When the system is in hibernation, Q1, Q3, and Q4 are in the OFF state, Q2 works. If the touch-screen is pressed down, firstly, Q1 and Q4 of MOS transistor comes into operation, 3.3V power supply will be added in the circuit which consists of X+ and X−. Meanwhile, closing Q2 and Q3 of MOS transistor, cutting the Y+ and Y−, then starting A/D conversion channel 1 (AIN1) of the processsor, the system will output component voltage by the touch between the circuit resistance and the touch-screen. Finally, with the A/D converter, the voltage becomes a

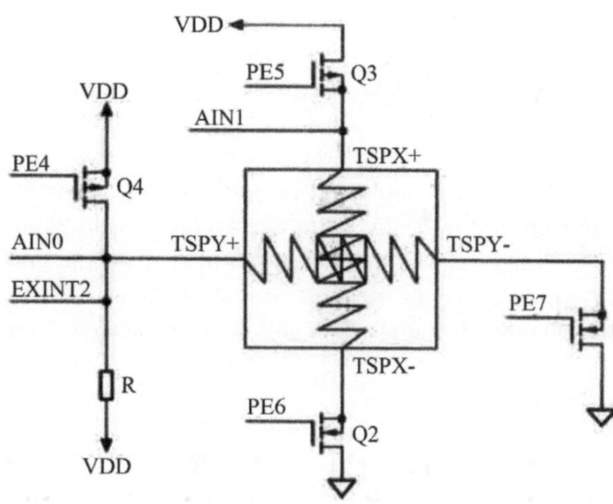

Figure 8. The connection interface of touch-screen and the processor.

Table 2. The determination of the X, Y directions.

direction	AD	N-MOS	P-MOS
X	AIN1	Q1(−) = 0 Q2(+) = 1	Q3(−) = 1 Q4(+) = 1
Y	AIN0	Q1(+) = 1 Q2(−) = 0	Q3(+) = 0 Q4(−) = 1

digital value, and the X-axis coordinate can be calculated. It shall also get the Y-axis coordinate. After the system obtains the coordinate value, it closes the Q1, Q3 and Q4, returns to the original state and waits for the next touch. The conversion values X, Y of coordinates can be expressed as:

$$\begin{aligned} X &= 240 * (Xmax - Xa)/(Xmax - Xmin) \\ Y &= 320 * (Ymax - Ya)/(Ymax - Ymin) \end{aligned} \quad (2)$$

In order to eliminate the conversion error as much as possible, this article makes use of multiple-point mean value method as follows, in the Formula (3):

$$\begin{aligned} Xa &= (X1 + X2 + \cdots + Xn)/n \\ Ya &= (Y1 + Y2 + \cdots + Yn)/n \end{aligned} \quad (3)$$

This system has been tested in the development environment with good results. With the touch-screen driver, the author compiles and transplants test procedures. When the touch-screen is pressed down, the coordinate values will be output to the terminal as **Figure 9**. What's more, this paper also transplants tslib and Qt4.6.3, and realizes the main interface design of airborne communication terminal. **Figure 9** also shows the result of this test. The touch-screen can be sensitive in the detection of the touch, and can send and receive data in real-time.

6. Conclusion

It is a point-to-point wireless communication system by using Samsung ARM9 core chip S3C2410 as a development platform. Main work includes the transplantation of Linux embedded system, design and implementation of the touch-screen driver, transplantation of a simple graphical interface based on QTE and point-to-point communication. What's more, procedures on the utilization of interrupt and multiple sampling average methods are to achieve precise coordinates positioning. Experimental results show that the touch-screen can work in airborne communication terminal with stable driver and convenient operation, and its future application is promising.

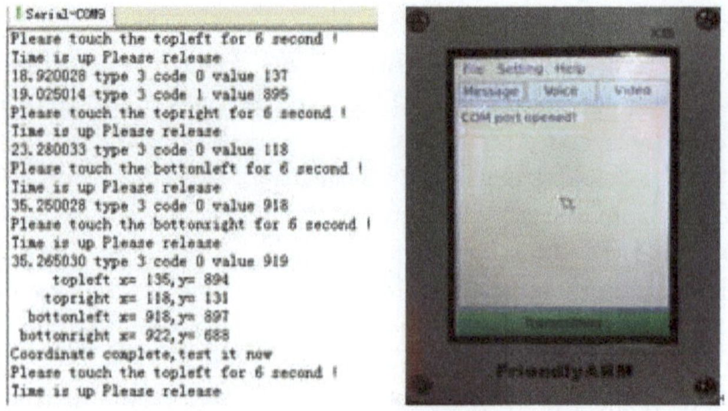

Figure 9. Result of test.

REFERENCES

[1] B. N. Xu and S. Hischke, "The Role of Ad Hoc Networking in Future Wireless Communications," *Proceedings of ICCT*, Beijing, 9-11 April 2003, pp. 1353-1358.

[2] Y. F. Li, "ARM Embedded Linux Device Driver Development," China Electric Power Press, Beijing, 2008, pp. 52-57.

[3] J. Corbet and Rubini, "Linux Equipment Driver," 3rd Edition, China Electric Power Press, Beijing, 2006.

[4] Z. R. Lin, "Embedded Linux Application Development Detailed Explanation," Mechanical Industry Press, 2005.

[5] Y. P. Sun, P. Peng and Y. Zhang, "Linux Transplantation Based on the Processor S3C2440," Electronic Measurement & Instruments (ICEMI), 2009, pp. 306-309.

[6] G. J. Feng, "The Design of Embedded Linux Driver," Tsinghua University Press, Beijing, 2008, pp. 196-207.

[7] H. J. Guo, Z. Wang and X. H. Wang, "Transplant of Linux and Embedded System of Boot Loader and LED Driver," *International Conference on Machine Vision and Human-Machine Interface*, Kaifen, 24-25 April, 2010.

[8] B. H. Song, "In-Depth Understanding of Linux Device Driver," People's Posts and Telecommunications Press, Beijing, 2008.

Tri-Band Fractal Antennas for RFID Applications

Abdelati Reha[1], Ahmed Oulad Said[2]

[1]Department of Telecommunications Engineering, Higher Institute of Applied Engineering (IGA), Marrakesh, Morocco; [2]Department of Electrical and Telecommunications Engineering, Royal Air Academy (ERA), Marrakesh, Morocco.

ABSTRACT

The RFID is a rapidly developing technology. It's used in many applications such as logistics, ticketing, security, employee attendance record and others. Also, fractal technology is used in many areas, and recently in antenna design because it allows making multi-band and wide-band antennas. In this paper, two tri-band fractal antennas are studied for the Radio Frequency Identification (RFID) applications using the Method of Moment (MoM). The first one is designed for the RFID readers and it operates at 3.85 GHZ, 5.80 GHZ and 8.12 GHZ. The second one is designed for the RFID Tags and it operates at 3.94 GHZ, 5.65 GHZ and 8.20 GHZ.

Keywords: Antenna; Fractals; RFID (Radio Frequency Identification); Multi-Band; Wide-Band

1. Introduction

RFID stands for Radio Frequency Identification, a term that describes any system of identification wherein an electronic device that uses radio frequency or magnetic field variations to communicate is attached to an item. The two most talked-about components of an RFID system are the tag, which is the identification device attached to the item we want to track, and the *reader*, which is a device that can recognize the presence of RFID tags and read the information stored on them. The reader can then inform another system about the presence of the tagged items. The system with which the reader communicates usually runs software that stands between readers and applications. This software is called RFID middleware [1].

In a typical RFID system [2], passive tags are attached to an object such as goods, vehicles, humans, animals, and shipments, while a vertical/circular polarization antenna is connected to the RFID reader. The RFID reader and tag can radio-communicate with each other using a number of different frequencies, and currently most RFID systems use unlicensed spectrum. The common frequencies used are low frequency (125 KHz), high frequency (13.56 MHz), ultra high frequency (860 - 960 MHz/2.45 GHz), and microwave frequency (3.6/3.9/5.8/5.9/8.2 GHZ [3]). The typical RFID readers are able to read (or detect) the tags of only a single frequency but multimode readers are becoming cheaper and popular which are capable of reading the tags of different frequencies [4].

Several calculation methods are adopted to solve the Maxwell equations and then study the performance of the antennas. Among these, two are widely used in simulation software:

1) The method of moments (MoM) is used among others in FEKO software (an abbreviation derived from the German phrase FEldberechnung bei Körpern mit beliebiger Oberfläche (Field computations involving bodies of arbitrary shape)) [5].

2) The method of finite integral (FIT) is used in the software CST Microwave Studio (Computer Simulation Technology Microwave Studio) [6].

In this work, we use the FEKO software.

2. Fractal Antennas

The fractal antennas use fractals and self-similar models. Several studies show that we can easily make compact, multi-band and wide-band antennas, suitable for miniaturization and various applications such as RFID, WLAN (Wireless Local Area Network), WIMAX (WorldWide Interoperability for Microwave Access), WIFI (WIreless FIdelity) [7-9].

3. Antennas Geometry

3.1. Readers Antennas

The reader antenna is a fractal antenna 2nd iteration (**Figure 1**).

The antenna is fed by a Coplanar Waveguide (CPW). The antenna, the feeding point and the ground plane are on the same plane (**Figure 2**). Several studies have used this mode of feeding [10-13].

The antenna is designed on an FR4 substrate (relative dielectric constant = 4.3, thickness = 1.6 mm, loss tangent = 0.025). The overall dimension of the substrate is 40 × 60 mm².

The dimensions of the antenna are: L_1 = 60 mm, L_2 = 40 mm, L_3 = 22 mm, H = 1.6 mm, H_1 = 23 mm, H_2 = 21 mm, H_3 = 10 mm, W = 2.2 mm, S = 1 mm, G = 0.5 mm, θ= 20°, α= 40° (**Figure 3**).

3.2. Tag Antenna

The Tag antenna is a fractal antenna 3rd iteration (**Figure 4**).

The antenna is fed by a Coplanar Waveguide (CPW). The antenna, the feeding point and the ground plane are on the same plane (**Figure 5**).

The antenna is designed on an FR4 substrate (relative dielectric constant = 4.3, thickness = 1.6 mm, loss tangent = 0.025). The overall dimension of the substrate is 30 × 90 mm².

The dimensions of the antenna are: L_1 = 90 mm, L_2 = 30 mm, L_3 = 10 mm, H = 1.6 mm, W = 2.4 mm, S = 1 mm, G = 1 mm (**Figure 6**).

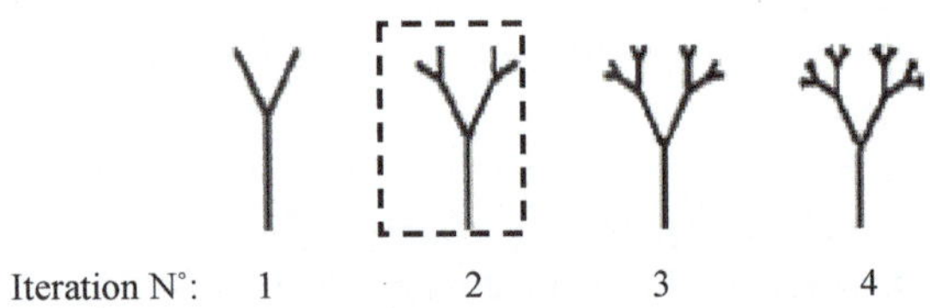

Figure 1. The first 4 iterations of the Tree-Fractal antenna.

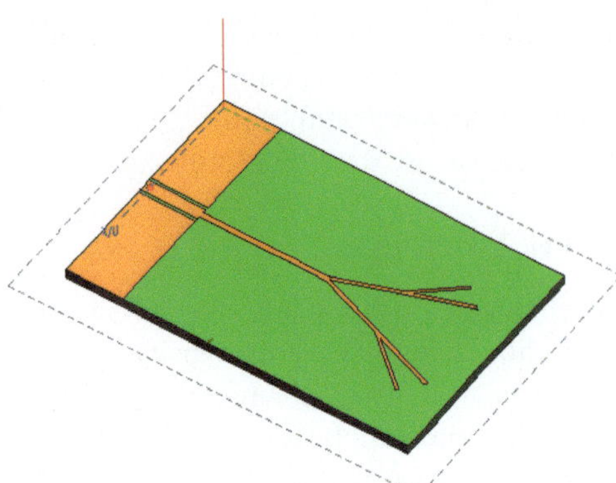

Figure 2. The Reader antenna fed by a coplanar waveguide (CPW).

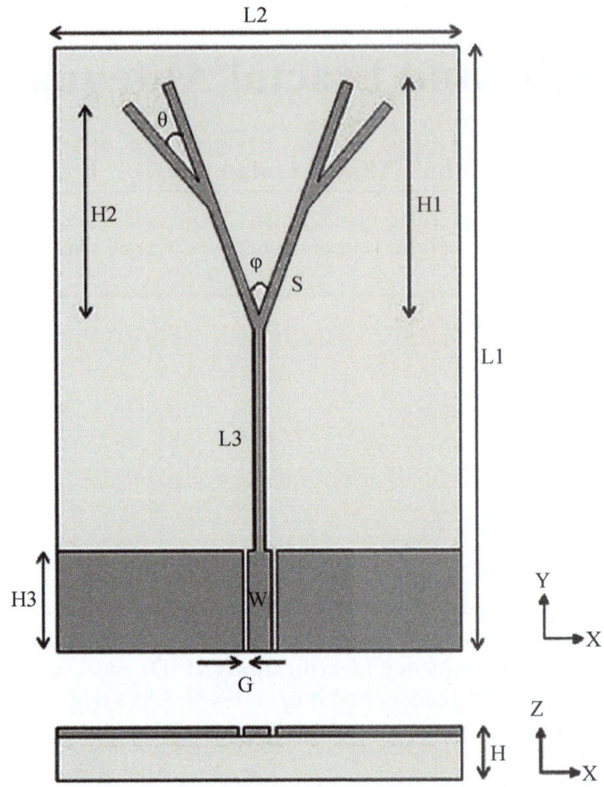

Figure 3. Dimensions of the Reader Antenna [3].

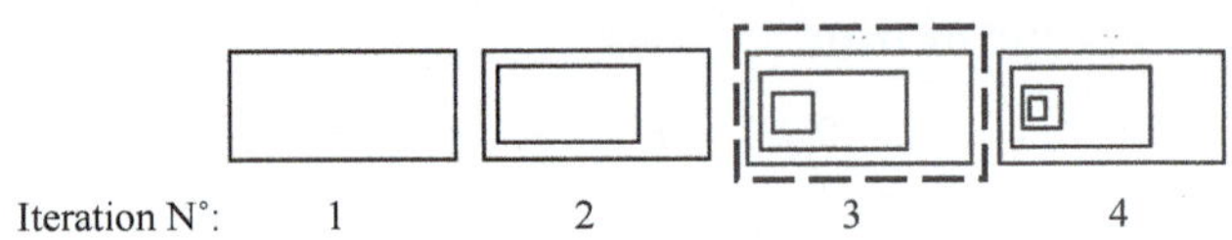

Figure 4. The first 4 iterations of the Rectangular-Fractal antenna.

Figure 5. The Tag antenna fed by a coplanar waveguide (CPW).

4. Results and Discussions

4.1. Reader Antenna

The analysis of the S_{11} parameter (**Figure 7**) shows that the Reader antenna operates at f_1 = 3.85 GHZ, f_2 = 5.8 GHZ and f_3 = 8.12 GHZ.

F_1 = 3.85 GHZ

Around this frequency, the S_{11} parameter shown by **Figure 8**, we note that the bandwidth is 305 MHZ (from 3.69 GHz to 4 GHZ).

We also note that the gain of this antenna is 2.6 dBi for $\varphi = 90°$ and $\varphi = 60°$ as shown by **Figure 9**.

F_2 = 5.8 GHZ

Around this frequency, the S_{11} parameter shown by **Figure 10**, we note that the bandwidth is 201 MHZ (from 5.743 GHZ to 5.944 GHZ)

We also note that the gain of this antenna is 1.7 dBi for $\varphi = 90°$ and $\varphi = 270°$ as shown by **Figure 11**.

F_3 = 8.12 GHZ

Around this frequency, the S_{11} parameter shown by **Figure 12**, we note that the bandwidth is 289 MHZ (from

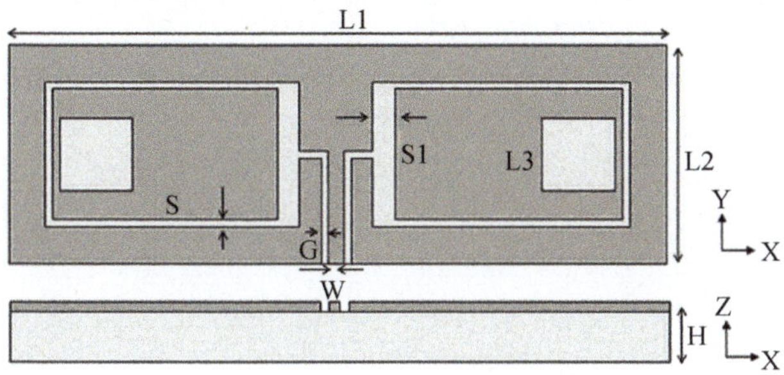

Figure 6. Dimensions of the Tag Antenna [3].

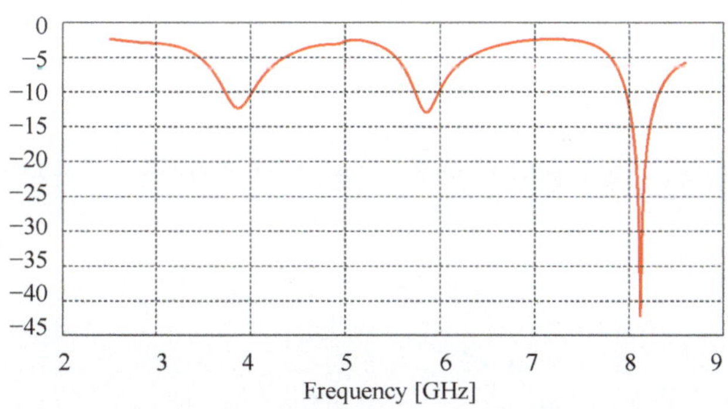

Figure 7. Simulated S_{11} versus frequency graph of the Reader antenna.

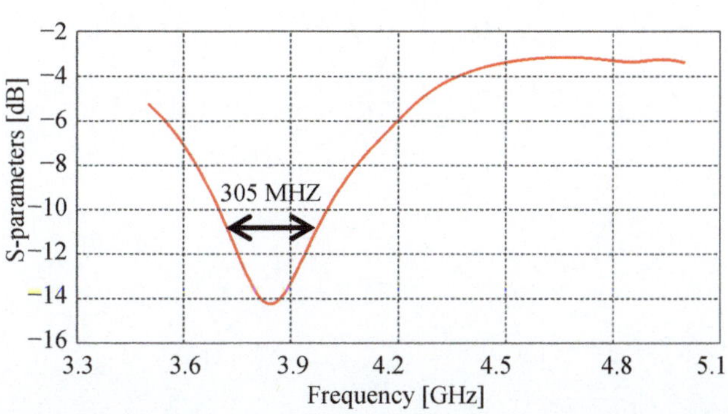

Figure 8. Evolution of S_{11} around the frequency f_1 = 3.85 GHZ.

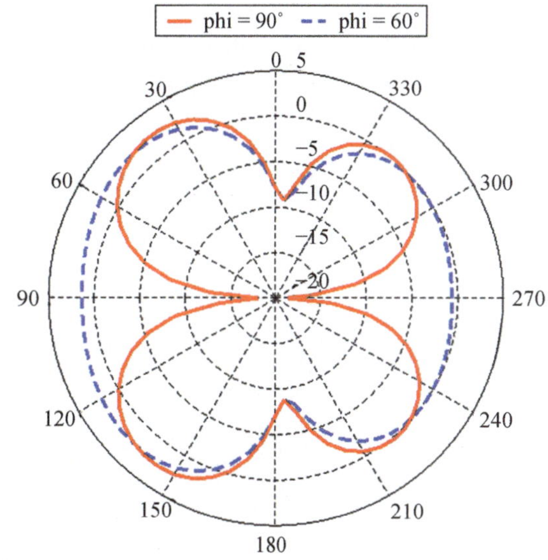

Figure 9. The gain of the antenna for f_1 = 3.85 GHZ (for $\varphi = 90°$ and $\varphi = 60°$) – G_{max} = 2.6 dBi.

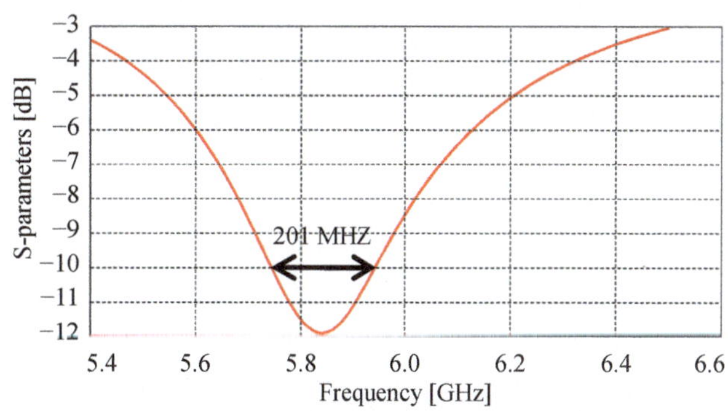

Figure 10. Evolution of S_{11} around the frequency f_2 = 5.8 GHZ.

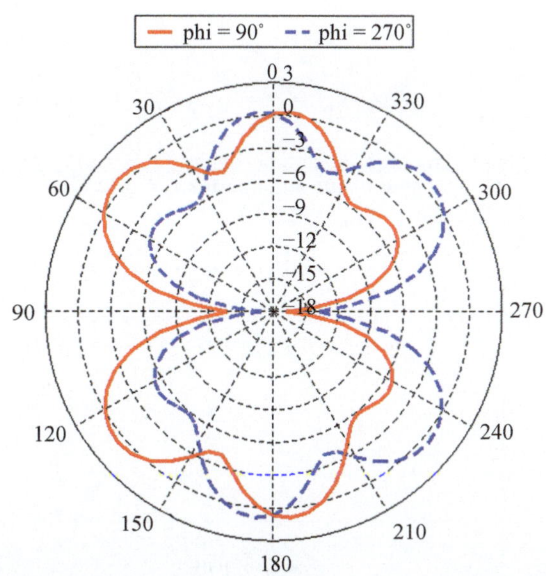

Figure 11. The gain of the antenna for f_2 = 5.8 GHZ (for $\varphi = 90°$ and $\varphi = 270°$) – G_{max} = 1.7 dBi.

7.989 GHZ to 8.278 GHZ).

We also note that the gain of this antenna is 2.7 dBi for $\varphi = 30°$ and $\varphi = 90°$ as shown by **Figure 13**.

4.2. Tag Antenna

The analysis of the S_{11} parameter (**Figure 14**) shows that the Tag antenna operates at f_1 = 3.94 GHZ, f_2 = 5.65 GHZ and f_3 = 8.2 GHZ.

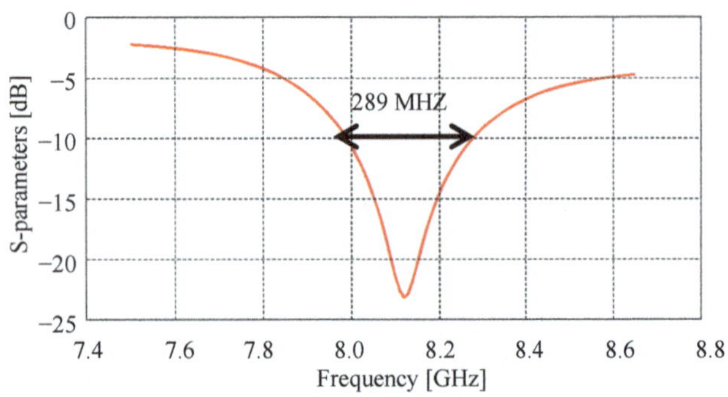

Figure 12. Evolution of S_{11} around the frequency f_3 = 8.12 GHZ.

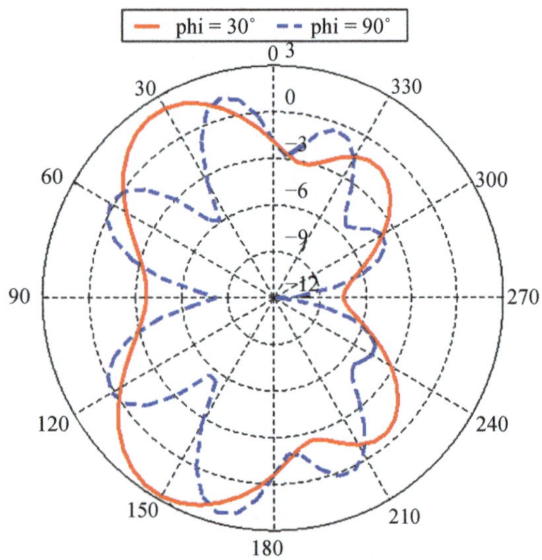

Figure 13. The gain of the antenna for f_3 = 8.12 GHZ (for φ = 30° and φ = 90°) – G_{max} = 2.7 dBi.

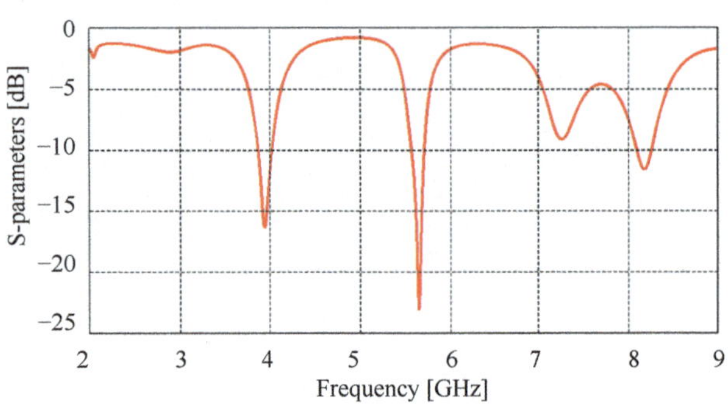

Figure 14. Simulated S_{11} versus frequency graph of the Tag antenna.

F_1 = 3.94 GHZ

Around this frequency, the S_{11} parameter shown by **Figure 15**, we note that the bandwidth is 165 MHZ (from 3.869 GHZ to 4.0348 GHZ).

We also note that the gain of this antenna is 3.8 dBi for $\varphi = 0°$ and $\varphi = 180°$ as shown by **Figure 16**.

F_2 = 5.65 GHZ

Around this frequency, the S_{11} parameter shown by **Figure 17**, we note that the bandwidth is 149 MHZ (from 5.564 GHZ to 5.713 GHZ).

We also note that the gain of this antenna is 1.25 dBi for $\varphi = 0°$ and $\varphi = 180°$ as shown by **Figure 18**.

F_3 = 8.2 GHZ

Around this frequency, the S_{11} parameter shown by **Figure 19**, we note that the bandwidth is 166 MHZ (from 8.101 GHZ to 8.267 GHZ).

We also note that the gain of this antenna is 2.65 dBi for $\varphi = 0°$ as shown by **Figure 20**.

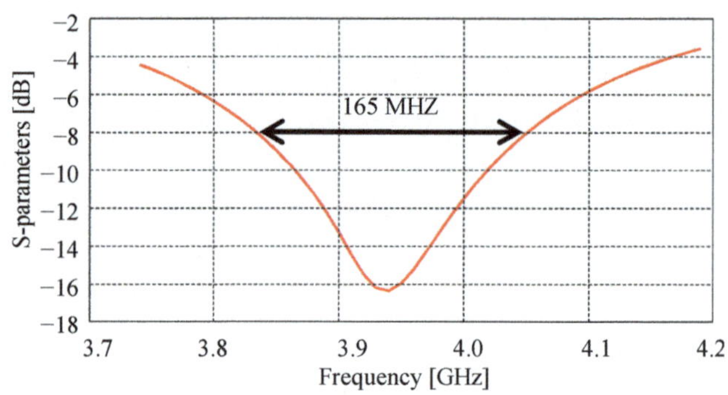

Figure 15. Evolution of S_{11} around the frequency f_1 = 3.94 GHZ.

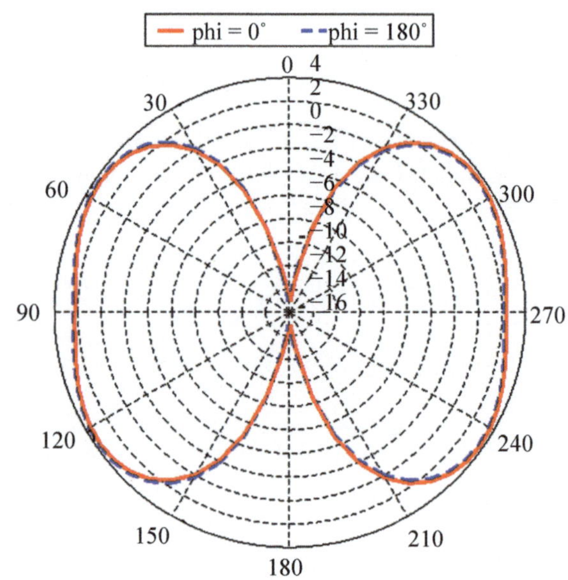

Figure 16. The gain of the antenna for f_1 = 3.94 GHZ (for φ = 0° and φ = 180°) – G_{max} = 3.8 dBi.

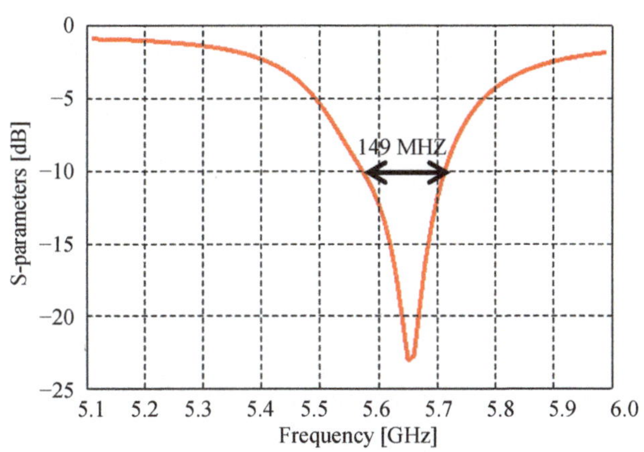

Figure 17. Evolution of S_{11} around the frequency f_2 = 5.65 GHZ.

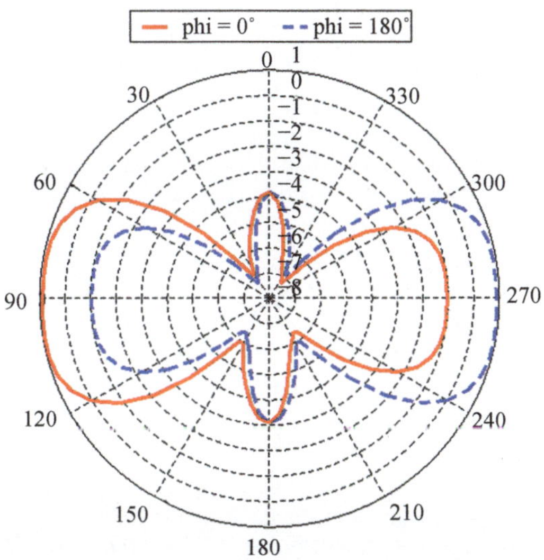

Figure 18. The gain of the antenna for f_2 = 5.65 GHZ (for φ = 0° and φ = 180°) – G_{max} = 1.25 dBi.

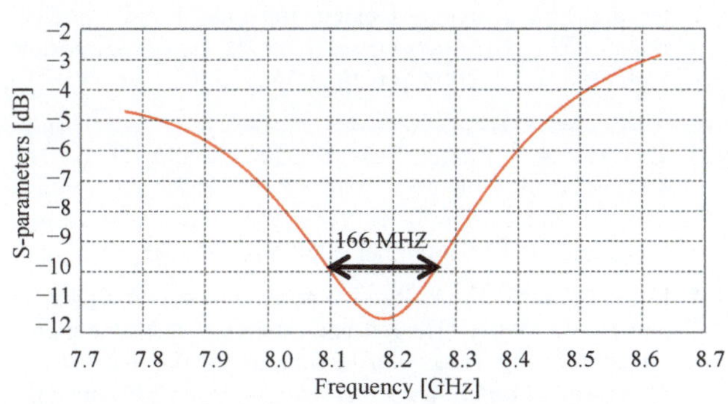

Figure 19. Evolution of S_{11} around the frequency f_3 = 8.2 GHZ.

5. Read Range of the Antennas

To calculate the read range of these antennas, we apply the First Transmission Equation (EQ1) which gives the relationship between the received power (Pr) as a function of the transmitted power (Pt), the antennas gain (Gt, Gr) and free space loss which can be expressed in terms of the wavelength (λ) and the distance between the antennas(R).

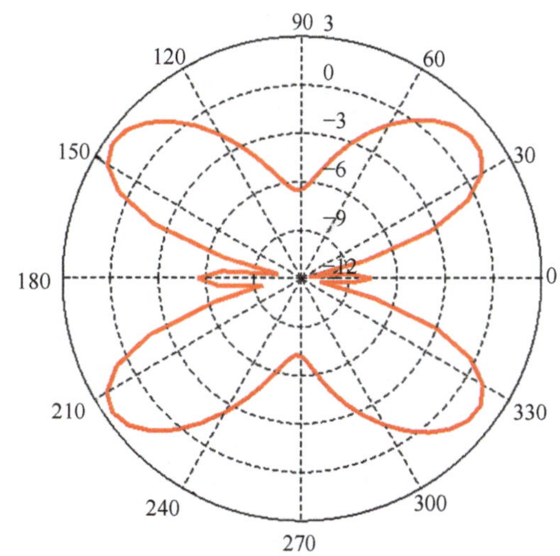

Figure 20. The gain of the antenna for f_3 = 8.2 GHZ (for φ = 0°) – G_{max} = 2.65 dBi.

$$\frac{P_r}{P_t} = G_t \cdot G_r \left(\frac{\lambda}{4 \cdot \pi \cdot R} \right)^2 \quad \text{(EQ1)}$$

$$R = \frac{c}{4 \cdot \pi \cdot f} \cdot \sqrt{\frac{P_t \cdot G_t \cdot G_r}{P_r}} \quad \text{(EQ2)}$$

$$\lambda = \frac{c}{f} \quad \text{(EQ3)}$$

For each one of the antennas, the read range calculation is done by taking a reference antenna of gain 3 dBi and minimum detectable power of 1 mW. The transmitting power is 4 W [3]. Therefore, we can build the table below (**Table 1**).

6. Comparison with other Methods of Calculation

This study has been done with The CST Microwave Studio which is based on the finite integral method (FIT) [3]. Our results are similar, but with small differences summarized in the **Table 2**.

Comparing these results, we note that:

- For the Reader, there is a small frequency lag of 0.25 GHZ ($f_{1\ FIT}$ = 3.6 GHZ and $f_{1\ MoM}$ = 3.85 GHZ)
- For the Tag, there is a small frequency lag of 0.25 GHZ ($f_{2\ FIT}$ = 5.9 GHZ and $f_{1\ MoM}$ = 5.65 GHZ)

7. Conclusions

The fractal antennas are preferred due to small size, light weight and multiband. This paper introduced RFID reader and tag antennas. The reader antenna exhibits a −10 dB bandwidth of 305 MHz at 3.85 GHz, 201 MHz at 5.8

Table 1. Read range antennas.

frequency (GHZ)	Antenna type	Gain of the antenna (dBi)	Read range (cm)
3.85	Reader	2.6	74.72
3.94	Tag	3.8	83.83
5.8	Reader	1.7	44.72
5.65	Tag	1.25	43.59
8.12	Reader	2.7	35.84
8.2	Tag	2.65	35.28

Table 2. Comparison of results between the two methods MoM and FIT.

frequency (GHZ)	Calculation Method (Simulation software)	Gain of the antenna (dBi)	Read range (cm)
	READER ANTENNA		
3.6	FIT (CST)	3.4	87.50
3.85	MoM (FEKO)	2.6	74.72
5.8	FIT (CST)	2.7	50.10
5.8	MoM (FEKO)	1.7	44.72
8.2	FIT (CST)	2.6	35.06
8.10	MoM (FEKO)	2.7	35.84
	TAG ANTENNA		
3.9	FIT (CST)	3.9	85.60
3.94	MoM (FEKO)	3.8	83.83
5.9	FIT (CST)	3.0	51.02
5.65	MoM (FEKO)	1.25	43.59
8.2	FIT (CST)	2.2	33.48
8.2	MoM (FEKO)	2.65	35.28

GHz, and 289 MHz at 8.12 GHz. The tag antenna exhibits a −10 dB bandwidth of 165 MHz at 3.94 GHz, 149 MHz at 5.65 GHz, and 166 MHz at 8.2 GHz. The maximum read range obtained by the reader antenna is 74.72 cm, and 83.83 cm by the tag antenna.

By simulating the antennas using the Method of Moment (MoM) and comparing the results with another study using the method of finite integral (FIT), we find that the results are similar, even if there are small lags that are tolerable.

REFERENCES

[1] H. Bhatt and B. Glover, "RFID Essentials," O'Reilly, Sebastopol, 2006.

[2] S. B. Miles, S. E. Sharma and J. R. Williams, "RFID Technology & Applications," Cambridge University Press, New York, 2011.

[3] C. Varadhan, J. K. Pakkathillam, M. Kanagasabai, R. Sivasamy, R. Natarajan and S. K. Palaniswamy, "Triband Antenna Structures for RFID Systems Deploying Fractal Geometry," *IEEE Antennas and Wireless Propagation Letters*, Vol. 12, 2013, pp. 437-440.

[4] G. O. Young, "Synthetic Structure of Industrial Plastics (Book Style with Paper Title and Editor)," In: J. Peters, Ed., *Plastics*, McGraw-Hill, New York, 1964, pp. 15-64.

[5] "FEKO 5.5 User's Manual," EM Software & Systems-S. A, 2009, pp. 1-1.

[6] B. Bieda and P. Słobodzian, "Efficiency of the IE-MoM Approach in the Analysis of Dielectric Bodies Embedded in a Cavity," 2010 *18th International Conference on Microwave Radar and Wireless Communications* (*MIKON*), Vilnius, 14-16 June 2010, pp. 1-4.

[7] A. Mehdipour, L. Rosca, A. Sabek, C. Trueman and S. Hoa, "Full Composite Fractal Antenna Using Carbon Nanotubes for Multiband Wireless Applications," *IEEE Antennas Wireless Propagation Letters*, Vol. 9, 2010, pp.891-895.

[8] A. Ismahayati, P. J. Soh and R. Hadibah, "Design and Analysis of a Multiband Koch Fractal Monopole Antenna," 2011 *IEEE International RF and Microwave Conference* (*RFM*), Seremban, 12-14 December 2011, pp. 58-62.

[9] D. H. Werner and S. Ganguly, "An Overview of Fractal Antenna Engineering Research," *IEEE Antennas and Propagation Magazine*, Vol. 45, No. 1, 2003, pp. 38-57.

[10] A. Mak, C. R. Rowell, R. D. Murch and M. Chi-Lun, "Compact Multiband Planar Antenna for 2.4/3.5/5.2/5.8-GHz Wireless Applications," *IEEE Antennas Wireless Propagation Letters*, Vol. 11, 2012, pp. 144-147.

[11] D. C. Chang, B. H. Zeng and J. Liu, "CPW-Fed Circular Fractal Slot Antenna Design for Dual-Band Applications," *IEEE Transactions on Antennas and Propagation*, Vol. 56, No. 12, 2008, pp. 3630-3637.

[12] J. H. Gu, S. S. Zhong, L. L. Xue and Z. Sun "Dual-Band Monopole Antenna for 2.45/5.8 GHz RFID Applications," 2008 *China-Japan Joint Microwave Conference*, Shanghai, 10-12 September 2008, pp. 133-135.

[13] B. D. Bala, M. K. A. Rahim, N. A. Murad, M. F. Ismail and H. A. Majid, "Design and Analysis of Metamaterial Antenna Using Triangular Resonator," 2012 *Asia-Pacific Microwave Conference Proceedings* (*APMC*), Kaohsiung, 4-7 December 2012, pp. 577-579.

Resource Optimisation for 3rd Generation Partnership Project (3GPP) Long Term Evolution OFDMA Downlink Interface Air

Ayoola Akinloye Akinniranye[1], Samson Adenle Oyetunji[2]

[1]MTN, Nigeria; [2]Federal University of Technology Akure, Akure, Nigeria.

ABSTRACT

This Paper describes the system model of the developed optimisation scheme which is implemented using a simulation package known as MATLAB. The 3rd Generation Partnership Project (3GPP) Long Term Evolution standard for the downlink air interface, Orthogonal Frequency Division Multiple Access (OFDMA) was adopted. Frequency Selective Rayleigh fading channel is considered as a model for real life scenario of the LTE Evaluation, operating in 5 MHz downlink bandwidth of OFDMA parameters as proposed in the standard. Simulation results show that, the inherent characteristics of a communication channel can be exploited and combined with the robustness of OFDMA to optimally allocate the radio resources. The optimal channel assignment of the dynamic allocation scheme exploited in this work meets the set target of high data rate and spectral efficiency.

Keywords: Communication Channels; Wireless Mobile Communications; Noise; OFDMA

1. Introduction

Wireless communication is enjoying a fast growth period in history which is coupled with technology improvements that permit its widespread deployment. That is the cellular concept developed by Bell Laboratories [1]. Mobile communication offers a full duplex communication using a radio to connect portable device to a dedicated Base station, which is then connected to a switching network. The first generation of mobile communication, known as Advanced Mobile Phone System (AMPS), was deployed in 1983 [1]. The second generation (2G) of mobile communication is known as Global System for Mobile communication (GSM) which was deployed in the 1990s [1] with 9.6 kbps data rate. The International Telecommunications Union (ITU) developed a plan in 1995 [2].

Called International Mobile Telecommunication 2000 (IMT-2000) implements a global frequency band [1,3] for the future mobile development. The third Generation system (3G) standard deployed in 21st century improves the data handling capacity with data rate of 64 kbps to 2 Mbps. A collaborative group of standards organisation and telecommunication companies called Third Generation Partnership Project (3GPP) was formed for enhanced versions to the 3G standard. Evolved from 3GPP standards in 2004 [4], it is the Release 8 version, which is known as Long Term Evolution (LTE) which targets to support high data rate of 100 Mbps for the downlink and 50 Mbps for the uplink with low delay, higher data rate, flexible bandwidth and optimised radio access and cell edge performance [5]. For the Downlink, Orthogonal Frequency Division Multiple Access (OFDMA) is considered for transmitting from Base Station (BS) otherwise known as Node B to the mobile user which is known as User Equipment (UE) while Single Carrier Frequency Division Multiple Access (SC-FDMA) is for the Uplink. This work focuses on the Downlink access. In a rapidly growing environment, overall system performance will depend on the ability to provide power and spectrum efficiency, adaptive to wireless fading and channel characteristics and support to changing UE traffics.

2. Wireless Channel Characteristics

Wireless channel is an unguided channel and signals not only contain the direct Line of Sight LOS waves; but also a number of signals as a result of diffraction, reflection and scattering as shown in **Figure 1**. This propagation

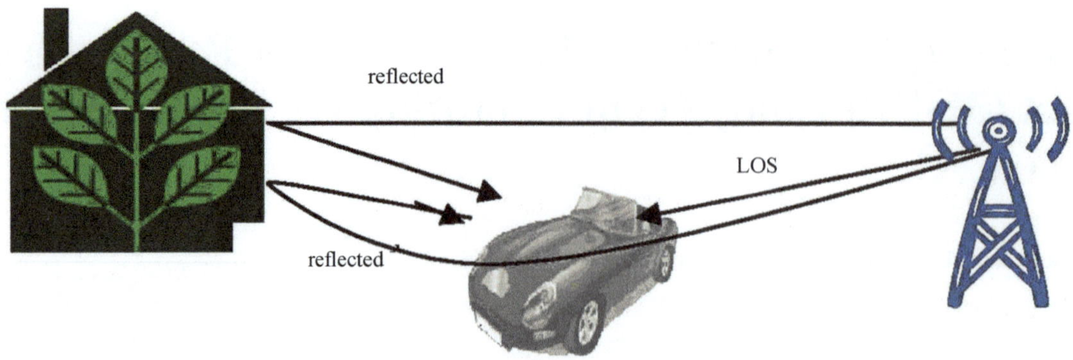

Figure 1. Multipath signal reception of a moving receiver.

type is termed Multipath [2] degrades the performance of the channel. Similarly, the channel may introduce Doppler Effect when the transmitter or receiver moves.

2.1. Additive White Gaussian Noise Channel

Additive White Gaussian Noise (AWGN) channel is a good model for the physical reality of channel, as long as the thermal noise at the receiver is the only source of disturbance [6]. The impairment this channel caused to signal is the addition of Gaussian distributed noise. Mathematically, it can be illustrated as:

$$r(t) = s(t) + n(t) \tag{1}$$

where $r(t)$ is the received signal, $s(t)$ is the transmitted signal and $n(t)$ is the noise.

2.2. Multi Path Fading Channels

An alternative class of channel used to model communication system is fading channels because mobile reception is harshly affected by multipath propagation which results in Fading or Inter-symbol.

Interference (ISI). This can be mathematically expressed as:

$$r(t) = s(t) * h(t) + n(t) \tag{2}$$

Time disperse signal are often affected by the delay spread. If the delay spread is less than the symbol period *Ts* or the signal bandwidth *BW* is less than the Coherence bandwidth, the signal channel is categorised as *Flat fading* which preserves of the spectral characteristics of the signal at the receiver [2]. In contrast, if signal bandwidth is more than the coherence bandwidth or delay spread is more than the symbol period, then the channel is categorised as *Frequency Selective fading* and leads to ISI which degrades the channel.

2.3. Channel Models

Andrea stated in [3] that deterministic channel models are rarely available. But to evaluate the performance of signals properly in fading channels, this work considered Flat and Frequency Selective fading channel and few of the models.

A. Rayleigh and Rician Fading Model

Rayleigh distribution model is often used for fading signal with infinite or large number of arrival paths at the same time whose gain are statistically independent and no dominant path [2]. The phase component of the channel gain is Gaussian distributed and equation 3 is its probability density function (PDF) as stated by Rappaport [1]:

$$p(r) = \begin{cases} \dfrac{r}{\sigma^2} e^{\left(-\frac{r^2}{2\sigma^2}\right)}, & 0 \le r \le \infty \\ 0, & r < 0 \end{cases} \tag{3}$$

where, σ is the RMS value of received signal before detection.

And according to [4], the average channel power is given by:

$$E[r] = 2\sigma^2 \tag{4}$$

Similar to the distribution properties of Rayleigh is the *Rician Distribution* model except for the presence of a dominant path with numerous weak paths. Inclusive in its pdf (equation 5 [2]) is the peak amplitude A of dominant signal and zero-order Bessel function I, of the first kind

$$p(r) = \begin{cases} \dfrac{r}{\sigma^2} e^{\left(-\frac{r^2 + A^2}{2\sigma^2}\right)} I_0\left(\dfrac{Ar}{\sigma^2}\right), & A \ge 0, r \ge 0 \\ 0, & r < 0 \end{cases} \tag{5}$$

B. Clarkes' Fading Model

The model assumes all multipath signals arrive at the same time in horizontal direction and when the mobile user moves, each path will experience a different Doppler shift. Hence, a uniform probability density function (PDF) of the rays is assumed and a Doppler effect is introduced due to respective movement of the receiver [7].

C. ITU Model

International Telecommunications Union published some generic test models that are commonly used in the communication industry. Depicted in [2] are the three common cases of the model-Indoor, Pedestrian and Vehicular. This is represented in **Table 1**. In this work, the interest is in the Channel B type of the Pedestrian model with 6 rays, median delay spread (750 *ns*) and 55% probability of occurrence in an outdoor to indoor environment. Each tap is modelled using Rayleigh fading distribution characterised by Clarkes' model. The spectrum defined [8] as:

$$S(f) \propto \frac{1}{\sqrt{1-\left(\frac{f}{f_d}\right)^2}} \quad \text{for } f \in -f_d, f_d \tag{6}$$

Assuming all the paths arrives at the same time and are uniformly distributed, the PSD, $S_h(f)$ is modelled as [2]:

$$\tilde{a}(t) = \sum_{i=0}^{N-1} a_i e^{j(2\pi f_i t + \theta_i)} \tag{7}$$

where, $f_i = f_d \cos\theta_i$

$$S_h(f) = \mathcal{F}\{R_h(\Delta t;\tau)\} = \begin{cases} \dfrac{P_{av}}{\sqrt{1-\left(\frac{f}{f_d}\right)^2}}, & |f| < f_d \\ 0, & |f| > f_d \end{cases} \tag{8}$$

where R_h is channel autocorrelation function, P_{av} is the average channel power, F_i is the Doppler shift in direction of travel for path θ_i and $\tilde{a}$ is the channel response in relation to Doppler shift.

3. Orthogonal Frequency Division Multiple Access

The multiple access scheme is based on Orthogonal Frequency Domain Multiplexing (OFDM) [9]. The multi-carrier access is achieved by assigning a group of subcarriers to a particular user as shown in **Figure 2**. A typical subframe is shown in **Figure 3**. OFDMA is inherently robust to time dispersion on the radio channel without recourse to complex receiver channel equalisation due to the combine use of narrow-band subcarrier transmission with cyclic prefix (CP) [10]. Basically, OFDMA is a suitable data access technique of transforming wideband frequency selective channel into flat fading narrow bands [11]. OFDM signals can be generated by Inverse Fast Fourier Transform (IFFT) and Fast Fourier Transform (FFT) implementation at Node B and UE. Thus, the basic OFDM signal $x(t)$ can be expressed during the time interval $0 \le t < T$ as [2];

$$x(t) = \sum_{k=0}^{N-1} X_k e^{-j2\pi k \Delta_f t} \tag{9}$$

where X_k is the complex modulated symbol carried in k_{th} subscriber in N size subcarriers. Δ_f is the subcarrier spacing and T is the symbol period.

As the downlink access standard prescribes [8], $\Delta_f = 15$ kHz This is made constant regardless of the transmission bandwidth simplify the implementation of 3G multimode terminal [2]. Hence, for an N size FFT, the sampling rate *fs* is given by:

$$f_s = N\Delta_f \tag{10}$$

Table 1. ITU pedestrian model [2].

Tap	Channel A Relative Delay (ns)	Channel A Average Power (dB)	Channel B Relative Delay (ns)	Channel B Average Power (dB)	Doppler Spectrum
1	0	0	0	0	Classic
2	110	−9.7	200	−0.9	Classic
3	190	−19.2	800	−4.9	Classic
4	410	−22.8	1200	−8.0	Classic
5			2300	−7.8	Classic
6			3700	−23.9	Classic

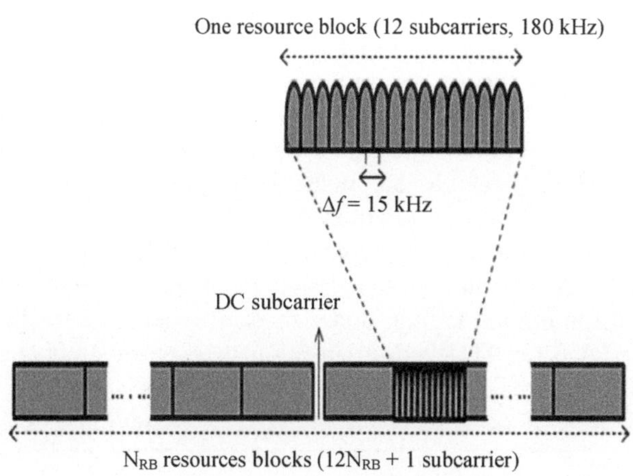

Figure 2. LTE downlink frequency-domain structure [6].

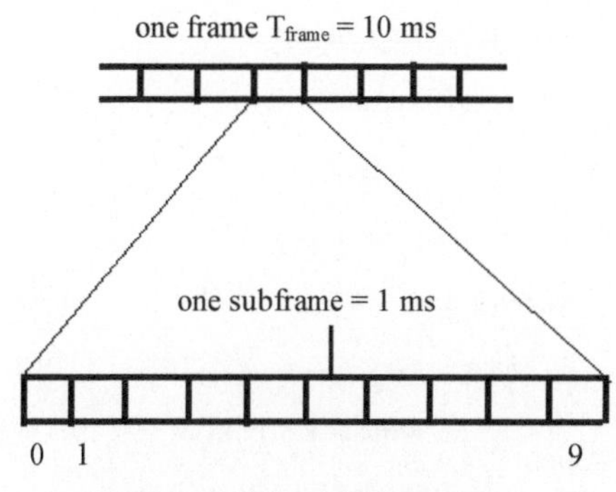

Figure 3. LTE subframe [2].

Also the bandwidth is made scalable by controlling the size of N, keeping the subcarrier spacing constant. Thus, with FFT size varying from 128 to 2048, downlink bandwidth range up to 20 MHz can be supported [12]. Similarly, two CPs, a short duration 4.69 μs and long duration 16.67 μs [13] are defined for flexibility of LTE system deployment. CP insertion means that the last part of the OFDM symbol is copied and attached to the beginning of the symbol. The CP length is chosen to be longer than the maximum delay spread and is carried out to retain the orthogonal properties between the subcarriers, in order to avoid inter-channel interference ICI.

4. Implementation of Resource Allocation for OFDMA

The main objective of this work is to determine an optimal channel assignment to achieve an effective resource optimisation scheme for 3GPP Long Term Evolution downlink's air interface which guarantees a designed coverage probability, high data rate, good spectral efficiency and ensure good quality of service. Allocation scheme has been modelled and presented. Resource optimisation is crucial to the overall performance of the proposed 3GPP LTE system. This methodology includes the system model of the developed optimisation scheme and as implemented using a simulation package known as MATLAB®. An initial model for all the modulation scheme is carried out in simple AWGN channel transmission, to test and evaluate the performance and detection of the modulated data as seen by the UE. Coherent detection is employed in the initial setup using Gray mapped signal constellation for all the modulation scheme and Maximum Likelihood Detection (MLD) at the receiver. Downlink transmission channel is characterised with multipath fading, dominance in frequency selectivity. A 6 ray ITU Pedestrian frequency selective channel B model (as shown in **Table 1**) is used to model the varying channel state. The signal arriving at the receiver is the sum of copies of the original signal with different delays and gain.

In the downlink of OFDMA systems, modulated data is converted from serial to parallel and mapped to different subcarriers. IFFT of the mapped data is carried out to convert the data into their corresponding time domain and the output signal are converted back to serial data called OFDM symbols. CP is attached to the beginning of the symbols (as guard interval, elimination of ISI and to enable circular convolution) before transmitting across the channel. At this point, a convolution with the channel is performed and Gaussian distributed noise n added. Subcarrier spacing in LTE is 15 kHz, CP length 4.69 μ or 16.67 μs and 1 ms time unit sub-frame (**Table 2**).

The description of OFDMA implementation above is in time domain, but OFDMA transmission in Frequency domain is exploited in this model making it relatively easier to implement frequency selective scheduling as 3GPP LTE standard prescribes. Therefore, OFDMA symbol implementation scheme can be described with the following analysis as described by **Figure 4**. Assume a Node B transmits symbols $s(t)$ which is passed through OFDM modulator yields OFDM symbols x as the output:

$$x = \mathcal{F}^{-1}\{s(t)\} \tag{11}$$

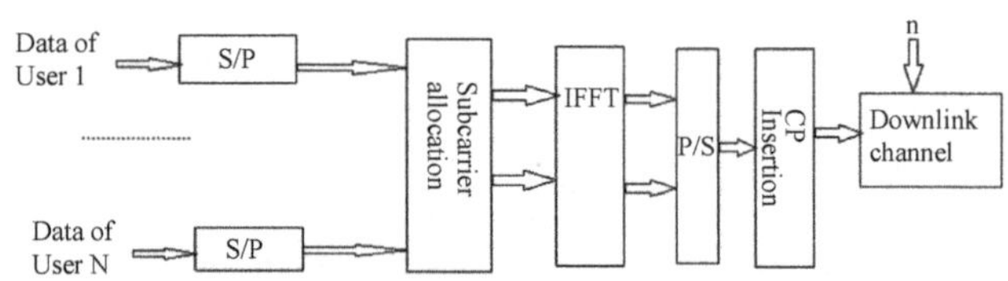

Figure 4. OFDMA downlink of 3GPP LTE [15].

Table 2. LTEOFDM parameters [14].

Parameter	Value					
Channel Bandwidth (MHz)	14	3	5	10	15	20
Number of Resource Blocks	6	5	25	50	75	100
Number of occupied subcarriers	72	180	300	600	900	1200
IDFT/FFT size	128	256	512	1024	1536	2048
Subcarrier spacing Δf (kHz)	15 (7.5)					
Sampling Rate (MHz)	1.92	3.84	7.68	15.36	23.04	30.72
Samples/slot	960	1920	3840	7680	11,520	15,360
CP sizes (μs) Normal CP = 15 kHz	5.21 (first symbols of the slot) 4.69 (other symbols of the slot) 7 symbols/slot					
CP sizes (μs) Extended CP = 15 kHZ	16.67 6 symbols/slot					

in the presence of noise n produce the received signal r at the receiver before demodulation is

$$r = h \otimes x + n \quad (12)$$

Demodulating r yields Sn

$$S_n = \mathcal{F}\{r\} \quad (13)$$

$$S_n = \mathcal{F}\{h \otimes x + n\} = \mathcal{F}\{h \otimes x\} + \mathcal{F}\{n\} \quad (14)$$

$$S_n = \mathcal{F}\{h\} \cdot \mathcal{F}\{x\} + \mathcal{F}\{n\} \quad (15)$$

$$S_n = \mathcal{F}\{h\} \cdot s(t) + \mathcal{F}\{n\} \quad (16)$$

where S_n are the received OFDMA symbols of n size FFT.

Therefore, frequency domain OFDMA modulation implementation overcomes the rigorous time domain implementation in. For correctness, the simulated OFDM model performance is compared with its expected performance of converting frequency selective fading channel to flat fading channel of the same distribution.

4.1. Optmisation Scheme for Resource Allocation

The main challenge is to develop a scheme that will increase the throughput. In this work optomisation of the resources becomes necessary by developing subscriber allocation capable of optmising the channel allocation. Two schemes are possible which are fixed and dynamic.

A. Subcarrier Allocation

In OFDMA systems, resources are mapped to the available subcarriers. The allocation strategies employed in assigning the subcarriers will affect the system performance in various ways such as data rate, power efficiency, spectrum efficiency and bandwidth flexibility. Two main assignment strategies is exploited in this work and presented below.

B. Fixed Subcarrier Allocation

Fixed allocation algorithm allocates a set of subcarriers to each User Equipment (UE) and no other UE can use them even if they are not in use. This is often viewed as being fair to all users and ensures that all users have a dedicated channel when the need arise. But this method amounts to waste of resources if UEs are in deep fade or when UE is not in need of the allocated subcarriers. Thus, the fixed allocation can be model as follows:

Assume Node B is scheduled to transmit N size subcarrier to k numbers of UEs, each user is allocated subcarrier sets M adjacent to each other. The algorithm is modelled as shown in (17) and (18):

$$S_k = n \in [s_k : k : s_N] \quad (17)$$

$$S_k = S_1, \cdots, S_k \quad (18)$$

where S_k is the set of subcarriers allocated to the kth user, n is the number of allocated subcarriers, S_k is the subcarrier index of the first subcarrier allocated to the kth user in the OFDM subcarrier stream of N size.

C. Dynamic Subcarrier Allocation

Since, subcarriers are currently spread across the channel band with equal spacing at implementation of OFDM, each OFDM symbol occupies a subcarrier (consisting of instantaneous subchannel, h_n for N size subcarriers). Recall that, 3GPP LTE specifies the use of CQI to estimate the UE performance per transmission subframe. Dynamic allocation relies on the knowledge of this estimate; hence the allocation scheme is modelled as follows:

Assume the Channel Quality of Information (CQI) of independent UE is known at Node B, which comprise of the same channel characteristics of the transmission channel but varies per transmission time. The varying CQI Hq for the kth UE can be model as random instantaneous channel gain $H_{q,k}$ by taking the frequency response of the randomly generated channel of the same delay profile as the transmission channel in n subcarrier transmission, which is expressed as:

$$H_{q,k\cdot t} = \sum_{n=1}^{F} \left| h_{q,k,n} \right|_t^2, \ f \text{ or } 0 : t : T. \quad (19)$$

where $h_{q,k,n}$ is the instanteneous subchannel extracted from k^{th} UE's quality indication at the instant time t of subcarrier n over $F = N$ subchannel of each UE (frequency response iteration of the randomly generated channel for each UE) within an OFDM data steam period T ($T = tN$). To apply CDS, in allocating resources to the UE with best CQI-essentially the best channel condition. The kth UE that get the nth subcarrier needs to maximize the following function:

$$n_k = \arg\max\left(H_{q,k}\right), T; n \in N. \quad (20)$$

where n_k is the subcarrier to be allocated to the kth user having $H_{q,k}$ that fulfils the condition above. Using the known Channel Quality of Information (CQI) from equation 20, Node B applies the nth subcarrier to the corresponding UE. Let $S_{k,n}$ denote the set of n subcarriers assigned to the kth UE. Thus, a (K X N) state matrix of subcarriers S and subchannels H is formed. If n is allocated to a k user, it returns a value and index the corresponding h_n in H, while returning false (0) for other users column-wise. Then, $S_{k,n}$ can easily be extracted row-wise. The objective of this dynamic allocation algorithm is to find a good allocation that will optimise the available resources. More subcarriers are allocated to best UE for optimal performance and a subcarrier can only be used by at most one user. Each subcarrier n as perceived by user k is subject to flat fading. Therefore, user capacity C_k, can be formulated and simplified from the description in equation in accordance with Shannon equation [5] as:

$$C_k = \max_{S_{k,n}} \sum_{n \subset S_{k,n}} \frac{BW}{N} \log_2 \left(1 + \frac{SNR|h|_{k,n}^2}{\frac{BW}{N}} \right) \quad (21)$$

As also described for a power constrained system, C_k will be subject to [16]

$$\sum_{k=1}^{K} \sum_{n \subset S_{k,n}} SNR \cdot N_0 \le P_{\max}. \quad (22)$$

where $P_{\max}$ is the maximum transmit power of the system, $|h|_{k,n}^2$ is the channel gain of subcarrier n of the kth UE. BW is the transmission channel bandwidth occupied by N size subcarriers $(n = 1, \cdots, N)$. SNR is the received Signal to noise power and N_0 is the noise power.

4.2. Simulation Results and Analysis

To evaluate the performance of the optimisation scheme, a Monte Carlo simulation approach of the model in MATLAB, using 32 uniformly distributed UE (*i.e.* 32 users) is carried out within a given cell. The channel model is as described for ITU and for 5 MHz channel bandwidth with $N = 512$ size FFT as illustrated in **Table 2** is implemented.

The numbers of users is made sufficiently large to exploit frequency selectivity gain and multiple of the user signals can undergo independent fading. For variations, CQI is estimated and all the subcarriers are allocated accordingly. OFDMA is expected to transform the channel from frequency selective Rayleigh fading to Flat fading. Provided below is the BER performance of simulated OFDMA in each modulation scheme as compared with a theoretical Rayleigh Flat fading.

Subcarrier Allocation Effects on Capacity

Since the downlink broadcast channel is characterised with fading, the transmission channel have independent random subchannel that changes over time [5]. Thus, the capacity of the downlink was estimated from which is known at the Node B according to the allocated subcarriers per UE as expressed in Equation (21). Capacity simulation results for 512 subcarriers adaptively allocated to 32 users over 5 MHz bandwidth.

The BER performances of the Bandpass modulation Techniques with OFDM in frequency selective fading channel were compared to the Bandpass modulation in flat fading channel. The results are presented in **Figure 5** to **Figure 8**. It was observed that the performance of OFDM in selective fading channel produced equivalent bandpass modulation performance in a flat fading channel. In addition the throughput rate increased with dynamic allocation scheme which responds to the UE request as shown in **Figure 9**. The spectral efficiency is dependent on the subcarriers allocation scheme adopted as established in **Figure 10**. The resources optimization shows the subcarrier allocation with efficient allocation of radio resources as determined by the choice of modu-

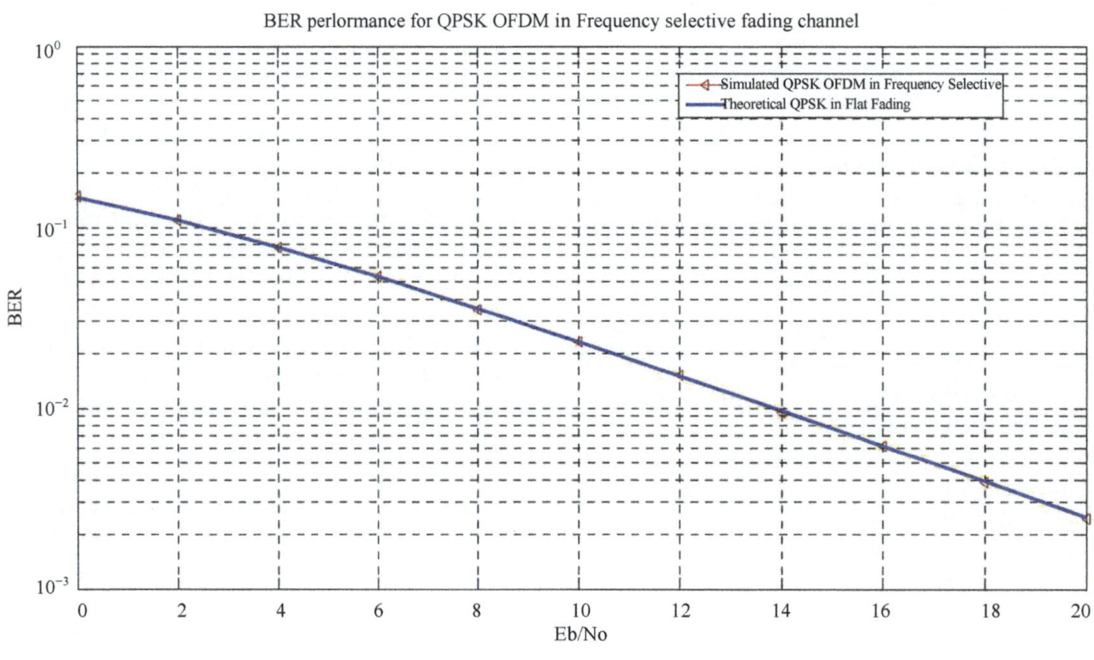

Figure 5. Comparison of BER performance of simulated QPSK OFDMA in frequency selective channel and theoretical QPSK in flat fading channel.

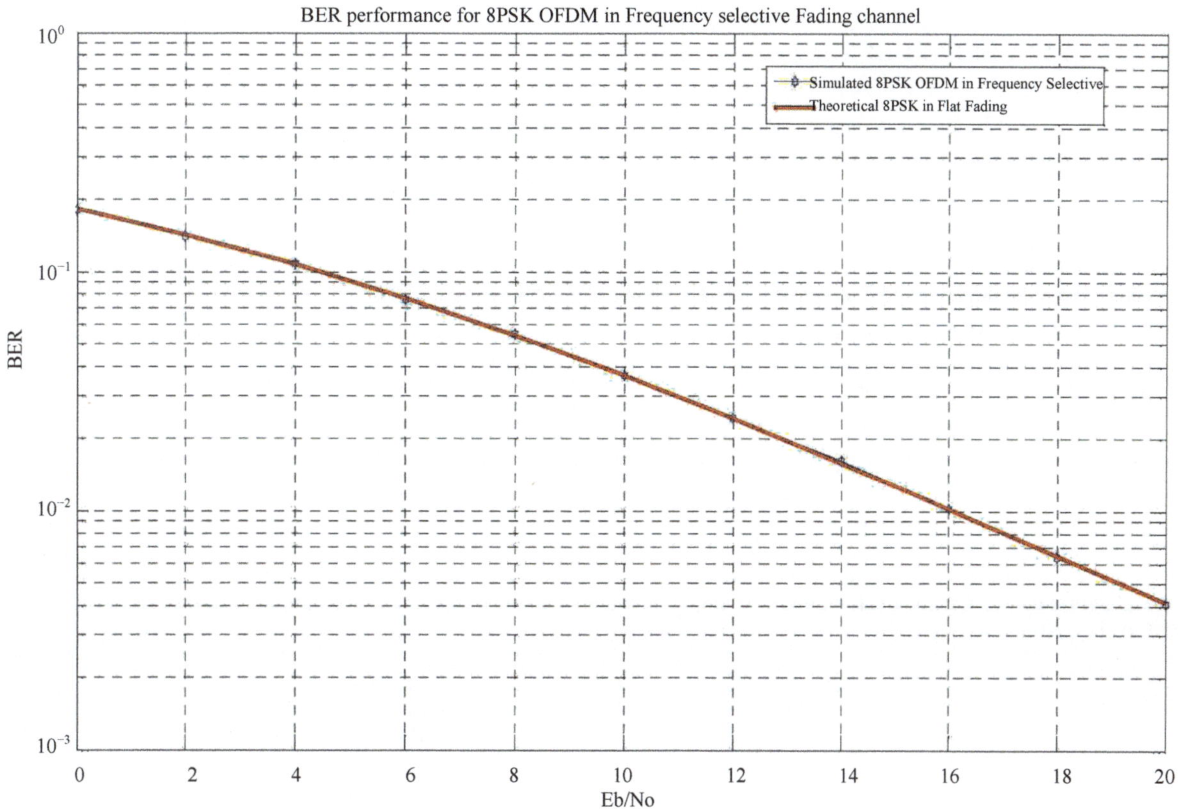

Figure 6. Comparison of BER performance of simulated 8PSK OFDMA in frequency selective channel and theoretical 8PSK in flat fading channel.

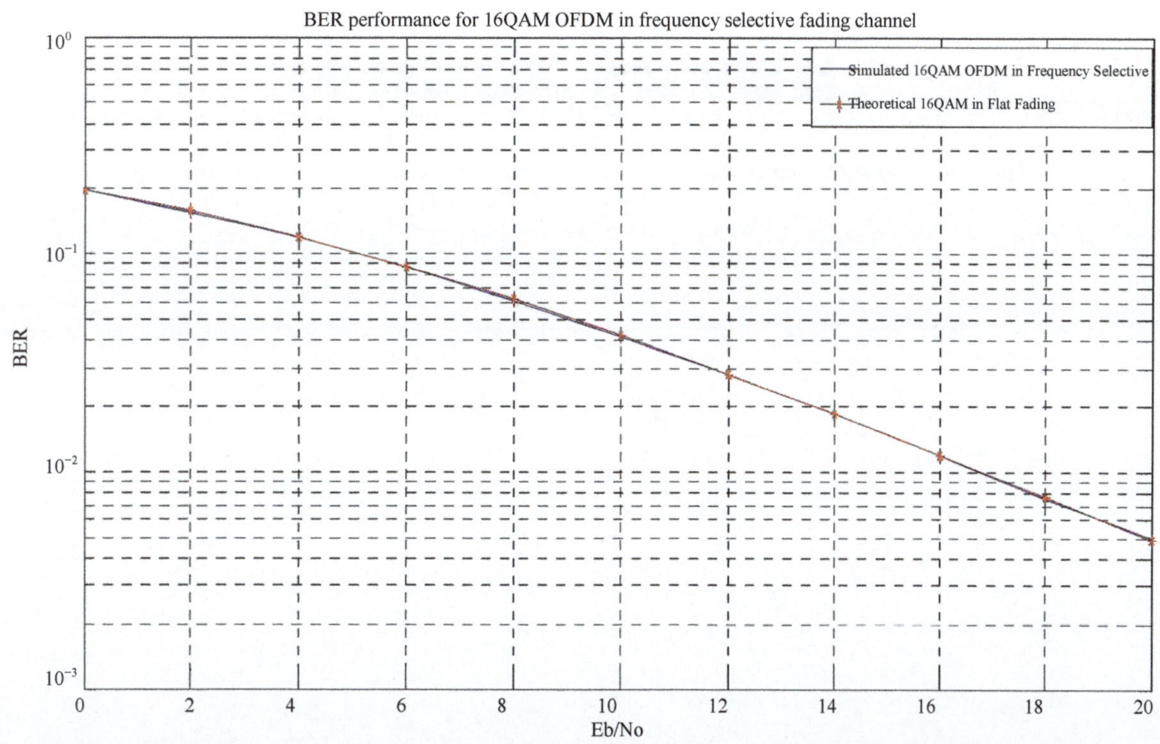

Figure 7. Comparison of BER performance of simulated 16QAM OFDMA in frequency selective channel and theoretical 16QAM in flat fading channel.

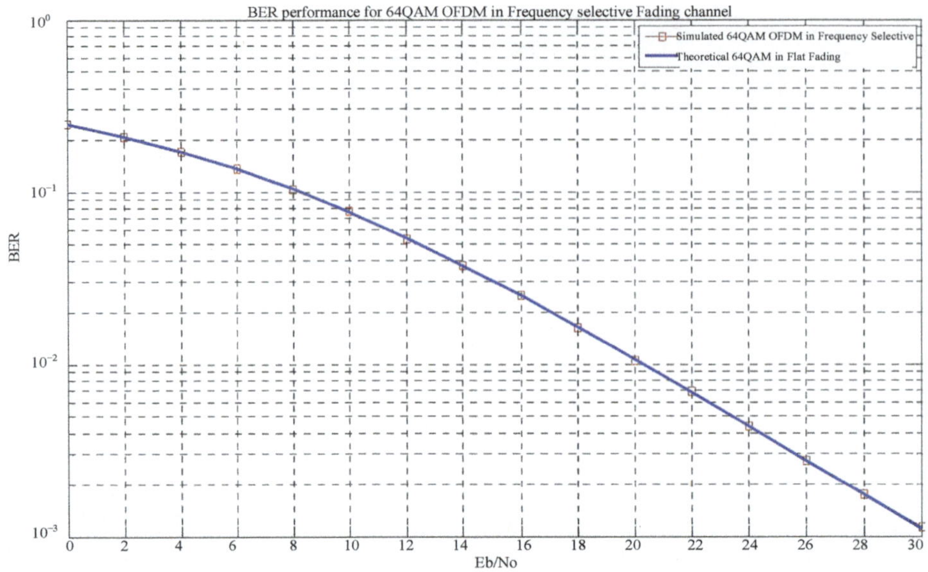

Figure 8. Comparison of BER performance of simulated 64QAM OFDMA in frequency selective channel and theoretical 64QAM in flat fading channel.

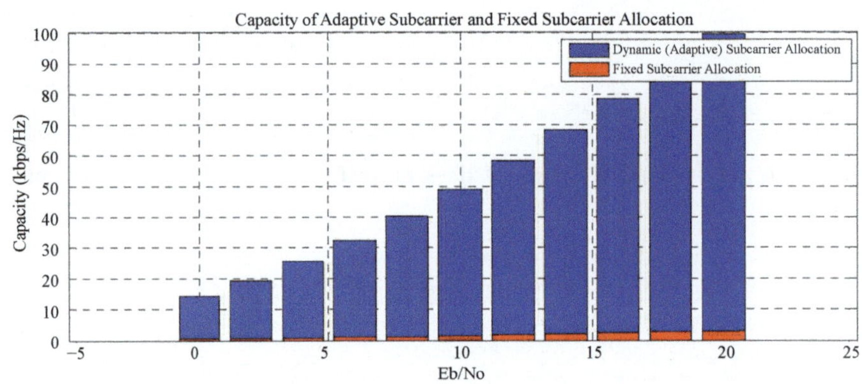

Figure 9. Grouped Spectral efficiency and Fixed Subcarrier Allocation vs Average SNR.

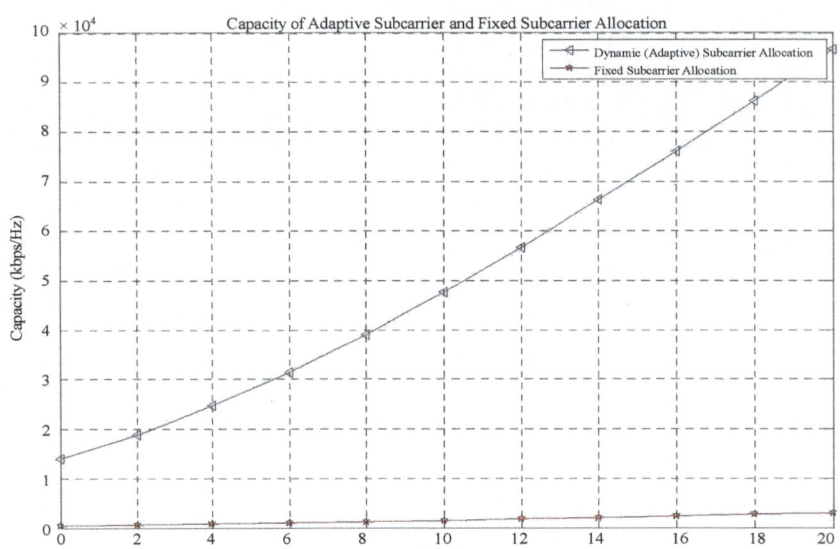

Figure 10. Spectral efficiency of adaptive and fixed subcarrier allocation vs SNR.

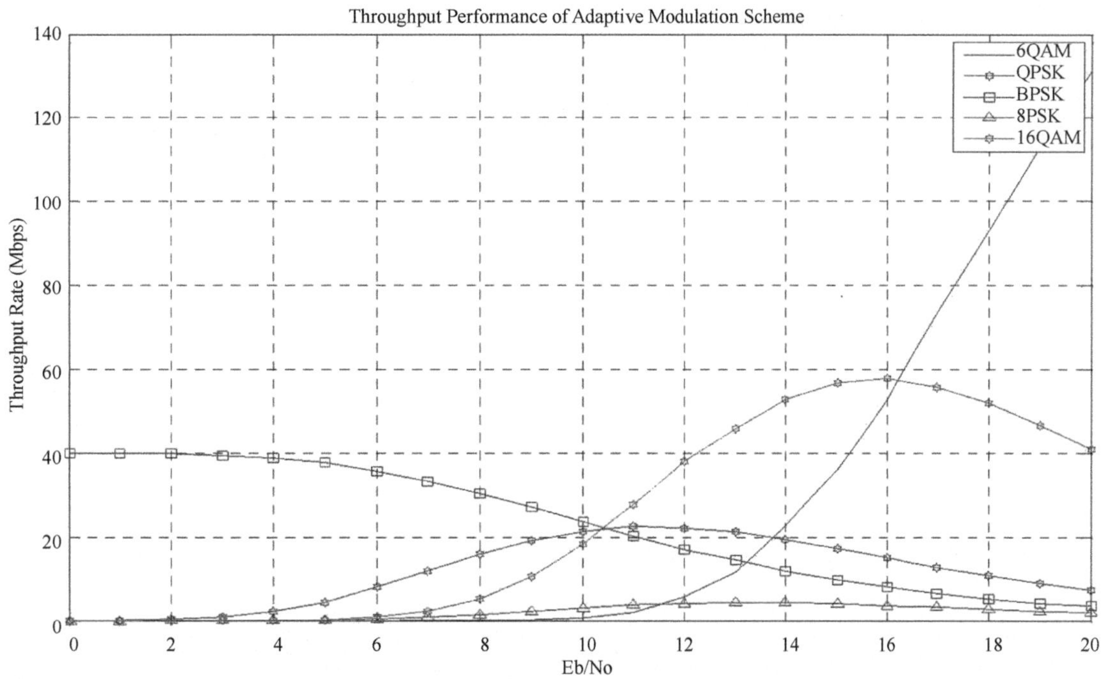

Figure 11. Comparison of subcarrier throughput rate with SNR.

lation. This is dynamically handled based on the channel quality information (**Figure 11**).

5. Conclusion

Efficient allocation of scarce radio resources is important for 3GPP LTE to provide high data rate transmission with improved spectral efficiency and guaranteed quality of service as the Next Generation network. Simulation results show that for an optimised resource allocation in the downlink, the inherent characteristics of a communi cation channel can be exploited and combined with the robustness of OFDMA to optimally allocate the radio resources. It was observed that optimal channel assignment of the dynamic allocation scheme exploited in this work is suitable for high data rate and spectral efficiency, and provides the required support for the increasing user demands encountered in the proposed 3GPP LTE downlink.

REFERENCES

[1] T. S. Rappaport, "Wireless Communications: Principle and Practice," 2nd Edition, Prentice Hall, Upper Saddle River, 2002.

[2] D. So, "Mobile and Wireless Communication," *Communications Engineering Lecture Notes*, University of Manchester, Manchester, 2010.

[3] A. Goldsmith, "Wireless Communications," Cambridge University Press, Cambridge, 2005.

[4] J. Berkmann, C. Carbonelli, F. Dietrich, C. Drewes and W. Xu, "On 3G LTE Terminal Implementation-Standard, Algorithm, Complexities and Challenges," *IEEE Wireless Communications and Mobile Computing Conference*, Crete Island, 6-8 August 2008, pp. 970-975.

[5] E. Dahlman, S. Parkvall, J. Skold and P. Beming, "3G Evolution: HSPA and LTE for Mobile Broadband," 1st Edition, Elsevier Academic Press, Amsterdam, 2007.

[6] H. Schulze and C. Luders, "Theory and Application of OFDM and CDMA," John Wiley, Hoboken, 2005.

[7] D. D. Bevan, V. T. Ermolayev, P. M. Grant, A. G. Flaksman and I. M. Averin, "Gaussian Channel Model for Macrocellular Mobile Propagation," *European Signal Processing Conference*, Antalya, 4-8 September 2005.

[8] 3GPP, "TS 36.104 v8.0 (Release 8 Series) BS Radio Transmission and Reception," 2010. http://www.3gpp.org/ftp/Specs/html-info/36-series.htm

[9] A. Furuskar, T. Jonsson and M. Lundevall, "The LTE Radio Interface-Key Characteristics and Performance," 19*th IEEE Personal, Indoor and Mobile Radio Communications*, Cannes, 15-18 September 2008, pp. 1-5.

[10] D. Astely, E. Dahlman, A. Furuskar, Y. Jading, M. Lindstrom and S. Parkvall, "LTE: The Evolution of Mobile Broadband," *IEEE Communications Magazine*, Vol. 47, No. 4, 2009, pp. 44-51.

[11] H. Schulze and C. Luders, "Theory and Application of OFDM and CDMA," John Wiley, Hoboken, 2005.

[12] S. Omar, A. Ancora and D. T. M. Slock, "Performance Analysis of General Pilot-Aided Linear Channel estima-

tion in LTE OFDMA Systems with Application to Simplified MMSE Schemes," *IEEE Personal, Indoor and Mobile Radio Communications*, Cannes, 15-18 September 2008, pp. 1-6.

[13] M. Sawahashi, E. Dahlman and K. Higuchi, "Evolved UTRA Technologies," *Enhanced Radio Access Technologies for Next Generation Mobile Communication*, pp. 217-276.

[14] S. Ahmadi, "An Overview of 3GPP Long-Term Evolution Radio Access Network," *New Directions in Wireless Communications Research*, Springer, Berlin, 2009.

[15] A. Haghighat, Z. Lin and G. Zhang, "Haar Compression For Efficient CQI Feedback Signalling in 3GPP LTE Systems," *IEEE Wireless Communications and Networking Conference*, Las Vegas, 31 March-3 April 2008, pp. 819-823.

[16] W. Rhee and J. M. Cioffi, "Increase in Capacity of Multiuser OFDM System Using Dynamic Subchannel Allocation," *IEEE 51st Vehicular Technology Conference Proceedings*, Vol. 2, Tokyo, 15-18 May 2000, pp. 1085-1089.

23

Estimation of Two-Dimensional Correction Factors for Min-Sum Decoding of Regular LDPC Code

Ahmed A. Hamad

Department of Electrical Engineering, University of Babylon, Babel, Iraq.

ABSTRACT

In this paper, two-dimensional (2-D) correction scheme is proposed to improve the performance of conventional Min-Sum (MS) decoding of regular low density parity check codes. The adopted algorithm to obtain the correction factors is simply based on estimating the mean square difference (MSD) between the transmitted codeword and the posteriori information of both bit and check node that produced at the MS decoder. Semi-practical tests using software-defined radio (SDR) and specific code simulations show that the proposed quasi-optimal algorithm provides a comparable error performance as Sum-Product (SP) decoding while requiring less complexity.

Keywords: LDPC Code; Sum-Product; Min-Sum; 2-D Correction Factors; Software-Defined Radio (SDR)

1. Introduction

Due to its exceptional error performance, low density parity check code (LDPC) [1] has received significant attention recently. It is adopted by many new generation communication standards, such as wireless LAN (IEEE 802.11n) [2], WiMax (IEEE 802.16e) [3] and DVB-S2 [4]. Although Sum-Product (SP) algorithm [5] provides a powerful tool for iterative decoding of LDPC codes, it requires a large hardware complexity. Sub-optimal algorithms like Min-Sum (MS) [6] can significantly reduce the hardware complexity of SP at the cost of performance degradation. In MS decoding, the complex computations at the check nodes are approximated by using simple comparison and summation operations. The sub-optimality of the MS decoding comes from the overestimation of check-to-bit and thereby, bit-to-check node messages, which lead to performance loss with respect to the SP decoding.

In many literatures like [7-11], various methods were proposed to elevate the performance degradation of the MS decoding which is considered as quasi-optimal algorithms. It has been noted that the degradation in error performance due to MS decoding can be compensated by linear post processing (normalization or offset) of the messages delivered by check and bit nodes.

In this study, a simple algorithm is presented to estimate the optimal correction factors for 2-D corrected MS decoding of regular LDPC codes. In a previous work [12], it has been shown that the error performance loss of SOVA compared to Log-MAP decoding can be compensated by scaling the a-posteriori LLR values of the information bits based on the fact that the latter is a growing estimate to the transmitted codeword, and the correction factors can be derived based on minimizing the mean-square difference (MSD) between them. A similar approach is applied to find the optimal correction factors which are utilized to scale the message passing from check to bit and bit to check nodes. Extensive simulations are carried out using the specifications of the LDPC adopted by the IEEE 802.11n standards. A significant improvement in error performance is achieved by using the modified MS decoder compared to the original one.

This paper is organized as follows. Section 2 briefly reviews the standard SP and MS decoding schemes. Section 3 presents the proposed 2-D corrected MS decoding. In Section 4, a simple SDR implementation for the modified and unmodified systems is demonstrated. Section 5 reports simulation and practical results and Section 6 concludes this paper.

2. Standard SP and MS Decoding Algorithms

The SP and its simplified version MS are soft-decision

decoders which make use of the channel information. They are developed to deal with log-likelihood values (LLR) of the received signal. As proposed by [13] efficient encoding of the information bits with a reasonable complexity is achieved to produce the codeword $c = \left[c_n \in \{0,1\} : n = 1 \text{ to } N\right]$. When a codeword is transmitted through an AWGN channel the binary vector c is first mapped into a transmitted signal vector $q = \left[q_n \in \{1,-1\} : n = 1 \text{ to } N\right]$ according to $q_n = 1 - 2c_n$. The received sequence is $y = [y_n] = q + w$, where the samples of the received signal y_n are given by

$$y_n = q_n + \omega_n \tag{1}$$

The vector $\omega = [\omega_n]$ are statistically independent Gaussian random variables with zero mean and two sided power spectral density of $N_0/2$. The received signals y_n have to be scaled by the channel reliability $L_c = 4R\frac{E_b}{N_o}$ before they deliver to the decoder. Here, the notations R and E_b define the information rate and the energy per information bit. Hence, $r_n = L_c y_n$ is the log-likelihood ratio (LLR) of bit n which is derived from the channel output y_n.

Suppose a regular binary (N, K) LDPC code has a bipartite Tanner graph which can be defined by the sparse parity check matrix $H = [H_{mn}]$. The set of neighbor check nodes of the bit node n is $A_n = \{m : H_{mn} = 1\}$. While, the set of neighbor bit nodes of the check node m, is $B_m = \{n : H_{mn} = 1\}$. The set of bits that participate in check node m except for bit n is denoted by $B_{m/n}$. On the other hand, the notation $A_{n/m}$ refers to the set of checks in which bit n participates except for check m. For each iteration i, the decoder calculates the following variables:

- E^i_{mn}: The LLR message of bit $\boldsymbol{n}$ which is passed from check node $\boldsymbol{m}$ to bit node $\boldsymbol{n}$ (extrinsic information).
- M^i_{mn}: The LLR message of bit $\boldsymbol{n}$ which is passed from bit node $\boldsymbol{n}$ to check node $\boldsymbol{m}$.
- M^i_n: The a-posteriori LLR of bit $\boldsymbol{n}$.

The standard SP [5], [14] decoding algorithm can be summarized by the following steps.

Step 1: Initialization: set $M^i_{mn} = r_n$ for each m and n, maximum number of iteration to i_{max}, $i = 1$ and failure = 0.

Step 2: Check node update (Horizontal step): For each m, n with $H_{mn} = 1$, compute

$$E^i_{mn} = 2\tanh^{-1}\left(\prod_{n' \in B_{m/n}} \tanh\left(\frac{M^{i-1}_{mn'}}{2}\right)\right) \tag{2}$$

Step 3: Bit node update (vertical step): For each m, n with $H_{mn} = 1$, compute

$$M^i_{mn} = r_n + \sum_{m' \in A_{n/m}} E^i_{m'n} \tag{3}$$

$$M^i_n = r_n + \sum_{m \in A_n} E^i_{mn} \tag{4}$$

Step 3: Tentative decision: Let denote the estimated codeword bits as

$$\hat{c}_n = \begin{cases} 0 & \text{if } M^i_n \geq 0 \\ 1 & \text{if } M^i_n < 0 \end{cases}$$

If $H\hat{c}^T = 0$ or $i = i_{\max}$ go to step 4, otherwise set $i = i + 1$ and go to step 2.

Step 4: Declare that $\hat{c}$ is the decoded codeword. If $H\hat{c}^T \neq 0$ and $i = i_{\max}$ then set failure = failure + 1.

The difference between MS and SP algorithm is the way by which MS decoder computes the check node update [6]. It is recognized that the term corresponding to the smallest E^i_{mn} dominates the product term in Equation (2) and so the product can be approximated by a minimum. Hence Equation (2) can be modified to

$$E^i_{mn} = \prod_{n' \in B_{m/n}} \operatorname{sgn}\left(M^{i-1}_{mn'}\right) \cdot \min_{n' \in B_{m/n}} M^{i-1}_{mn'} \tag{5}$$

The product of sign and minimum operations are much simpler to implement and faster in time processing than the product of hyperbolic tan and its inverse. These advantages spot a light on the MS decoding algorithm and make it more feasible and preferable in practical implementations.

3. Proposed Corrected MS Decoder

Although the MS algorithm greatly reduces the decoding complexity for implementation, it suffers from a significant deterioration in its error performance compared to SP decoding. Many researches state that the correlation of the extrinsic information with the original a priori bit LLR (r_n) is what prevents the resulting a-posteriori probabilities from being exact. In line with the foundation that explored in [12], it is suggested that we scale the marginal LLR values E^i_{mn} and M^i_{mn} by the correction factors a^i and b^i to make the a-posteriori probabilities as close as possible from the transmitted codeword, thereby reducing the probability of error. The corrected values $\acute{E}^i_{mn}$ and $\acute{M}^i_{mn}$ are given by the following equations:

$$\acute{E}^i_{mn} = a^i E^i_{mn} \tag{6}$$

$$\acute{M}^i_{mn} = b^i M^i_{mn} \tag{7}$$

It is possible to derive the correction factor a^i by minimizing the mean square difference $\delta(a)$ between the a-posteriori probability M^i_n and the mapped codeword $\boldsymbol{q}$. Assume that $V^i = \left[M^i_n\right]$, then

$$\delta(a^i) = E\left[\left(q - a^i\, sgn(V^i)\right)^2\right] \quad (8)$$

$E[\cdot]$ denotes the expected value. $\delta(a^i)$ may be considered as a measure of how efficient the algorithm is with the proposed modification. It is obvious that to get better suboptimal decoding, the parameter a^i should be found to minimize $\delta(a^i)$. The value of a^i is simply found by derive $\delta(a^i)$ with respect to a^i and equating the result to zero.

$$\frac{d\delta(a^i)}{da^i} = a^i E\left[\left(\mathrm{sgn}(V^i)\right)^2\right] - E\left[\mathrm{sgn}(V^i) * q\right] = 0 \quad (9)$$

Considering that $\left(\mathrm{sgn}(V^i)\right)^2 = 1$, the value of a^i is given by

$$a^i = E\left[\mathrm{sgn}(V^i) \cdot q\right] \quad (10)$$

In similar manner, the value of b^i can be found by minimizing the MSD between the modulated codeword and the vector representing the vertical sum of M^i_{mn}. Suppose that

$$U^i_n = \sum_{m \in A_n} M^i_{mn} \quad (11)$$

and let the vector $U^i = \left[U^i_n\right]$, then

$$b^i = E\left[\mathrm{sgn}(U^i) \cdot q\right] \quad (12)$$

It is worth mentioning that both Equations (10) and (12) need to be computed to the transmitted codeword q and this is not available at the receiver in real-world systems. It is suggested to solve this problem in three different ways. First, it is possible to compute the values of a^i and b^i offline for different channel characteristics and store their values in memory to be used latter by the decoder online. This will reduce the proposed modification to simple multipliers. Second, pilots and headers are known signals to the decoder, so they offer online estimation for the correction factors. Third, from the simulation of different systems, it is observed that the values of a^i and b^i have nearly stationary values for certain number of iterations, code parameters and channel characteristics. Thereby sending few known codewords to the receiver at the start of or in between transmission is sufficient to compute the correction factors with accepted accuracy.

Actually, for certain code H and channel characteristics (s) (e.g., E_b/N_o), the correction factors are function of the iteration's number (i) and the number of simulated frames (t), (*i.e.*, $a^i_{s,t}$ and $b^i_{s,t}$). The average values of $a^i_{s,t}$ over all simulated codewords is given by

$$a^i_s = E_t\left[a^i_{s,t}\right] \quad (13)$$

Further, if the average is done over all iterations, we get

$$a_s = E_i\left[a^i_s\right] \quad (14)$$

And finally, the averaging over all simulated values of E_b/N_o produces a constant

$$a = E_s\left[a_s\right] \quad (15)$$

Different simulation tests are carried out to demonstrate the effect of averaging presented in Equations (13)-(15) on the error performance of LDPC.

4. Implementation by Software-Defined Radio

To be close from the implementation issues of the real-world wireless communications, semi-practical tests to some of the simulated systems are experienced using the MATLAB data acquisition toolbox and PC sound card. Real-world passband signals for some of the modified and unmodified coded systems are transmitted and received via room-to-room audio sender operating in the 2.4 GHz industrial, scientific and medical (ISM) radio band. The acquired signal from the PC sound card can be directed to the RAM and/or the hard disk to be processed later by the SDR receiver. The SDR receiver takes consider of all the required DSP algorithms like modulation/demodulation, pulse shaping filtering, synchronization and signal to noise power ratio (SNR) estimation in MATLAB code.

Figure 1 depicts simple Graphical User Interface (GUI) panel that used as SDR receiver.

The panel offers different options that can be changed to suit the desired test like codeword length (N), coding rate (R), maximum number of iterations ($i_{\max}$), baud rate and oversampling value. It is also possible to change the order and roll-off factor of the square root raised cosine filter which is used in pulse shaping process for the transmitted and received signals.

Figure 2 shows the block diagram of the proposed SDR system. While, **Figure 3** depicts a picture for the set of SDR test system.

In asynchronous data transmission, the frame and bit

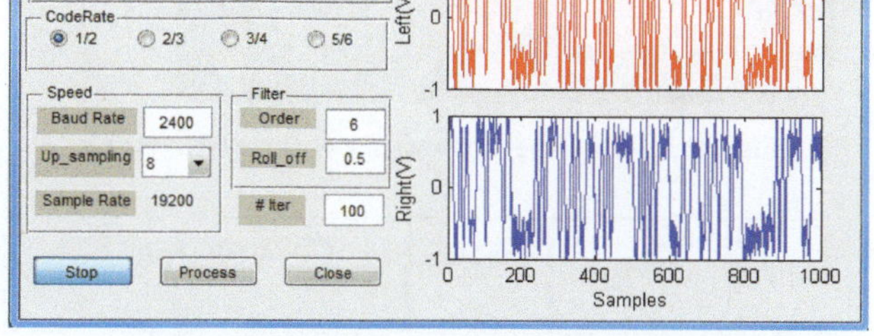

Figure 1. GUI panel used for SDR receiver.

synchronization are considered as a challenging problem. In this study, a simple frame construction is considered to cope with this problem as shown in **Figure 4**. In the front of any transmitted frame a header is generated by one or more Barker code(s) of thirteen bits in length. The length of the header determines the capability of the receiver to mark the starting point of the received frame and reduces the probability of miss detection. However, the quality of the channel is the main factor in determining the header length.

Intuitively, the data followed the header is random for the receiver which leads to a mismatch between the actual and reference header at the receiver due to filtering process. This is manifested especially at the junction between the header and data fields. Since the header signal is utilized to estimate the SNR, this undesirable distortion will reduce the accuracy of estimation. Therefore, three bits (e.g., 101) are appended by the header bits. The padded bits should be omitted during the SNR estimation to get rid of this distortion. **Figure 5** illustrates a section from the transmitted frame at the junction between header and data, received noisy signal, and the output of square raised cosine filter. The estimated E_b/N_o at the receiver is around 15 dB for this case.

The SNR estimation technique presented here originated from a method for measuring channel distortion errors in wideband telemetry systems [15], in which the noise in a signal at a point in a system is defined as the mean-square error (MSE) between the actual signal and a desired signal at that point. The estimated SNR is given by

$$SNR = \frac{R_{DA}^2(\tau_m)}{P_D P_A - R_{DA}^2(\tau_m)} \tag{16}$$

where P_D and P_A represent the average powers in the reference (desired) header and the actual (received) header, respectively. The maximum correlation $R_{DA}(\tau_m)$ between the actual and the desired header signals indicates the system time delay (τ_m).

Figure 6 demonstrates the correlation between the reference and actual headers. The maximum correlation $R_{DA}(\tau_m)$ marks the beginning of the received frame. It is worth mentioning that the abscissa of the BER performance graphs for the practical tests presented in section 5.2 is the average of the estimated SNR over all transmitted frames.

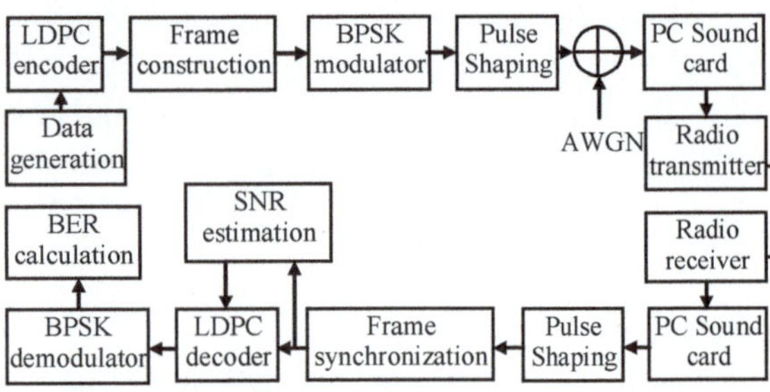

Figure 2. Block diagram of the proposed SDR system.

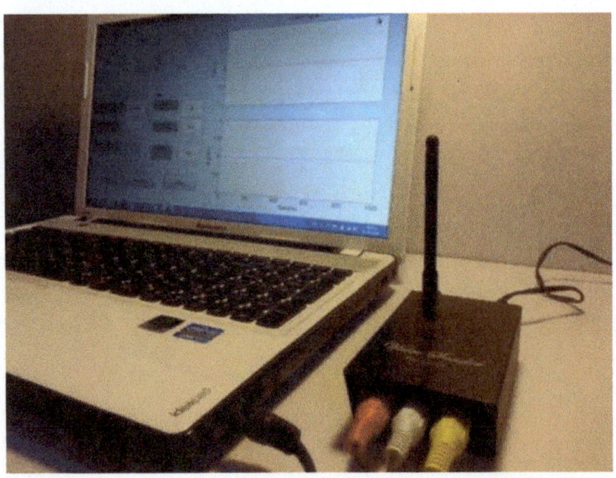

Figure 3. The experimental set of the SDR system.

Header Multiple of 13 bits barker code	3 padding bits	Codeword *N* bits

Figure 4. Frame construction.

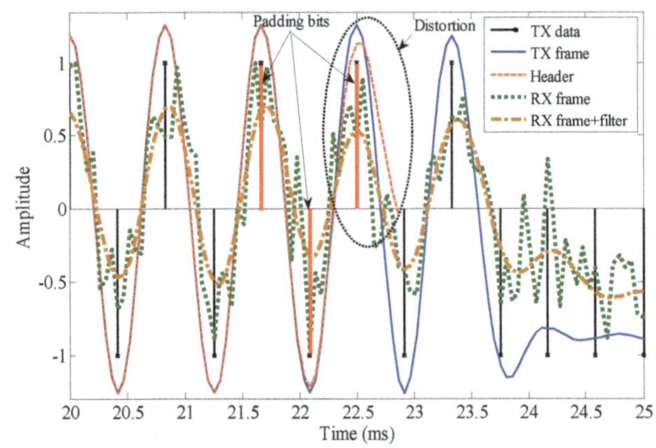

Figure 5. Transmitted data, transmitted pulse shaped, received noisy and filtered signals (part of).

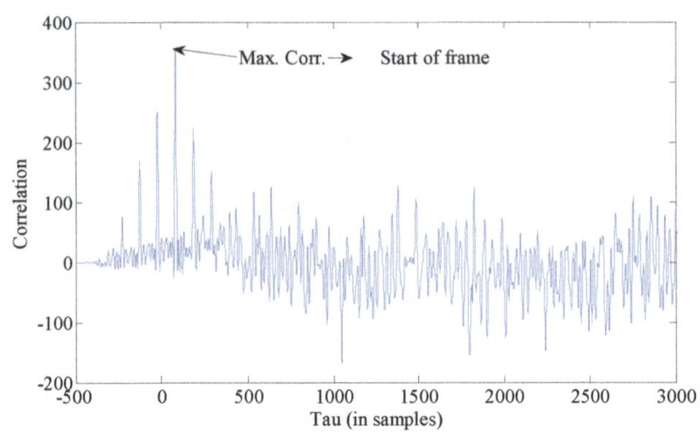

Figure 6. Correlation between desired and actual headers.

5. Simulation and Practical Results

5.1. Simulation Results

Simulations are carried out over the AWGN channel with BPSK modulation for the proposed 1-D and 2-D corrected MS decoding for LDPC code which is adopted by the standard IEEE 802.11n [2]. **Figure 7** presents the BER performance of regular (1944, 1296) rate 2/3 LDPC with the decoding algorithms; SP, original MS, 1-D MS-(a_s^i), 1-D MS-(b_s^i) and 2-D MS-(a_s^i, b_s^i). The systems in **Figure 7** (excluding 1-D MS-(b_s^i)) are retested using the regular (648, 324) rate 1/2 LDPC and depicted in **Figure 8**. It is obvious that the performance gap between the proposed 2-D MS-(a_s^i, b_s^i) and the SP decoders is only of 0.1 dB and 0.04 dB at BER = 10^{-4} for the (1944, 1296) and (648, 324) codes, respectively.

Table 1 illustrates the values of the correction factors a_s, b_s, a, b and the mean value of a and b, *i.e.*, $E(a,b)$, for the (1944, 1296) rate 2/3 LDPC code.

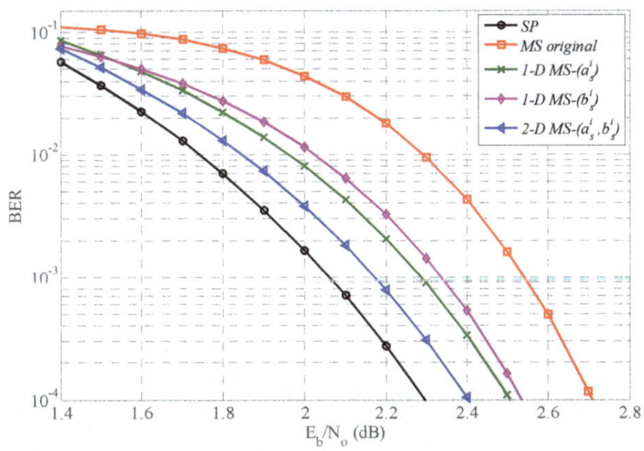

Figure 7. BER performance of SP, original MS, 1-D MS-$\left(a_s^i\right)$, 1-D MS-$\left(b_s^i\right)$ and 2-D MS-$\left(a_s^i, b_s^i\right)$ decoding algorithms for regular (1944, 1296) LDPC, $i_{\max} = 100$.

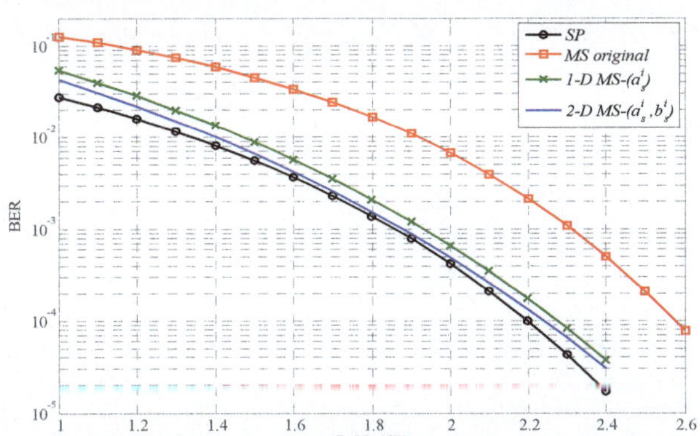

Figure 8. BER performance of SP, original MS, 2-D MS-$\left(a_s^i, b_s^i\right)$ and 1-D MS-$\left(a_s^i\right)$ decoding algorithms for regular (648, 324) LDPC, $i_{\max} = 100$.

It is interested to note that the values of a_s and a_s are very close to each other, so it is possible to use the same correction factor at the bit and check node for more reduction in complexity.

Figure 9 demonstrates the effect of applying the average of correction factors over all iterations and further over the whole tested range of E_b/N_o. It is clear that the averaging of the correction factors over all iterations (a_s, b_s) results a deterioration in error performance, but only about 0.02 dB.

Applying a and b instead of a_s, and b_s in MS decoding gives a slight improvement especially at high SNR. This interested result urged on testing the proposed scheme by applying a single correction factor $E(a,b)$ = 0.889 at the bit and check nodes for the whole range of SNR and along all decoding iterations. The 2-D MS-$E(a,b)$ decoding scheme shows a reduction in performance in comparison with other schemes at low E_b/N_o. After E_b/N_o=2 dB the BER performance of this scheme outperforms the 2-D MS-(a_s, b_s) and 2-D MS-(a, b) systems and becomes even better than the 2-D MS-$\left(a_s^i, b_s^i\right)$ scheme after E_b/N_o=2.3 dB.

Figure 10 presents the probability of failure for the (1944, 1296) LDPC, for SP, original MS, and 2-D MS-a_s^i, b_s^i decoding algorithms. Probability of failure can be

Table 1. Correction factors (a_s, b_s), their average values (a, b) and the average of a and b of regular (1944, 1296) LDPC.

Correction factors	$\frac{E_b}{N_o}$ (dB)				a	b	$E(a,b)$
	1.4	1.8	2.2	2.6			
a_s	0.8526	0.8867	0.9137	0.9363	0.8973	0.9007	0.899
b_s	0.8604	0.8910	0.9150	0.9363			

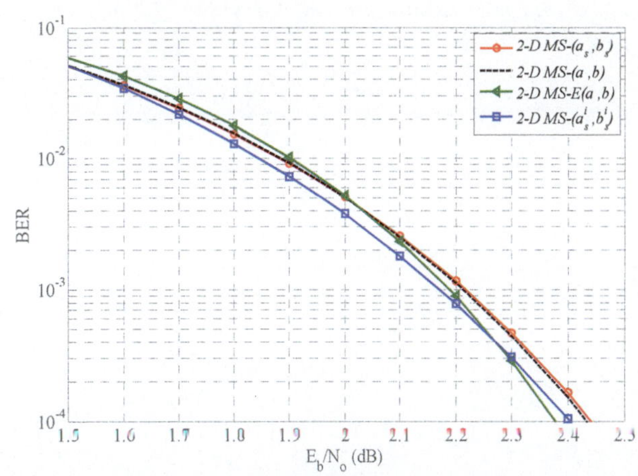

Figure 9. BER performance of 2-D MS-$\left(a_s^i, b_s^i\right)$, 2-D MS-$\left(a_s^i, b_s^i\right)$, 2-D MS-($a$, b) and 2-D MS-$E(a,b)$ decoding algorithms for regular (1944, 1296) LDPC, $i_{\max} = 100$.

defined as the probability of the decoder to reach the maximum number of iterations $i_{\max}$ without satisfying the parity check equations, *i.e.*, $H\hat{c}^T \neq 0$. **Figure 11** shows the average number of iteration for each simulated system in **Figure 10**. It is clear from **Figures 10** and **11** that the modified MS decoder demonstrates a comparable performance to the optimum SP decoder.

Figure 12 presents the variation of a_s^i and b_s^i over 100 iterations with different values of E_b/N_o. **Figure 13** demonstrates the values of a_s^i over the simulation of 50 and 500 frames. It is clear that the two curves are very close to each other which mean that it is possible to estimate the correction factors online with few known received frames without losing a lot of time in this process.

5.2. Practical Results

Experiments of wireless indoor transmission (utilizing the SDR approach explained in Section 4) are carried out to prove the effectiveness of the proposed systems. The tests are running over approximately the same environments. The transmitter and the receiver are fixed inside a 4 m × 4 m room during the experiments. The data is transmitted with a bit rate of 2400 bit/sec using BPSK modulation. The filter order and roll-off factor at both transmitter and receiver are adjusted to 6 and 0.5 respectively.

Figure 14 presents the resulted BER performance of original MS, 2-D MS-$\left(a_s^i, b_s^i\right)$, and 2-D MS-*E*(*a*, *b*) decoding algorithms of the regular (1944, 1296) LDPC code. It is noted that the BER performance of the proposed 2-D MS-$\left(a_s^i, b_s^i\right)$, decoder outperform the original MS by about 1.1 dB measured at BER = 10^{-3}. Reducing the complexity by using one correction factor $E(a,b) = 0.8816$ for both bit and check messages results in reduction of performance by about 0.12 dB in compare with the 2-D MS-$\left(a_s^i, b_s^i\right)$ decoding. **Table 2** illustrates the correction factors (a_s, b_s), their average values (a, b) and the mean value of a and b, $E(a, b)$ that acquired from the practical test of the regular (1944, 1296) LDPC code.

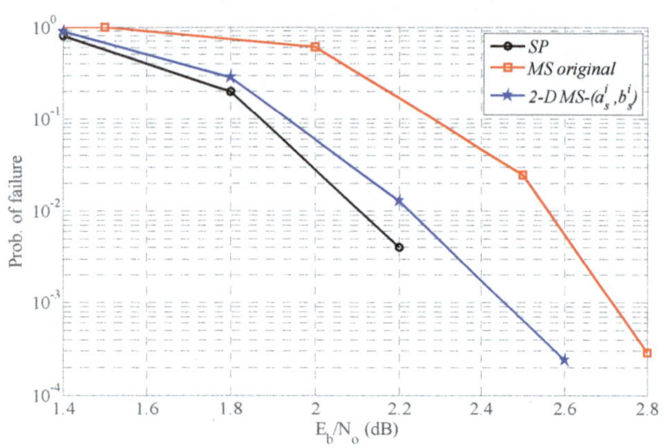

Figure 10. Probability of failure of SP, original MS and 2-D MS-$\left(a_s^i, b_s^i\right)$ algorithms for regular (1944, 1296), 2/3 rate LDPC.

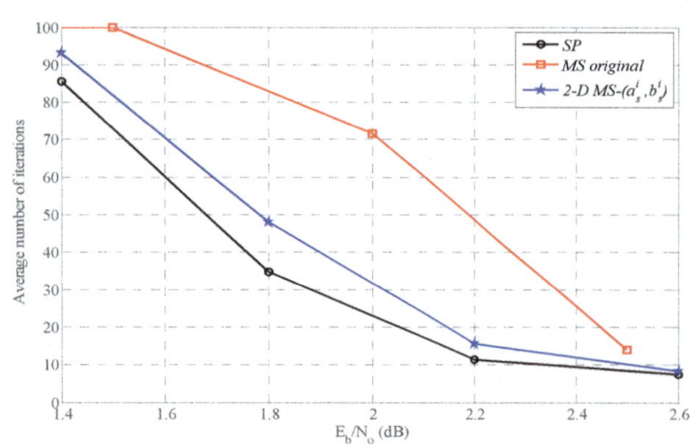

Figure 11. Average number of iteration of SP, original MS and 2-D MS-$\left(a_s^i, b_s^i\right)$ algorithms for regular (1944, 1296), 2/3 rate LDPC.

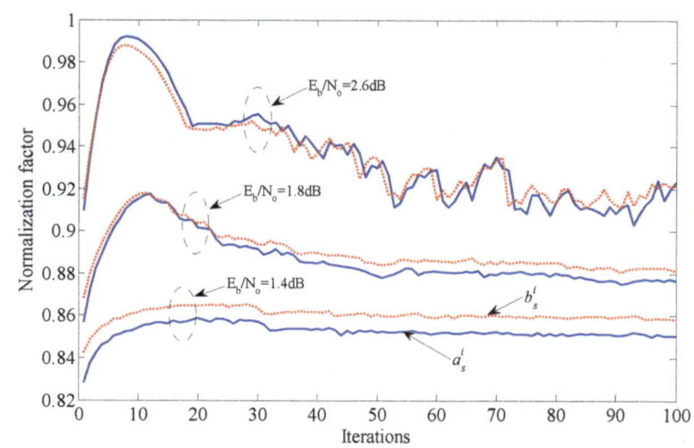

Figure 12. The correction factors a_s^i and b_s^i over 100 iterations with different $\frac{E_b}{N_o}$.

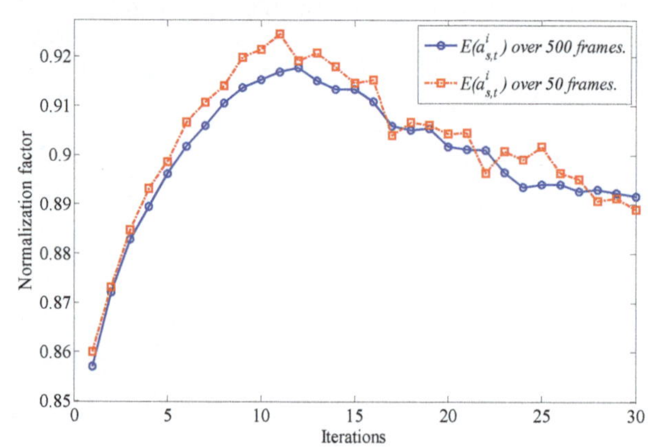

Figure 13. The average of correction factor $a_{s,t}^i$ over t = 50 and 500 transmitted frames of regular (1944, 1296), 2/3 LDPC at $\frac{E_b}{N_o} = 2.2$ dB.

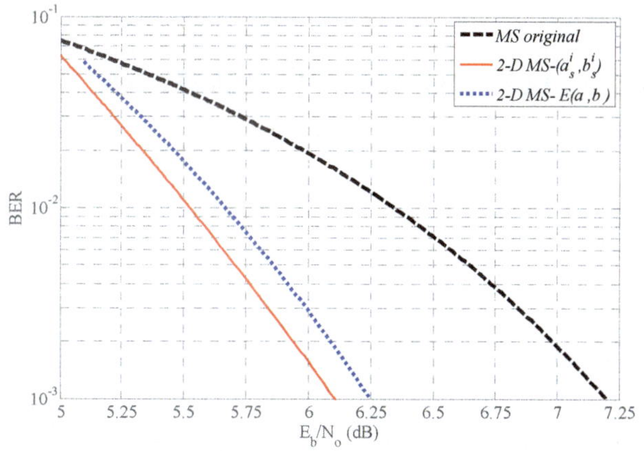

Figure 14. BER performance of original MS, 2-D MS-(a_s^i, b_s^i), and 2-D MS-*E* (*a*, *b*) decoding algorithms of the regular (1944, 1296) LDPC code.

Table 2. Correction factors (a_s, b_s) and their average values (a, b) for practical test of the regular (1944, 1296) LDPC code.

Correction factors	Estimated $\frac{E_b}{N_o}$ (dB)				a	b	$E(a, b)$
	4.9721	5.2444	5.7512	6.1188			
a_s	0.8421	0.8777	0.8959	0.9023	0.8795	0.8837	0.8816
b_s	0.8500	0.8807	0.8994	0.9045			

6. Conclusion

In this paper, a 2-D corrected MS decoding has been presented to improve the performance of standard MS decoding. Simulation and practical tests show that 2-D corrected MS provides a good performance versus complexity tradeoff for decoding regular LDPC codes. In comparison with the standard SP, the proposed method requires considerably less complexity while introducing small performance degradation, especially if a single correction factor for both check and bit nodes is considered. Further analysis and hardware can be done for irregular LDPC codes.

REFERENCES

[1] R. G. Gallager, "Low-Density Parity Check Codes," *IRE Transactions on Information Theory*, Vol. 8, No. 1, 1962, pp. 21-28.

[2] IEEE 802.11n, "Wireless LAN Medium Access Control and Physical Layer Specifications: Enhancement for Higher Throughput," IEEE P802.11n/D1.0, 2006.

[3] IEEE 802.16e, "Air Interface for Fixed and Mobile Broadband Wireless Access Systems," IEEE P802.16e/D12 Draft, 2005.

[4] European Telecommunications Standards Institude (ETSI), "Digital Video Broadcasting (DVB) Second Generation Framing Structure for Broadband Satellite Applications," EN 302 307 v1.1.1, 2005.

[5] N. Wiberg, "Codes and Decoding on General Graphs," Ph. D. Thesis, Linkoping University, Linkoping, 1996.

[6] M. P. C. Fossorier, M. Mihaljevic and H. Imai, "Reduced Complexity Iterative Decoding of Low-Density Parity Check Codes Based on Belief Propagation," *IEEE Transactions on Communications*, Vol. 47, No. 5, 1999, pp. 673-680.

[7] J. Chen and M. P. Fossorier, "Near Optimum Universal Belief Propagation Based Decoding of Low Density Parity Check Codes," *IEEE Transactions on Communications*, Vol. 50, No. 3, 2002, pp. 406-414.

[8] J. Zhang, M. Fossorier, D. Gu and J. Zhang, "Two-Dimensional Correction for Min-Sum Decoding of Irregular LDPC Codes," *IEEE Communications Letters*, Vol. 10, No. 3, 2006, pp. 180-182.

[9] V. Savin, "Self-Corrected Min-Sum Decoding of LDPC Codes," ISIT 2008. *IEEE International Symposium on information Theory*, 2008. *ISIT* 2008, Toronto, 6-11 July 2008, pp. 146-150.

[10] H. Wei, J. G. Huang and F. F. Wu, "A Modified Min-Sum Algorithm for Low-Density Parity-Check Codes," 2010 *IEEE International Conference on Wireless Communications, Networking and Information Security* (*WCNIS*), Beijing, 25-27 June 2010, pp. 449-451.

[11] X. F. Wu, Y. Song, L. Cui, M. Jiang and Ch. Zhao, "Adaptive-Normalized Min-Sum Algorithm," 2010 *2nd International Conference on Future Computer and Communication* (*ICFCC*), Wuhan, 21-24 May 2010, pp.V2-661-V2-663.

[12] A. A. Hamad, "Performance Enhancement of SOVA Based Decoder in SCCC and PCCC Schemes," *Scientific Research Magazine, Wireless Engineering and Technology*, Vol. 4, No. 1, 2013, pp. 40-45.

[13] T. J. Richardson and R. L. Urbanke, "Efficient Encoding of Low-Density Parity-Check Codes," *IEEE Transactions on Information Theory*, Vol. 47, No. 2, 2001, Article ID: 638456.

[14] D. J. C. MacKay, "Good Error-Correcting Codes Based on Very Sparse Matrices," *IEEE Transactions on Information Theory*, Vol. 45, No. 2, 1999, pp. 399-431.

[15] T. H. Shepertycki, "Telemetry Error Measurements Using Pseudo-Random Signals," *IEEE Transactions on Space Electronics and Telemetry*, Vol. 10, No. 3, 1964, pp. 111-115.

Performance Assessment for LTE-Advanced Networks with Uniform Fractional Guard Channel over Soft Frequency Reuse Scheme

Mahammad A. Safwat[1], Hesham M. El-Badawy[1], Ahmad Yehya[2], Hosni El-Motaafy[3]

[1]Network Planning Department, National Telecommunication Institute, Cairo, Egypt; [2]Department of Electrical Engineering, Al-azhar University, Cairo, Egypt; [3]Department of Electronics and Computer Engineering, HIT, Egypt.

ABSTRACT

Dropping probability of handoff calls and blocking probability of new calls are two important Quality of Service (QoS) measures for LTE-Advanced networks. Applying QoS for Cell edge users in soft frequency reuse scheme in LTE system is a challenge as they already suffer from limited resources. Assigning some resources for handover calls may enhance dropping probability but this is in price of degradation in the blocking probability for new calls in cell-edge. Uniform Fractional Guard Channel (UFGC) is a call admission policy that provides QoS without reserving resources for handover calls. In this paper, the performance of Soft Frequency Reuse (SFR) in presence of Uniform Fractional Guard Channel (UFGC) will be investigated using queuing analysis. The mathematical model and performance metrics will be deduced in this assessment. The impact of UFGC will be evaluated in edge and core part separately. Then the optimal value for the parameter of UFGC will be obtained to minimize the blocking probability of new calls with the constraint on the upper bound on the dropping probability of handoff calls.

Keywords: Soft Frequency Reuse; Queuing; Uniform Fractional Guard Channel

1. Introduction

In cellular mobile systems, spectrum available for communication is limited. Hence, efficient utilization of the scarce spectrum allocated for cellular communication is a major challenge in cellular system design [1]. This limitation means that the radio resources have to be reused as much as possible in order to support a numerous number of simultaneous calls that may arise in many typical mobile networks. By reusing frequency more and more times, a scarce spectrum can be used many times and an efficient utilization for the spectrum is achieved. Many frequency reuse schemes have been introduced in the literature [2-4]. A comparison of different frequency reuse schemes is introduced in [5] in terms of outage probability, network throughput, spectral efficiency, and average cell edge user Signal to Interference plus Noise Ratio (SINR). Soft frequency Reuse has been introduced as one of the most promising frequency reuse schemes and it is introduced to be used in LTE-Advanced networks [6]. The SFR scheme divides UEs within each cell into two groups, cell core UEs and cell edge UEs, depending on which type of bandwidth they have access to. Cell-edge users are confined to cell-edge Resource blocks (RBs) while cell-core users can access the cell-core RBs and can also access the cell-edge RBs but with less priority than cell-edge users.

Handover is a key element in wireless cellular networks in order to sustain the provided QoS to the users and to support users' mobility. There are two QoS parameters in these networks: new call blocking probability and handover call dropping probability. The probability of assigning no channel to handover call is defined as handover call dropping probability P_D. The probability of assigning no channel to new call is defined as new call blocking probability P_B. There is a trade-off between P_B and P_D. Call Admission Control (CAC) schemes are some strategies to keeping this parameters under desired level. From user point of view, blocking a new call is less annoying than dropping a handover one [7].

So Handover prioritization schemes result in a decrease of handover dropping probability and in an increase of new call blocking probability that, in turn, reduces the total admitted traffic. The concept of these strategies is to reserve a number of channels called guard channels (GC) exclusively for handovers [1]. More handover attempts can be admitted by buffering process [8].

When applying this strategy in soft frequency reuse, the blocking probability of new call at edge part will increase dramatically as they already suffer from resources availability. Providing suitable QoS for handover users while keeping blocking probability fairly acceptable, is a challenge especially, users in the edge part of the cell. In order to overcome this problem different schemes have been proposed: fractional guard channel (FGC), limited fractional guard channel (LFGC) and the uniform fractional guard channel (UFGC) schemes. General call admission FGC schemes have been studied in [9,10] and are used to improve the call blocking probability. This policy depends on acceptance of new call with a certain probability that depends on the current channel occupancy and acceptance of handover calls as long as channels are available. [11,12] explained LFGC and UFGC schemes, which are particular examples of FGC scheme. LFGC scheme controls communication service quality by effectively varying the average number of reserved channels by a fraction of one where as UFGC accepts new calls with an admission probability independent of channel occupancy.

In [13], an analytical approach for the performance analysis and optimization of FGC and UFGC in wireless cellular networks is developed for which different types of calls are prioritized based on a channel reservation scheme.

In [9], an algorithm for finding the optimal value of new call acceptance probability is deduced; this algorithm minimizes the blocking probability of new calls with the constraint on the upper bound on dropping probability of handover calls.

The current work assists the performance of SFR with UFGC in terms of blocking and dropping probability. In addition, using an algorithm in [12], [14] an optimal acceptance probability β^*_{opt} is obtained to provide minimum new call blocking probability in presence of constrained handover dropping probability P_h.

In [15], a good model for representing SFR using queuing analysis is introduced, this model is modified to suit UFGC, and a set of linear balance equations are deduced from the queuing model. A successive over relaxation (SOR) as an iterative algorithm is used to solve these balance equations to get steady state probabilities.

This paper is organized as follows: In section II, the system model for SFR with UFGC is presented. An iterative algorithm to get steady state probability is introduced in section III, and Numerical results and analysis are provided in section IV. Finally, conclusion is presented in section V.

2. Model Description

A homogeneous multi-cellular system is assumed that has the same traffic patterns. This allows considering only one cell for performance study and all other cells catch the interaction through handoff call arrival process.

A two dimension Markov chain is used to model SFR with UFGC considered. Horizontal axis stands for the number of RBs used by cell-core users and vertical axis represents the number of RBs used by cell-edge users.

2.1. Model Assumptions

In this paper the following assumption are considered:

- The basic resource element considered in this paper is the physical resource block (PRB) which spans both frequency and time dimensions.
- N is the number of available PRBs that can be used for transmission in each transmission time interval (TTI) in the cell. The maximum number of PRB that can be assigned to the edge-users and core-users is E and C respectively; the ratio of cell-edge PRBs to the total number of PRBs each cell is η, so $E = \eta N$ where $E + C = N$.
- Users are uniformly distributed in a cell. A new call follows a Passion process with the mean arrive rate λ. The effective distance between users to eNodeB in a cell is the dominant parameter to be considered. The target cell can be modeled by two queues with the mean arrival rates $\lambda_c = \xi_c \lambda$ and $\lambda_e = \xi_e \lambda$ respectively, where ξ_c represents the ratio of cell-core area to the whole cell area, while ξ_e represents the ratio of cell-edge area to the whole cell area. λ_h is call arrival rate for handover calls.
- The FG policy uses a vector $\beta = [\beta_0, .\beta_k, \cdots, \beta_{N-1}]$ to accept the new calls, where $0 \leq \beta_k \leq 1$ (for $k = 0, \cdots, N-1$). This policy accepts new calls with probability β_k when k channels are busy. The UFGC can be obtained from FG by setting $\beta_k = \beta^*$.
- The cell-edge PRB is available for both of cell-edge users (with acceptance probability β^*) and handover users and if there are none of them, it can be occupied by cell-core users.
- A cell-edge user may be blocked if all cell-edge PRBs in the cell is occupied by cell-edge users or handover users. A cell-core user may be blocked if there is no more cell-core PRBs or cell-edge PRBs in target cell.
- An ongoing handover call may be dropped if all cell-edge PRBs in the target cell is occupied by cell-edge users or handover users.
- System may force the cell-core call which has already connected to the networks to be terminated if the cell-

core call has occupied cell-edge PRBs and a new cell-edge user initialized a new call simultaneously or an ongoing handover call entered the cell.

- The service rate of a cell-core user, a cell-edge user and handover users are negative exponentially distributed with μ and for simplicity it is assumed to be equal for three users.

2.2. Queuing Model

The queuing model is used in order to tackle and investigate the contention problem. This problem is a raised due to the limitation of the available radio resources. **Figure 1** explains the state diagram in SFR with UFGC. In this figure, the cell is characterized as two dimensional Markov chain. Horizontal axis stands for the number of PRBs used by cell-core users and is denoted by i while vertical axis stands for the number of PRBs used by cell-edge users or handover users and is denoted by j. Then, a two dimensional state space Γ can be defined as:

$$\Gamma=\{(i,j)\mid 0\le i\le N, 0\le j\le E, i+j\le N\} \quad (1)$$

Let $\Pi(i,j)$ be the steady state probability for a valid state $(i,j)\in\Gamma$.

The steady state probabilities should satisfy the normalization constraint.

$$\sum_{(i,j)\in\Gamma}\prod(i,j)=1 \quad (2)$$

In the following, based on the state diagram shown, the set of global balance equations are introduced:

For the state $(i,j)=(0,0)$

$$\left(\lambda_c+\beta^*\lambda_e+\lambda_h\right)\pi(0,0)=\mu\pi(1,0)+\mu\pi(0,1) \quad (3)$$

For states $1\le i<N; j=0$

$$\begin{aligned}&\left(\lambda_c+\beta^*\lambda_e+\lambda_h+i\mu\right)\pi(i,0)\\&=\lambda_c\pi(i-1,0)+(i+1)\mu\pi(i+1,0)+\mu\pi(i,1)\end{aligned} \quad (4)$$

For states $1\le j<E;\ i=0$

$$\begin{aligned}&\left(\lambda_c+\beta^*\lambda_e+\lambda_h+j\mu\right)\pi(0,j)\\&=\left(\beta^*\lambda_e+\lambda_h\right)\pi(0,j-1)+(j+1)\mu\pi(0,j+1)\\&\quad+\mu\pi(1,j)\end{aligned} \quad (5)$$

For states $1\le i<N; 1\le j<E$

$$\begin{aligned}&\left(\lambda_c+\beta^*\lambda_e+\lambda_h+i\mu+j\mu\right)\pi(i,j)\\&=\lambda_c\pi(i-1,j)+(i+1)\mu\pi(i+1,j)\\&\quad+\left(\beta^*\lambda_e+\lambda_h\right)\pi(i,j-1)+(j+1)\mu\pi(i,j+1)\end{aligned} \quad (6)$$

For the state $(i,j)=(0,E)$;

$$\begin{aligned}&\left(\lambda_c+E\mu\right)\pi(0,E)=\left(\beta^*\lambda_e+\lambda_h\right)\pi(0,E-1)\\&\quad+\mu\,\pi(1,E)\end{aligned} \quad (7)$$

For the states $(i,j)=(N,0)$;

$$\left(\beta^*\lambda_e+\lambda_h+N\mu\right)\pi(N,0)=\lambda_c\pi(N-1,0) \quad (8)$$

For states $1\le i<C; j=E$

$$\begin{aligned}&\left(\lambda_c+i\mu+E\mu\right)\pi(i,E)\\&=\lambda_c\pi(i-1,E)+(i+1)\mu\pi(i+1,E)\\&\quad+\left(\beta^*\lambda_e+\lambda_h\right)\pi(i,E-1)\end{aligned} \quad (9)$$

For states $C<i<N,\ i+j=N$;

$$\begin{aligned}&\left(\beta^*\lambda_e+\lambda_h+i\mu+j\mu\right)\pi(i,j)\\&=\lambda_c\pi(i-1,j)+\left(\beta^*\lambda_e+\lambda_h\right)\pi(i,j-1)\\&\quad+\left(\beta^*\lambda_e+\lambda_h\right)\pi(i+1,j-1)\end{aligned} \quad (10)$$

For the state $(i,j)=(C,E)$;

$$\begin{aligned}&\left(C\mu+E\mu\right)\pi(C,E)\\&=\lambda_c\pi(C-1,E)+\left(\beta^*\lambda_e+\lambda_h\right)\pi(C,E-1)\\&\quad+\left(\beta^*\lambda_e+\lambda_h\right)\pi(C+1,E-1)\end{aligned} \quad (11)$$

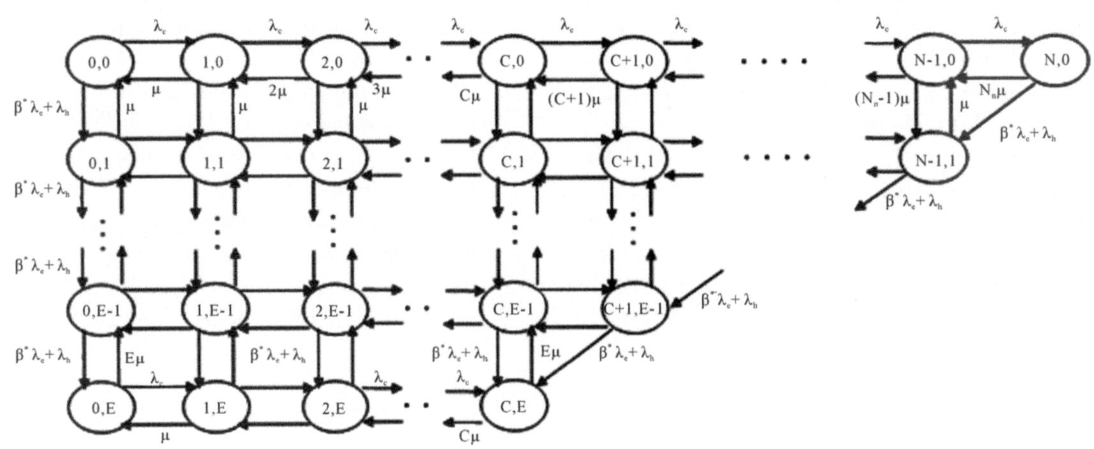

Figure 1. The state diagram of SFR with UFGC.

3. Assessment Criteria

In this section, the proposed queuing model will solved using successive over relaxation method to get steady state probability for each state. Performance metrics can be obtained from this steady state probability.

3.1. Successive over Relaxation

The successive over relaxation is an iterative method [15], [16] that used to solve the set of linear equation; the successive over-relaxation (SOR) is a variant of the Gauss-Seidel method for solving a linear system of equations, resulting in faster convergence.

In this method, a new set of equations, called SOR equations, are deduced from balance equation, the left hand side of these equations is a new value of steady state probability which is obtained iteratively using previous value for steady state probability on the right hand side. The speed of convergence is determined by relaxation factor ω, the choice of relaxation factor is not necessarily easy, and depends upon the properties of the coefficient matrix. For symmetric, positive-definite matrices it can be proven that $0 < \omega < 2$ will lead to convergence, but we are generally interested in faster convergence rather than just convergence.

The set of SOR equation is obtained as follow:

For the state $(i,j)=(\mathbf{0},\mathbf{0})$

$$\pi^{(k)}(0,0) = (1-w)\pi^{(k-1)}(0,0)+w\left(\lambda_c+\beta^*\lambda_e+\lambda_h\right)^{-1}\left(\mu\pi^{(k-1)}(1,0)+\mu\pi^{(k-1)}(0,1)\right) \tag{12}$$

For states $1\le i<N;\ j=0$

$$\pi^{(k)}(i,0) = (1-w)\pi^{(k-1)}(i,0)+w\left(\lambda_c+\beta^*\lambda_e+\lambda_h+i\mu\right)^{-1}\left(\lambda_c\pi^{(k)}(i-1,0)+(i+1)\mu\pi^{(k-1)}(i+1,0)+\mu\pi^{(k-1)}(i,1)\right) \tag{13}$$

For states $1\le j<E;\ i=0$

$$\pi^{(k)}(0,j) = (1-w)\pi^{(k-1)}(0,j)+w\left(\beta^*\lambda_e+\lambda_h+\lambda_c+j\mu\right)^{-1}\left(\left(\beta^*\lambda_e+\lambda_h\right)\pi^{(k)}(0,j-1)+(j+1)\mu\pi^{(k-1)}(0,j+1)+\mu\pi^{(k-1)}(1,j)\right) \tag{14}$$

For states $1\le i<N;1\le j<E$ (see Equation (15) below)

For the state $(i,j)=(0,E)$;

$$\pi^{(k)}(0,E)=(1-w)\pi^{(k-1)}(0,E)+w\left(\lambda_c+E\mu\right)^{-1}\left(\left(\beta^*\lambda_e+\lambda_h\right)\pi^{(k)}(0,E-1)+\mu\pi^{(k-1)}(1,E)\right) \tag{16}$$

For the state $(i,j)=(N,0)$;

$$\pi^{(k)}(N,0)=(1-w)\pi^{(k-1)}(N,0)+w\left(\beta^*\lambda_e+\lambda_h+N\mu\right)^{-1}\left(\lambda_c\pi^{(k)}(N-1,0)\right) \tag{17}$$

For states $1\le i<C;j=E$

$$\pi^{(k)}(i,E) = (1-w)\pi^{(k-1)}(i,E)+w\left(\lambda_c+i\mu+E\mu\right)^{-1}\left(\lambda_c\pi^{(k)}(i-1,E)+(i+1)\mu\pi^{(k-1)}(i+1,E)+\left(\lambda_h+\beta^*\lambda_e\right)\pi^{(k)}(i,E-1)\right) \tag{18}$$

For states $C<i<N;i+j=N$

$$\pi^{(k)}(i,j) = (1-w)\pi^{(k-1)}(i,j)+w\left(\beta^*\lambda_e+\lambda_h+i\mu+j\mu\right)^{-1}\left(\left(\beta^*\lambda_e+\lambda_h\right)\pi^{(k)}(i,j-1)+\left(\beta^*\lambda_e+\lambda_h\right)\pi^{(k)}(i+1,j-1)+\lambda_c\pi^{(k)}(i-1,j)\right) \tag{19}$$

For the state $(i,j)=(C,E)$;

$$\pi^{(k)}(C,E) = (1-w)\pi^{(k-1)}(C,E)+w(C\mu+E\mu)^{-1}\left(\lambda_c\pi^{(k)}(C-1,E)+\left(\beta^*\lambda_e+\lambda_h\right)\pi^{(k)}(C,E-1)+\left(\beta^*\lambda_e+\lambda_h\right)\pi^{(k)}(C+1,E-1)\right) \tag{20}$$

Figure 2 explains the SOR algorithm which comprises three main parts. First, all equations are started using initial valid state probability distribution and convergence criteria and relaxation factor. Second, we iterate SOR equations aforementioned until the steady probability distribution satisfies the convergence condition or iterations exceed 1000. This equation should solve in sequence as some of it values depends on the values from the former equation. Finally, performance metrics can be obtained if the algorithm could acquire the steady state probability corresponding with the convergence condition [15].

$$\pi^{(k)}(i,j)=(1-w)\pi^{(k-1)}(i,j)+w\left(\lambda_c+\beta^*\lambda_e+\lambda_h+i\mu+j\mu\right)^{-1}\left(\lambda_c\pi^{(k)}(i-1,j)+(i+1)\mu\pi^{(k-1)}(i+1,j)+\left(\lambda_e+\beta^*\lambda_h\right)\pi^{(k)}(i,j-1)+(j+1)\mu\pi^{(k-1)}(i,j+1)\right) \tag{15}$$

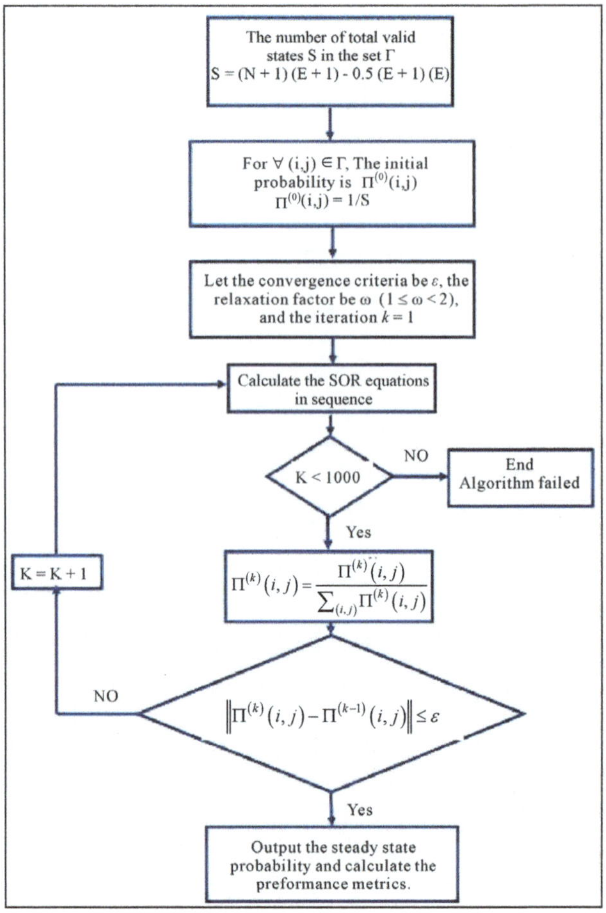

Figure 2. Flow chart for SFR algorithm.

3.2. Performance Metrics

The performance of the current model will be evaluated by determine of blocking probability (P_B) as well as dropping probability (P_D). Cell blocking probability is the probability that a new arriving cell-core user and a cell-edge user are blocked. Let ψ_{bc} and ψ_{be} be the subsets of states where a new arriving cell-core user and a cell-edge user are blocked respectively.

$$\Psi_{bc} = \left\{\prod(i,j) \in \Psi_{bc}; C \le iN, j = N - i\right\}$$

$$\Psi_{be} = \left\{\prod(i,j) \in \Psi_{be1}; 1 \le i < E, 1 \le j \le N\right\} \cup \left\{\prod(i,j) \in \Psi_{be2}; i = E, 1 \le j \le C\right\} \quad (21)$$

Then the blocking probability is calculated as [12]:

$$P_B = \sum_{(i,j)\in\Psi_{bc}} \xi_c \prod(i,j) + \sum_{(i,j)\in\Psi_{be1}} \left(1-\beta^*\right) \xi_e \prod(i,j) + \sum_{(i,j)\in\Psi_{be2}} \xi_e \prod(i,j) \quad (22)$$

Finally let ψ_d be the subsets of states where the system forces to terminate the ongoing handover call.

$$\Psi_d = \left\{\prod(i,j) \in \Psi_d; i = E, 1 \le j \le C\right\} \quad (23)$$

Then the cell dropping probability is calculated as:

$$P_D = \sum_{(i,j)\in\Psi_d} \xi_e \prod(i,j) \quad (24)$$

3.3. Optimization of New Call Acceptance Probability β^*_{opt}

The objective of this part is to obtain the optimal value of *new call acceptance probability* β^*_{opt}. This will minimize the new call blocking probability subjected to a hard constraint on handover dropping probability or minimize P_B such that $P_D \le P_h$.

The presented algorithm in [12] will be used to find β^*_{opt}. This algorithm is given in **Figure 3** and can be described as follows. At the beginning, the algorithm considers two cases: the first case when all channels are exclusively used for handover calls. If the exclusive use of channels for handover calls does not satisfy the level of QoS, then the number of allocated channels to the cell is not sufficient and the algorithm terminates; the second case when all channels are shared between handover and new calls. If the complete sharing satisfies the level of QoS, then the algorithm considers that β^*_{opt} =1 and the algorithm terminates; otherwise the algorithm searches for the optimal value of β^*_{opt}. The search method used in this algorithm is binary search.

4. Numerical Analysis and Results

In this section, the performance of SFR scheme in presence of UFGC is analyzed and evaluated using queuing model. The aforesaid performance metric of blocking

Let $\beta^*_{lower} = 0$ & $\beta^*_{lower} = 1$
If $P_D\left(\beta^*_{upper}\right) \le P_h$ then
$\beta^*_{opt} = \beta^*_{upper}$
else if $P_D\left(\beta^*_{lower}\right) \ge P_h$ then
$\beta^*_{opt} = \beta^*_{lower}$
end if
while $\left(\beta^*_{upper} - \beta^*_{lower}\right) \ge 0.0001$ & $\beta^*_{opt} \notin \{0,1\}$ do
Let $\beta^*_{opt} = \left(\beta^*_{upper} + \beta^*_{lower}\right)/2$
If $\left(P_D\left(\beta^*_{opt}\right)\right) > P_h$ then
$\beta^*_{upper} = \beta^*_{opt}$
else
$\beta^*_{lower} = \beta^*_{opt}$
end if
end while
return β^*_{opt}

Figure 3. Algorithm for obtaining optimal value of new call acceptance probability β^*_{opt}.

probability P_B and dropping probability P_D is used for evaluation. The effect of new call acceptance probability β^* and handover arrival rate λ_h in the blocking and dropping probability is studied. For more depicted analysis, the blocking probability is studied in cell-edge and cell-core separately.

The queuing model parameters for the presented results are as follow: the available PRB in the cell (N) is 48; the mean service period (μ) is 90 seconds. The SOR parameters are $\omega = 1.05$, $\varepsilon = 10^{-5}$ and $k = 1000$.

4.1. The Blocking Probability P_B

Figures 3 to **6** illustrate the effect of handover arrival rate λ_h and new call acceptance probability β^* on blocking probability for both of cell-edge and cell-core part.

The SFR parameters setting for the presented results are: $\xi = 1/2$ and $\eta = 1/3$. These parameters are in consistence with Huawei proposal [6]. For practical consideration and system sustainability, the values of handover arrival rate λ_h are chosen to be less than the arrival rate of new call for cell-edge user λ_e.

The new call acceptance probability β^* is 0.5 in the obtained results in **Figures 3** & **4** and the handover arrival rate λ_h is 0.25 λ_e in the obtained results in **Figures 5** & **6**.

The effect of handover arrival rate λ_h on blocking probability of cell-edge and cell-core users is introduced in **Figures 4** & **5** respectively.

As illustrated in **Figures 4** & **5**, the increasing of λ_h will leads to have more blocking for new calls. This may be happened as a result of the increasing of radio resources competition by the handover traffic.

These results may be used as an evidence for the stability and sustainability of the proposed model.

The system behavior in **Figure 4** is more degraded than that in **Figure 5**. This may be explained due to the shortage of the edge to resources that are allocated for new and handover call. The effect on the core is less than that in the edge due to there are many opportunities for the allocation of new calls.

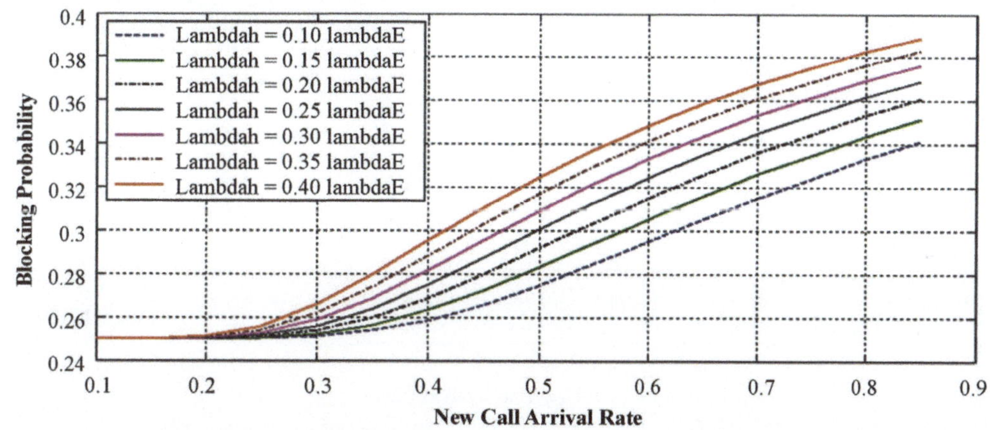

Figure 4. Blocking probability for cell-edge user with different λ_h.

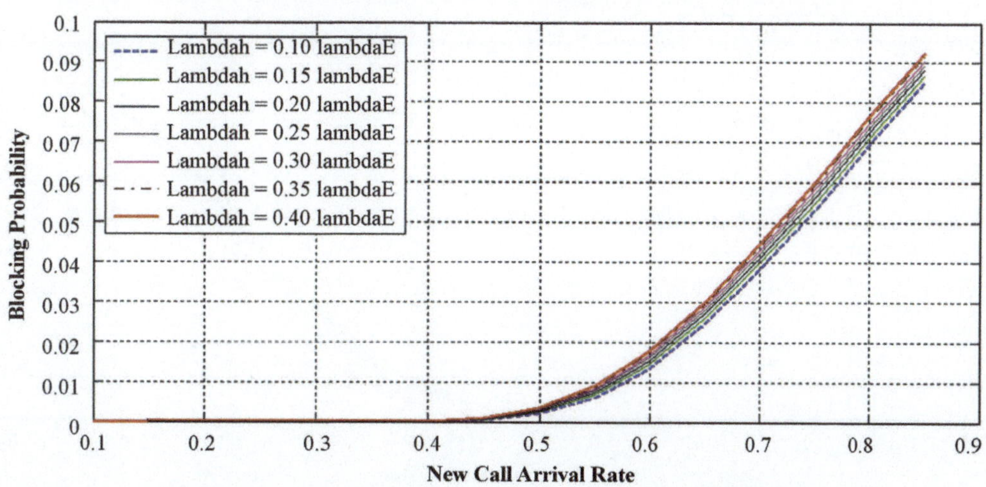

Figure 5. Blocking probability for cell-core user with different λ_h.

Figure 6 explains the noticeable effect of new call acceptance probability β^* in blocking probability for cell-edge users and it is compared with the system without UFGC.

It can be noticed that the increase of blocking probability when using UFGC as a results of reducing the resources permitted to new call request.

It is clear from **Figure 6** that the blocking probability decreases while β^* is increased as more resources are admitted for new calls. In addition, It can be noticed that the impact of β^* on blocking probability differs with regard to traffic load. This can be interpreted as follow:

In heavy traffic region there are more and more new call requests, so more resources are occupied, hence blocking probability for more new calls will increase. So the blocking probability in this region is due to increaseing of arrival rate of new call rather than the effect of acceptance probability. On the other hand, when new call arrival rate is becoming light, there is a little call requests and a lot of available resources. Consequently, the blocking (if happened) will be occurred as a result of the lack of the available resources admitted for new calls.

So the dominant factor that affect on blocking is traffic load. In heavy traffic region blocking will be mainly influenced by λ. whereas, in the light traffic region block will be controlled mainly by β^*.

Figure 7 explains the impact of β^* on blocking probability for cell-core users. This impact is a result of accessing of the core users the edge resources when there are no edge or handover users (some sort of resource borrowing from edge to serving the core customers). This impact is limited in comparison with cell-edge case; this is due to blocking of new call in cell-core part is occurred as a result of full usage of the resources. Consequently, the effect of acceptance probability is limited. On the contrary of **Figure 6**, the blocking probability in **Figure**

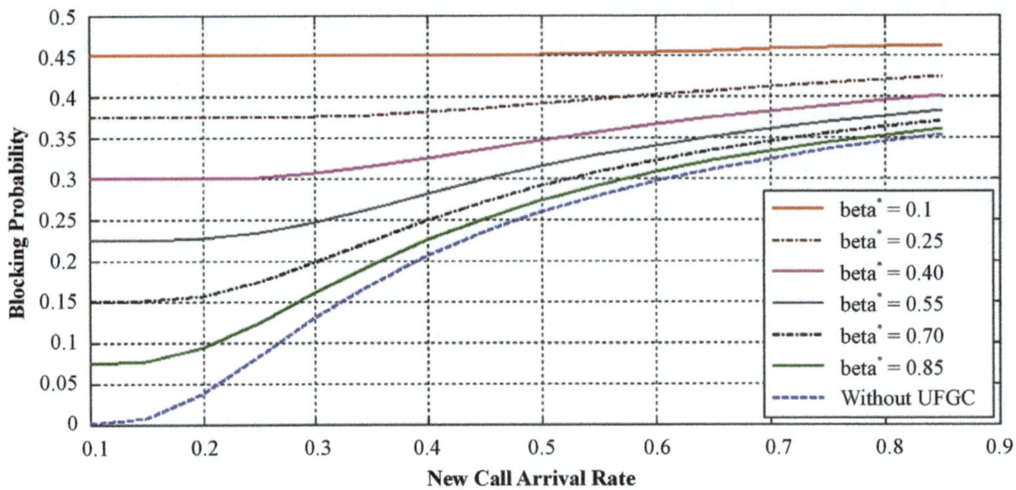

Figure 6. Blocking probability for cell-edge user with different β^*.

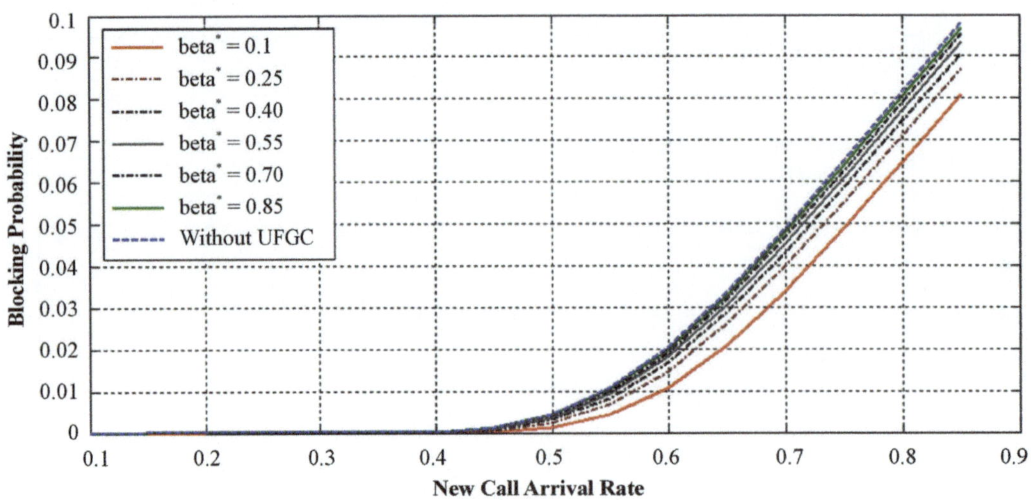

Figure 7. Blocking probability for cell-core user with different β^*.

7 increases as β^* is increasing. This is because of new call arrival rate is always greater than handover arrival rate, consequently with large values of β^*, the system is permitted to admit new call requests more than handover requests. This leads to more overall call requests and more resources occupation and probability of blocking increases. On the other hand **Figure 6** shows the improvement of blocking probability for cell-core users when using UFGC. This occurred as a result of some resources is admitted for handover call which have lower rate. So the overall service request decreases and provides cell-core users more available resources. But, without UFGC whole resources is admitted for new call of edge users, this leads to higher overall service requests. So the system starves to serve cell-core users.

4.2. The Dropping Probability P_D

In the following subsection, the effect of handover arrival rate λ_h as well as new call acceptance probability β^* on dropping probability P_D are investigated. These effects are presented in **Figures 8** and **9** respectively.

The new call acceptance probability β^* is 0.5 in the obtained results in **Figure 8** and the handover arrival rate λ_h is 0.25 λ_e in the obtained results in **Figure 9**.

Figure 8 explains the effect of handover arrival rate on dropping probability. It is shown that by having more handover request, more resources are occupied. Then, the probability for serving handover calls is reducing.

The impact of new call acceptance probability β^* is illustrated in **Figure 9**. Large values of acceptance probability means more new call admitted than handover call.

This leads to increasing in the overall arrival rate, so the system will be fully occupied faster. Consequently, handover call will suffer from more dropping due to the limitation of the available resources.

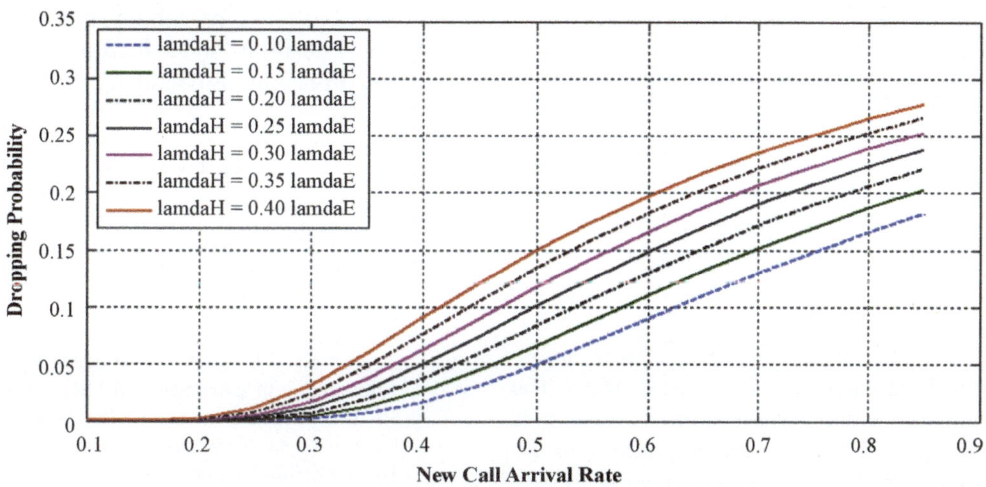

Figure 8. Dropping probability with different λ_h.

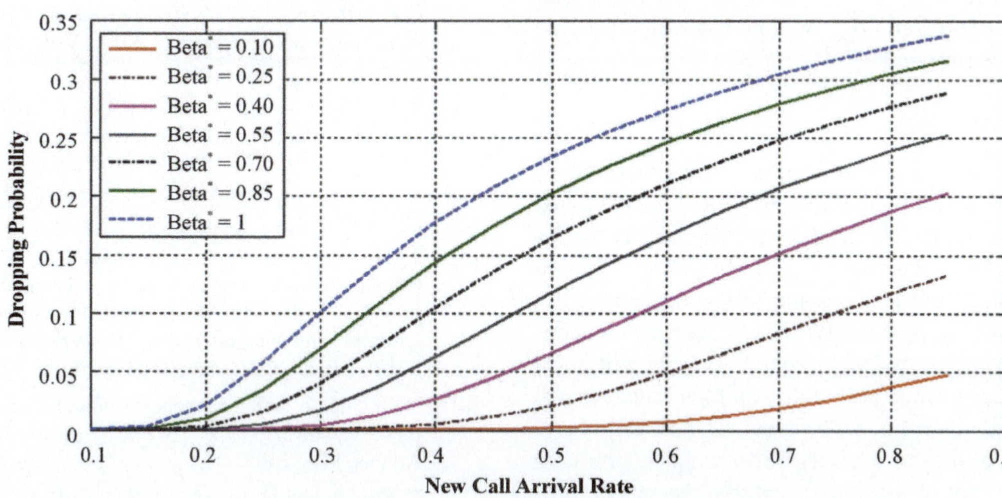

Figure 9. Dropping probability with different β^*.

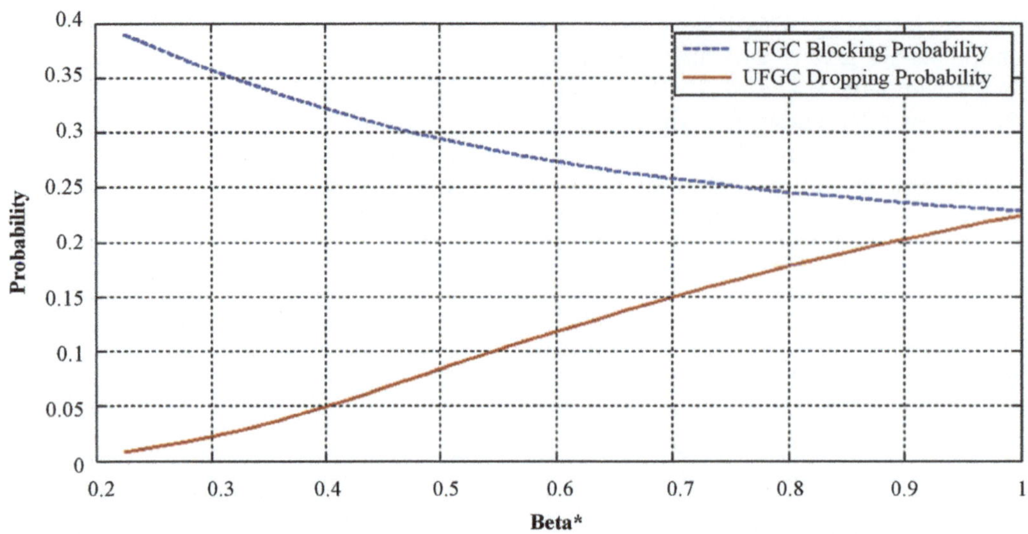

Figure 10. Blocking and dropping probability as a function of β^{*}_{opt}.

4.3. Optimization of New Call Acceptance Probability β^{*}_{opt}

In order to find out the most optimized value of acceptance probability β^{*}_{opt} that provides hard constraint of dropping probability P_h, we will follow **Figure 3** algorithm.

Figure 10 is illustrating the optimized values of β^{*}_{opt} that provide a hard constraint $P_D \le P_h$ (for $0.008 \le P_h \le 0.2$). Following algorithm of **Figure 3**, β^{*}_{opt} may be determined satisfying the condition $P_D \le P_h$. At each point of β^{*}_{opt}, the equivalent value of dropping probability P_D and blocking probability P_B are determined. The new call arrival rate λ and handover arrival rate λ_h is taken to be 0.5 and 0.05 respectively. So, the shown results in **Figure 10** may be used as design curve to give the minimum blocking probability for a given P_h. As an example, the required P_h to be taken as 0.05 (as a QoS measure), the optimized β^{*}_{opt} will be 0.4 which provide blocking probability of 0.33.

5. Conclusion

In this paper, the effect of uniform fractional guard channels in resources availability of SFR is investigated using queuing model. A steady state probability is deduced using successive over relaxation method. Through numerical results and analysis the following can be deduced: the blocking probability of edge users can be controlled by varying new call acceptance probability. On the hand the acceptance probability and handover arrival rate has a tremendous impact on blocking probability and dropping probability of edge users. In addition, the core users affected, but to a lesser extent, by acceptance probability and handover arrival rate. Finally, the optimal value of acceptance probability that provides minimum blocking probability under hard constraint on handover dropping probability is obtained.

REFERENCES

[1] V. Goswami and P. K. Swain, "Analysis of Finite Population Limited Fractional Guard Channel Call Admission Scheme in Cellular Networks," *Procedia Engineering*, Vol. 30, No. 2012, pp. 759-766.

[2] X. Zheng and B. Walke, "Frequency Reuse Techniques for Attaining Both Coverage and High Spectral Efficiency in OFDMA Cellular Systems," 2010 *IEEE Wireless Communications and Networking Conference* (*WCNC*), Sydney, 18-21 April 2010, pp. 1-6.

[3] C. Lei and D. Yuan, "Generalizing and Optimizing Fractional Frequency Reuse in Broadband Cellular Radio Access Networks," *EURASIP Journal on Wireless Communications and Networking* 2012, Vol. 2012, 2012, pp. 1-15.

[4] M. L. Qian, *et al.*, "Inter-Cell Interference Coordination through Adaptive Soft Frequency Reuse in LTE Networks," 2010 *IEEE Wireless Communications and Networking Conference* (*WCNC*), Shanghai, 1-4 April 2012, pp. 1618-1623.

[5] T. Novlan, J. G. Andrew, S. Illsoo, R. K. Ganti and A. Ghosh, "Comparison of Fractional Frequency Reuse Approaches in the OFDMA Cellular Downlink," 2010 *IEEE Global Telecommunications Conference* (*GLOBECOM* 2010), Miami, 6-10 December 2010, pp. 1-5.

[6] "R1-050507, Soft Frequency Reuse Scheme for UTRAN LTE," *Huawei 3GPP TSG RAN WG1 Meeting No.*41, Athens, May 2005.

[7] Z. Firouzi and H. Beigy, "A New Call Admission Control Scheme Based on New Call Bounding and Thinning II

Schemes in Cellular Mobile Networks," *IEEE International Conference on Electro/Information Technology*, 2009, Windsor, 7-9 June 2009, pp. 40-45.

[8] M. Jain, G. C. Sharma and R. Mittal, "Performance Analysis of Prioritized Call Admission Control Schemes for Integrated Traffic in Wireless Network," *International Journal of Electronics Communication and Computer Engineering*, Vol. 4, No. 1, 2013, pp. 283-291.

[9] R. Ramachandran, D. Towsley and R. Nagarajan, "On Optimal Call Admission Control in Cellular Networks," *Wireless Networks*, Vol. 3, No. 1, 1997, pp. 29-41.

[10] J. Vazquez-Avila, F. A. C. Cruz-Perez and L. Ortigoza-Guerrero, "Performance Analysis of Fractional Guard Channel Policies in Mobile Cellular Networks," *IEEE Transactions on Wireless Communications*, Vol. 5, No. 2, 2006, pp. 301-305.

[11] H. Beigy and M. R. Meybodi, "Uniform Fractional Guard Channel Policy," *Proceedings of the 6th SCI'*2002, Vol. 15, Orlando, July 2002.

[12] H, Beigy and M. R. Meybodi, "A New Fractional Channel Policy," *Journal of High Speed Networks*, Vol. 13, No. 1, 2004, pp. 25-36.

[13] H. Kim, *et al.*, "An Analytical Approach to the Analysis of Guard-Channel-Based Call Admission Control in Wireless Cellular Networks," *Journal of Applied Mathematics*, Vol. 2012, 2012, Article ID: 676582.

[14] H. Beigy and M. R. Meybodi, "Adaptive Uniform Fractional Channel Algorithms," *Iranian Journal of Electrical and Computer Engineering*, Vol. 3, No. 1, 2004, pp. 47-53.

[15] S. Li, X. Wen, Z. Liu, Z. Wei and Y. Sun, "Queue Analysis of Soft Frequency Reuse Scheme in LTE-Advanced," *2nd International Conference on Computer Modeling and Simulation*, Vol. 1, Sanya, 22-24 January 2010, pp. 248-252.

[16] Y. Zhang, Y. Xiao and H. H. Chen, "Queueing Analysis for OFDM Subcarrier Allocation in Broadband Wireless Multiservice Networks," *IEEE Transactions on Wireless Communication*, Vol.7, No. 10, 2008, pp. 3951-3961.

Suppression of Log-Normal Distributed Weather Clutter Observed by an S-Band Radar

Shuji Sayama[1], Seishiro Ishii[2]

[1]Department of Communications Engineering, National Defense Academy, Yokosuka, Japan; [2]Department of Information Networks, Sendai National College of Technology, Sendai, Japan.

ABSTRACT

We have observed weather clutter containing targets (ships) using an S-band radar with a frequency 3.05 GHz, a beam-width 1.8°, and a pulsewidth 0.5 μs. To investigate the weather clutter amplitude statistics, we introduce the Akaike Information Criterion (AIC). We have found that the weather clutter amplitudes obey the log-normal, Weibull, and log-Weibull distributions with the shape parameters of 0.308 to 0.470, 4.42 to 4.51, and 15.91 to 16.44, respectively, for small data within the beam width of an antenna. We have proposed the log-normal/CFAR circuit modified a Cell-Averaging (CA) LOG/CFAR circuit. It is found that weather clutter is suppressed with improvement of 51.58 dB by log-normal/CFAR. As a result, we have showed that weather clutter observed by S-band radar does not obey the Rayleigh distribution and our log-normal/CFAR circuit has an effect on suppression of clutter and detection of target, while conventional LOG/CFAR circuit does not. In addition, if our circuit can be realized, we will have an advantage economically.

Keywords: CFAR; S-Band Radar; Weather Clutter; Log-Normal Distribution

1. Introduction

It is important for a marine radar to detect targets of ships. The radar displays azimuth and range to targets by polar coordinates on the round screen called PPI (Plan Position Indicator). But there is a noise called clutter besides targets in return signal. Such noise makes detection of targets difficult. To detect targets of ships in such clutter, anticlutter techniques have been improved [1].

Recently, many anticlutter techniques have been developed by using polarization, phase and so on. Because most of marine radars use only amplitude, such anticlutter techniques cannot be applied to these radars. But anticlutter technique by using amplitude has been hardly researched. Therefore, it is important to research anticlutter technique by using amplitude.

For example, a Constant False Alarm Rate (CFAR) circuit suppresses a false alarm probability low and constantly by using statistical characteristics of clutter amplitude. Here the false alarm probability is the probability that a clutter signal above a threshold level is misjudged as a target signal [2].

One of the CFAR circuit is the LOG/CFAR circuit making use of a logarithmic amplifier. When Rayleigh distributed clutter is passed through the LOG/CFAR circuit, this circuit keeps the variance of clutter output and the false alarm probability constant, that is, CFAR is obtained. But it has been found that clutter amplitude obeys non-Rayleigh distribution with advance of a high-resolution radar [3].

For example, we observed ground clutter which is the reflected waves from town and hill using an S-band radar having a frequency 3.05 GHz, a beam width 1.8°, and a pulse width 1.0 μs. It was discovered that the ground clutter amplitudes obey the compound log-normal and compound log-Weibull distributions with the shape parameters of $\sigma_1 = 0.217$ to 0.429, $\sigma_2 = 0.028$ to 0.410, and $c_1 = 9.36$ to 11.20, $c_2 = 17.00$ to 23.81, respectively, for small data within the beam width of an antenna [4].

Sea clutter which is the reflected waves from sea waves was measured using an X-band radar with a frequency 9.41 GHz, a beam width 1.0°, and a pulse width $0.5\ \mu s$. We found that the sea clutter amplitudes obey the log-normal, Weibull, log-Weibull, and generalized gamma distributions with the shape parameters of 0.467,

2.86 to 3.39, 9.37 to 10.76 and 2.52 to 4.18, respectively, for small data within the beam width of an antenna [5].

We observed weather clutter which is the reflected waves from rain and fog using an L-band long-range air-route surveillance radar (ARSR) having a frequency 1.3 GHz, a beamwidth 1.2°, and a pulsewidth $3.0\ \mu s$. It was discovered that the weather clutter amplitudes obey the Weibull distribution with the shape parameters of 2.07 for entire data and the Weibull, log-Weibull, and K-distributions with the shape parameters of 1.73 to 2.43, 10.60, and 5.13 to 50.93, respectively, for small data within the beam width of an antenna [6].

Moreover, sea ice clutter was measured using a millimeter wave radar with a frequency 34.86 GHz, a beamwidth 0.25°, and a pulsewidth 30 ns. We found that the sea ice clutter amplitudes obey the log-Weibull distribution with the shape parameters of 2.97 for entire data and 2.69 to 3.15 for small data within the beam width of an antenna [7].

The Rayleigh distribution is identical to the Weibull distribution for $c = 2$ [8] and indicates the K-distribution in the region $\nu = \infty$ [9-12], so it is found that various clutter observed by various radars obey non-Rayleigh distribution.

If non-Rayleigh distributed clutter is passed through the LOG/CFAR circuit, then the false alarm probability is not kept constant by a wrong threshold level. And then targets have been embedded in clutter which cannot be suppressed or have been erased on a PPI screen. Therefore, it is necessary to devise a new CFAR circuit for suppression of non-Rayleigh distributed clutter.

In the following, we have observed weather clutter data containing targets of ships using an S-band radar with a frequency 3.05 GHz, a beamwidth 1.8°, and a pulsewidth 0.5 μs. We investigate the weather clutter amplitudes by using the Akaike Information Criterion (AIC) [13] and four probability distribution models. It is found that weather clutter amplitudes obey non-Rayleigh distribution. Therefore, we propose a new log-normal CFAR circuit modified a Cell-Averaging (CA) LOG/CFAR circuit. We apply this circuit and conventional LOG/CFAR circuit to the observed data, it is found that this circuit has an effect on suppression of clutter and detection of target, while conventional LOG/CFAR circuit does not. In addition, if our circuit can be realized, we will have an advantage economically.

The paper is organized as follows: In Section 2, we indicate radar system and describe weather clutter observed by this system. Explanations of four probability distribution models and AIC and result of distribution estimation are given in Section 3. We propose the log-normal/CFAR circuit modified a CA LOG/CFAR circuit and apply this circuit to weather clutter in Section 4.

2. Observations of Weather Clutter by an S-Band Radar

As mentioned in Section 1, various clutter make detection of targets difficult in a radar. The most considerable clutter is weather clutter in an S-band marine radar. Because X-band radars installed together in many ships cannot be used in rain. Therefore, we observed weather clutter data containing targets of ships using an S-band radar. Here, we explain the radar used for observation and the radar system used for record of data. And we illustrate the observed data.

2.1. Radar System

S-band radar used for observation is installed at the National Defense Academy in Yokosuka-City. The S-band radar characteristics are given in **Table 1**.

The radar system block diagram is shown in **Figure 1**. The antenna rotation signal (AZP) and the head marker (AZR) were always sent from the radar to the azimuth control unit. The sampling start signal was sent from the azimuth control unit to the personal computer (IBM PC) through the PS/2 terminal or the USB terminal. Then, the video signal from the radar was digitized in 8-bit (256 levels) by the A/D converter. The 8-bit digitized data were transferred to the hard disk drive through the bus controller and were displayed in the monitor at the same time. The region of the observed data recorded was set up by the azimuth control unit. According to this setup, the azimuth direction and the radial direction were controlled by the trigger signal from the radar. The sampling rate of A/D convert was set by the personal computer and was 20 MHz in this time.

Table 1. S-band radar parameters.

TOKIMEC BR-3440MA-S314	
Frequency	3.05 GHz
Transmitted peak power	30 kW
Pulsewidth	0.5 μs
Pulse-repetition frequency	1600 Hz
Antenna length	14 ft
Antenna scan rate	22 rpm
Horizontal antenna beamwidth	1.8°
Vertical antenna beamwidth	25°
Polarization	horizontal
Receiver	linear
Intermitted frequency	60 MHz
Intermitted bandwidth	5 MHz

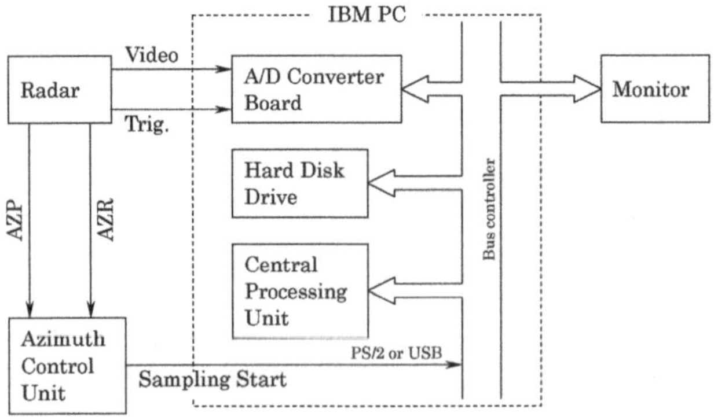

Figure 1. Radar system block diagram.

2.2. Observed Data

The observed data is shown in **Figure 2**. We observed it at 18:46 December 11, 2009. The observation was made in area including the Uraga Suido Traffic Route off Cape Futtu, Chiba Prefecture. On that day, it was rain. The precipitation was 3.5 mm/h. The wind was in the north, the wind velocity was 7 m/s. Because it was foggy, when the ships entered the Uraga Suido Traffic Route, they gave the fog alarms. The observed data was from 25.2° to 47.7° in the azimuth direction and from 3.91 km to 5.81 km in the radial direction. The observed data was a region of 256 range sweeps in azimuth and 256 range bins in radial interval. The data points totaled 65,536. The reflection amplitude at each point was quantized to 256 levels. The observed data is represented in gray scale as 256 gradation from the black of amplitude 0 to the white of 255.

Weather clutter from rain and fog looks white in the figure. Two targets surrounded by circles of white broken line are a passenger ship SARUBIA MARU with a ship length 121 m in upper left and a cargo ship HOEGH AMERICA with a ship length 200 m in lower right. The information about ship names and lengths were gotten from the AIS (Automatic Identification System). In the figure, the vertical lines are radar interference with other radars having the same frequency. Many ships pass through the Uraga Suido Traffic Route while they operate radars having the same frequency. Therefore, such a radar interference occurs.

3. Distribution Estimation of Weather Clutter

To estimate a distribution model obeying clutter amplitude, we apply some models to observed data and compare "goodness" of models. In this paper, we use four probability distribution models, the log-normal, Weibull, log-Weibull, and K-distributions, and introduce the Akaike Information Criterion (AIC). Here, we explain the probability density functions and the properties of these four models and summarize the AIC.

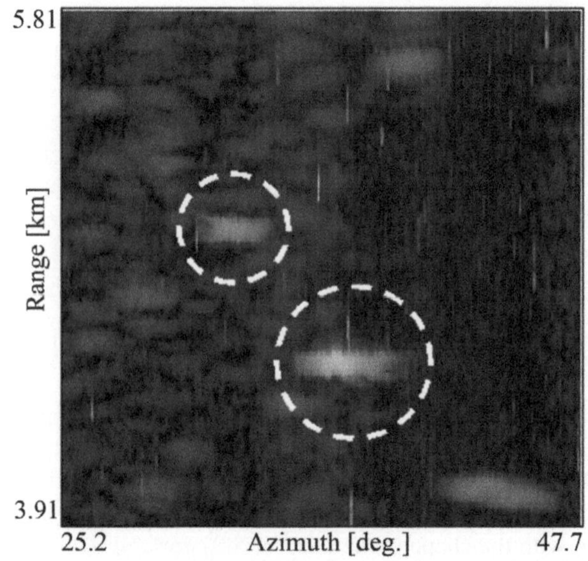

Figure 2. Observed data.

3.1. Distribution Estimation Models

The log-normal distribution is written as follows:

$$p_{LN}(x) = \frac{1}{\sqrt{2\pi}\sigma x}\exp\left[-\frac{(\ln x - \mu)^2}{2\sigma^2}\right] \quad (x \ge 0, \sigma > 0) \quad (1)$$

Here x is the amplitude of the return signals, μ is an average of $\ln x$ (a scale parameter) and σ is the standard deviation of $\ln x$ (a shape parameter). The character of this distribution is its long tail. Recently, to develop a high-resolution radar, return signals accompany many spikes. It has been observed that the clutter amplitude has a longer tail [8].

The Weibull distribution was proposed by Prof. Walodi Weibull at the Swedish Royal Institute of Technology in 1939 [14]. This distribution is written as follows:

$$p_W(x) = \frac{c}{b}\left(\frac{x}{b}\right)^{c-1}\exp\left[-\left(\frac{x}{b}\right)^c\right] \quad (x \ge 0, b > 0, c > 0) \quad (2)$$

Here b is a scale parameter and c is a shape parameter. It is important that the Weibull distribution includes the exponential and Rayleigh distributions. That is, for $c = 1.0$ and 2.0 in Equation (2), the Weibull distribution is identical to the exponential and Rayleigh distributions, respectively [8].

We proposed a new log-weibull distribution as the probability distribution model for clutter [4,15]. The log-Weibull distribution is written as follows:

$$p_{LW}(x) = \frac{c}{bx}\left(\frac{\ln x}{b}\right)^{c-1}\exp\left[-\left(\frac{\ln x}{b}\right)^c\right] \quad (x \ge 0, b > 0, c > 0) \quad (3)$$

Here b is a scale parameter and c is a shape parameter. The log-Weibull distribution has the advantage of both the log-normal and Weibull distributions. That is, the log-Weibull distribution has a long tail and is flexible in its shape.

The K-distribution was proposed by E. Jakeman and P. N. Pusey in 1976 [9]. This distribution is written as follows:

$$p_K(x)=\frac{4h^{\nu+1}}{\Gamma(\nu)}x^{\nu}K_{\nu-1}(2hx)\ (x\ge 0,h>0,\nu>0) \quad (4)$$

Here h is a scale parameter and ν is a shape parameter. $\Gamma(x)$ is the gamma function and $K_\nu(x)$ is the ν th modified Bessel function. ν has been found to lie in the region $\nu=\infty$, indicating Rayleigh distributed clutter. The smaller the value of ν becomes, the longer the tail of the K-distribution becomes and the larger the gap between the Rayleigh and K-distributions becomes [9-12]. The K-distribution includes the modified Bessel function. Therefore, the K-distribution is more difficult than the other three distributions in its calculation.

3.2. The Akaike Information Criterion (AIC)

Now we assume that the M numbers of the independent values $\{x_1,x_2,\cdots,x_M\}$ are observed. We consider applying a model $p(x|\theta)$ to these observed data (θ are parameters of the probability density function of this model). The logarithmic likelihood $L(\theta)$ of this model is defined as

$$L(\theta)=\sum_{i=1}^{M}\ln\{p(x_i|\theta)\}. \quad (5)$$

We obtain $L(\theta_0)$ (the maximum likelihood estimation) which gives the largest $L(\theta)$ and select the $p(x_i|\theta_0)$ (the maximum likelihood model). Considering some given probability distribution models, we select the model which gives the largest logarithmic likelihood, and we obtain the nearest model to the true probability distribution. The expectation for data x of the average logarithmic likelihood of the maximum likelihood model corresponds to the expectation average logarithmic likelihood. The maximum likelihood is considered as an estimation of the expectation average logarithmic likelihood. On detailed investigation, we find the bias that the maximum likelihood tends to be larger than the true expectation average logarithmic likelihood. We investigate the relationship between the extent of the bias and the number of parameters included in the model. It is found that

$$\begin{aligned}&(\text{maximum logarithmic likelihood})\\&-(\text{number of parameters included in the model})\end{aligned} \quad (6)$$

is approximately an unbiased estimation of the expectation average logarithmic likelihood. According to historical circumstances, Equation (6) multiplied by –2 is called AIC [13]. That is,

$$\begin{aligned}\text{AIC}&=-2\times(\text{maximum logarithmic likelihood})\\&+2\times(\text{number of parameters included in the model})\end{aligned} \quad (7)$$

The AIC is not the Kullback-Leibler entropy [16] but the expectation average logarithmic likelihood estimation. The AIC does not depend on each value of data and is determined by the true probability distribution of data, the number of data, and the model. Therefore, the value of the AIC itself is not significance, the differences in the values of the AIC are important. The significant differences are larger than unity from the relationship between the AIC and the entropy. The model which yields the smallest AIC (MAIC; minimum AIC estimation) is regarded as the best one [17].

3.3. Result of Distribution Estimation

In general, return signals have strong correlation and the same properties within the beam width of an antenna. Thus we divide the observed data into small data within the horizontal beam width of an antenna 1.8°. We estimate each small data by using the AIC and four probability distribution models.

Here, a range sweep number (0 - 255) is assigned to 256 points in azimuth direction. From the relation between the horizontal beam width of an antenna and the range sweep number, the horizontal beam width of an antenna 1.8° is correspond to the range sweep number 20 points. The observed data in **Figure 2** is divided into every 20 points like strips. The number of small data is 12 and the number of its data points is 5120.

We summarize the parameters and the values of the AIC for different range sweep numbers in **Table 2**. The smallest AIC (MAIC) are indicated by underline. The numbers of the MAIC of the log-normal, Weibull, log-Weibull, and K-distributions are 8, 2, 2, and 0, respectively, from 12 small data within the beam width of an antenna. The log-normal distribution is best fit to data for 8 small data more than half of the entire data. Moreover, the log-normal distribution is best fit to data for 5 small data of 6 small data containing targets (range sweep 60 - 179).

In **Table 2**, we have found that the weather clutter amplitudes observed by an S-band radar obey the log-normal, Weibull, and log-Weibull distributions with the shape parameters of 0.308 to 0.470, 4.42 to 4.51, and 15.91 to 16.44, respectively, for small data within the beam width of an antenna. If clutter obeyed the Rayleigh distribution, clutter amplitude should obey the Weibull distribution for $c=2.00$. But, there is no such small data. It is found that all small data obey non-Rayleigh distribution.

Table 2. Result of distribution estimation for different range sweep numbers.

range sweep	log-normal distribution			Weibull distribution			log-Weibull distribution			K-distribution		
numbers	μ	σ	AIC	b	c	AIC	b	c	AIC	h	ν	AIC
0 - 19	4.17	0.303	44964.4	74.16	3.80	44656.7	4.30	16.44	44517.3	0.131	84.70	47479.0
20 - 39	4.24	0.285	45109.0	78.71	4.51	44042.3	4.36	19.42	44064.8	0.130	93.89	47958.7
40 - 59	4.27	0.286	45382.0	81.13	4.42	44602.7	4.39	19.03	44616.7	0.120	84.91	48290.1
60 - 79	4.14	0.308	44830.3	72.61	3.28	45412.2	4.27	14.37	45059.4	0.113	61.04	47388.2
80 - 99	4.12	0.379	46743.3	73.70	2.69	47328.8	4.28	11.74	46948.5	0.113	63.82	48110.5
100 - 119	4.13	0.321	45168.4	71.94	3.76	44606.7	4.27	15.91	44547.4	0.138	87.94	47242.2
120 - 139	4.14	0.393	47374.0	76.86	2.28	48908.3	4.32	10.26	48247.6	0.113	71.21	49062.9
140 - 159	3.97	0.470	47426.6	67.21	2.00	48688.2	4.18	8.60	48048.1	0.058	14.77	48534.7
160 - 179	3.83	0.465	45949.9	58.40	2.19	46699.5	4.04	8.89	46265.9	0.113	42.01	46768.7
180 - 199	3.67	0.422	43278.8	48.60	2.28	44373.0	3.86	9.02	43793.1	0.179	71.47	44524.2
200 - 219	3.62	0.457	43642.9	47.86	1.85	45666.9	3.83	7.51	44691.1	0.050	5.62	45205.0
220 - 239	3.79	0.431	44747.9	55.28	2.20	45968.7	3.99	8.99	45363.1	0.113	37.34	46022.9

Examples of small data obeying the log-normal, Weibull, and log-Weibull distributions are illustrated in **Figures 3-5**, respectively. In any case, the distribution estimation models are best fit to data.

4. Log-Normal/CFAR Circuit Modified a Cell-Averaging LOG/CFAR Circuit

It has been discovered that weather clutter observed by an S-band radar obeys non-Rayleigh distribution. It is necessary to devise a new CFAR circuit in order to suppress such clutter and detect targets. If we can realize the new CFAR circuit by modifying the LOG/CFAR circuit which had been already put to practical use, we will have an advantage economically. It has been found that the log-normal distribution is best fit to data for many small data containing targets in Section 3.3. Therefore, we propose a new log-normal CFAR circuit modified a Cell-Averaging (CA) LOG/CFAR circuit.

The log-normal CFAR circuit modified a CA LOG/CFAR circuit is shown in **Figure 6**. This circuit can be realized by only adding a circuit calculating a mean-squared value to a conventional CA LOG/CFAR circuit. Here, it is logically explained how this circuit maintains CFAR for log-normal distributed clutter.

4.1. The CFAR Maintenance in Log-Normal Distributed Clutter

In **Figure 6**, if the clutter signal x obeying the log-normal distribution in Equation (1) is passed through an idealized logarithmic amplifier, the output signal y is

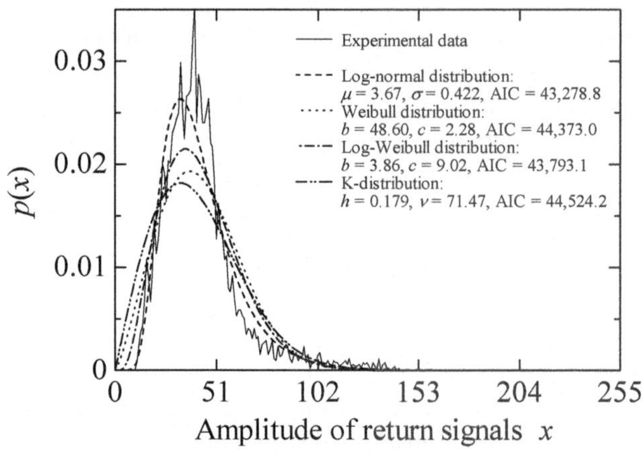

Figure 3. Log-normal distribution is best fit to data in range sweep 180 - 199.

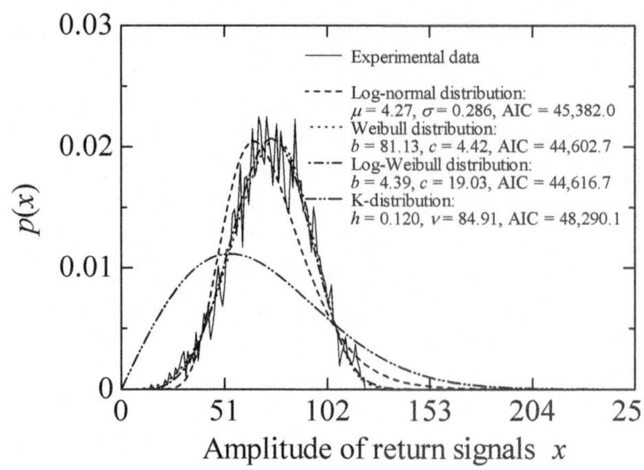

Figure 4. Weibull distribution is best fit to data in range sweep 40 - 59.

represented by

$$y = \ln x \tag{8}$$

y is put in a delay line, the mean signal level of y is calculated, $\langle y \rangle$ is

$$\langle y \rangle = \int_0^\infty y p_{LN}(x)\mathrm{d}x = \mu \tag{9}$$

Subtracting $\langle y \rangle$ from y, the output signal z passed through an antilogarithmic amplifier is represented by

$$z = \mathrm{e}^{y-\langle y \rangle} = \frac{x}{\mathrm{e}^{\mu}} \tag{10}$$

And Equation (1) is transformed into the probability

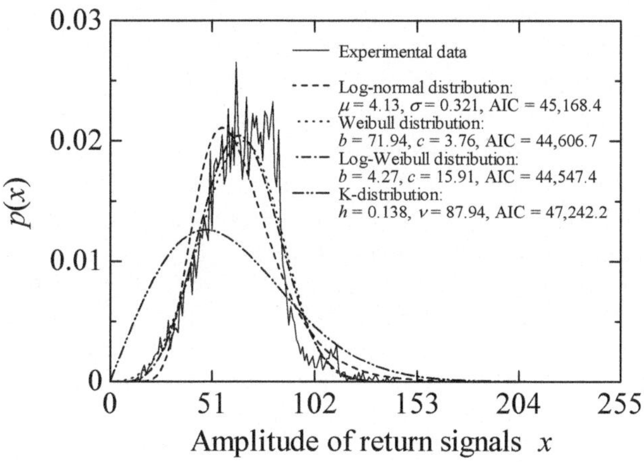

Figure 5. Log-Weibull distribution is best fit to data in range sweep 100 - 119.

density function of z by using Equation (10). $p_{LN}(z)$ is given by

$$p_{LN}(z) = \frac{1}{\sqrt{2\pi}\sigma z}\exp\left(-\frac{\ln z}{\sqrt{2}\sigma}\right)^2 \tag{11}$$

When a threshold level T_h is set to the signal z obeying the distribution represented by Equation (11), since the signal of $z < T_h$ becomes $z = 0$ by a comparator, the false alarm probability P_N is given by

$$P_N = \int_{T_h}^\infty p_{LN}(z)\mathrm{d}z = \frac{1}{2} - \frac{1}{2}\mathrm{erf}\left(\frac{\ln T_h}{\sqrt{2}\sigma}\right) \tag{12}$$

Here, erf( ) is the error function. P_N is independent of μ and depends on σ. That is, P_N changes by the value of σ and is not constant. CFAR is not maintained.

But, in Equation (12), if the value of σ has been already obtained, T_h can be set in order to keep P_N constant according to the value of σ. CFAR can be maintained. To obtain σ, by using a new additional circuit calculating a mean-squared value, the mean-squared signal level $\langle y^2 \rangle$ becomes

$$\langle y^2 \rangle = \int_0^\infty y^2 p_{LN}(x)\mathrm{d}x = \sigma^2 + \mu^2 \tag{13}$$

Now, $\langle y \rangle$ is squared and subtracted from $\langle y^2 \rangle$. The variance of y is given by

$$V(y) = \langle y^2 \rangle - \langle y \rangle^2 = \sigma^2 \tag{14}$$

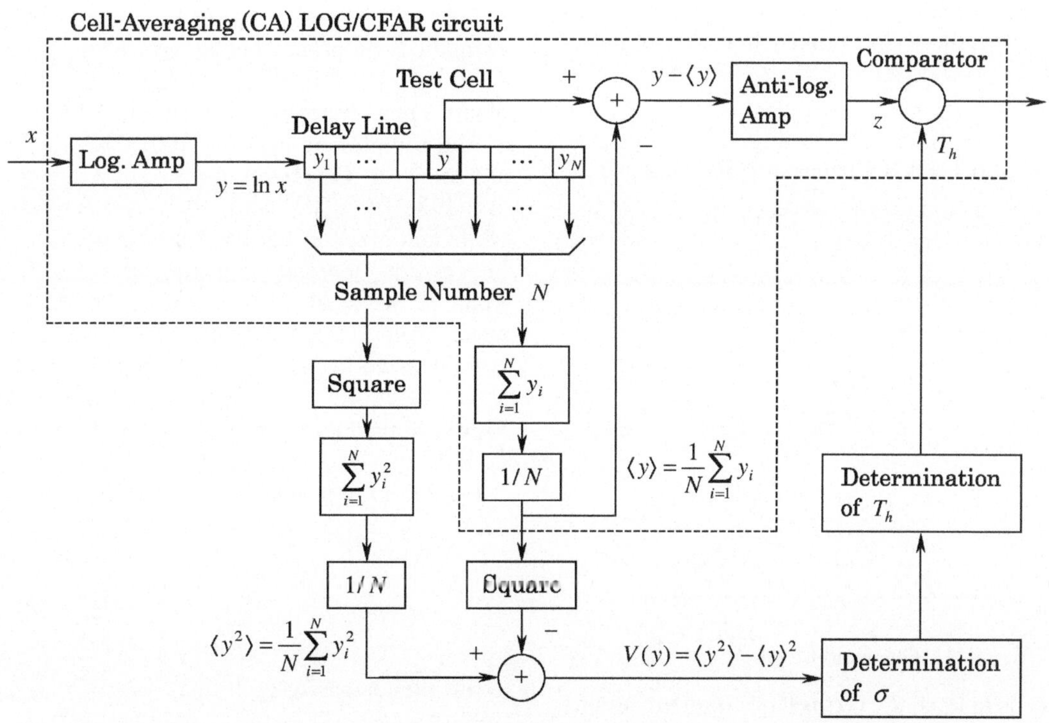

Figure 6. Log-normal/CFAR circuit modified a Cell-Averaging (CA) LOG/CFAR circuit.

$V(y)$ is independent of μ and depends only on σ. That is, σ can be determined by calculating $V(y)$. Therefore, if $\langle y^2 \rangle$ can be calculated besides $\langle y \rangle$ calculated by a CA LOG/CFAR circuit, σ can be calculated from $V(y)$. If the comparator decides whether there is target or not by T_h set according to σ, P_N is kept constant. That is, it is possible to maintain CFAR for the log-normal distributed clutter.

T_h for obtaining aimed P_N is determined by substituting σ calculated from Equation (14) into Equation (12). But, Equation (12) corresponds to the case of a sample number $N = \infty$ in **Figure 6**. When the CA LOG/CFAR circuit is used practically, N is finite. Moreover, N becomes a comparatively small value, because a radar is operated almost in real time. It is impossible to use Equation (12) representing P_N for $N = \infty$. Therefore, random numbers obeying the log-normal distribution were generated by the multiplicative congruential method on a computer. The random numbers were passed through the CA LOG/CFAR circuit in **Figure 6**. On every values of σ, we set T_h to the output z. P_N were calculated as the probability of z exceeding T_h. σ were calculated at intervals of 0.001, and T_h were calculated down to two decimal places. Each of the number of generated random numbers were 100,000,000.

Example is shown in **Figure 7**. The random numbers obeying the log-normal distribution with $\sigma = 0.329$ were generated. The relation between T_h and P_N were calculated for $N = 2$, 4, 8, 16, 32, 64, and 128. For example, P_N is assumed to be 10^{-3}. T_h is determined as an intersection in curve and broken line. T_h is determined 3.47, 3.12, 2.94, 2.85, 2.81, 2.79, and 2.78, respectively.

4.2. Log-Normal/CFAR Processing Results

We applied the log-normal/CFAR circuit modified a CA

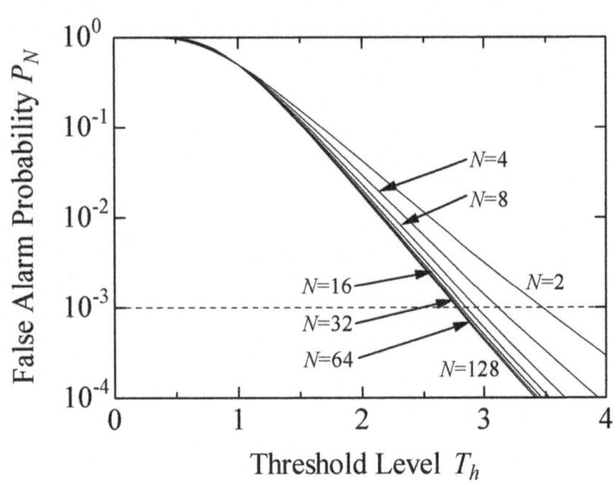

Figure 7. Threshold level T_h versus false alarm probability P_N $(\sigma = 0.329)$.

LOG/CFAR circuit to the observed data in **Figure 2**. First, to evaluate the log-normal/CFAR circuit quantitatively, we introduce the target-to-clutter ratio. The target-to-clutter ratio (T/C) is defined as

$$(T/C) = 10\log\frac{s^2}{\langle x^2 \rangle} \tag{15}$$

Here, s and x are the signals of target and clutter, respectively [18].

When the observed data was actually processed by this circuit for the sample number $N = 2$ and 4, two and four consecutive signals of same values were occasionally put in both sides of the test cell on the delay line. The variance became $V(x) = 0$, that is, $\sigma = 0$. The threshold level T_h could not be calculated. For $N = 8$ and 16, the values of σ occasionally became extremely big or small values which were completely different from the results of distribution estimation in Section 3.3. (For example, the maximum and minimum values of σ were 0.008 and 0.855 for $N = 8$, respectively.) Therefore, for $N = 2$, 4, 8, and 16, it is of no practical use.

Next, for $N = 32$, 64, and 128, since the number of data is 65,536, the false alarm probability P_N is set to 10^{-1}, 10^{-2}, and 10^{-3}. We calculate the target-to-clutter ratio (T/C) from Equation (15) and summarize the ratio (T/C) in **Table 3**. The unit is dB. In **Table 3**, (T/C) cannot be calculated for $N = 32$, $P_N = 10^{-2}$ and 10^{-3} and $N = 64$, $P_N = 10^{-3}$, because clutter can be suppressed but the targets cannot be detected by log-normal/CFAR processing. (T/C) of the observed data is 7.85 dB. Therefore (T/C) is improved on all log-normal/CFAR processing except them. A maximum of (T/C) is 59.43 dB for $N = 128$, $P_N = 10^{-3}$. Thus we obtain a maximum improvement of 51.58 dB.

The samples of log-normal/CFAR processing image are shown in **Figures 8** and **9**. These are processed for $N = 128$, $P_N = 10^{-2}$ and 10^{-3}, respectively. Here, as white and black in **Figure 2** are reversed, the processed images are represented in gray scale as 256 gradation from the white of amplitude 0 to the black of 255. To make easily to see, a maximum of signal after processing is normalized to 255. Circles of black broken line correspond to circles of white broken line in **Figure 2**. In **Figure 8**, the targets are completely detected. The meas-

Table 3. (T/C) of log-normal/CFAR processed results.

		sample number N		
		32	64	128
false alarm probability P_N	10^{-1}	21.02	24.34	25.86
	10^{-2}	-	33.63	42.10
	10^{-3}	-	-	59.43

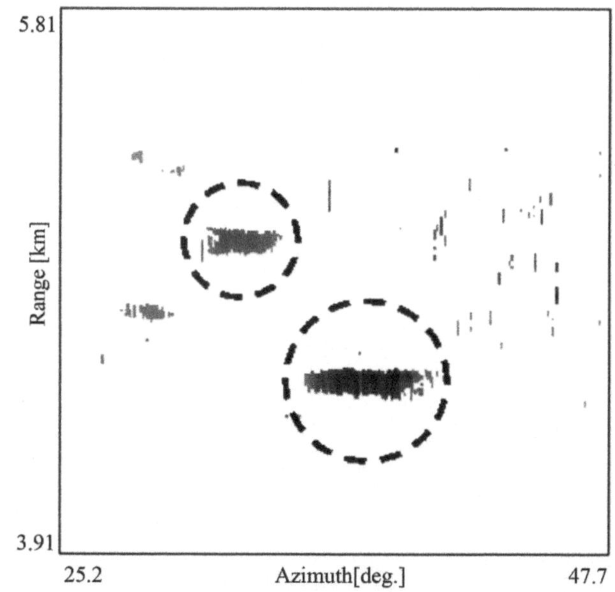

Figure 8. Log-normal/CFAR processed image ($N = 128$, $P_N = 10^{-2}$).

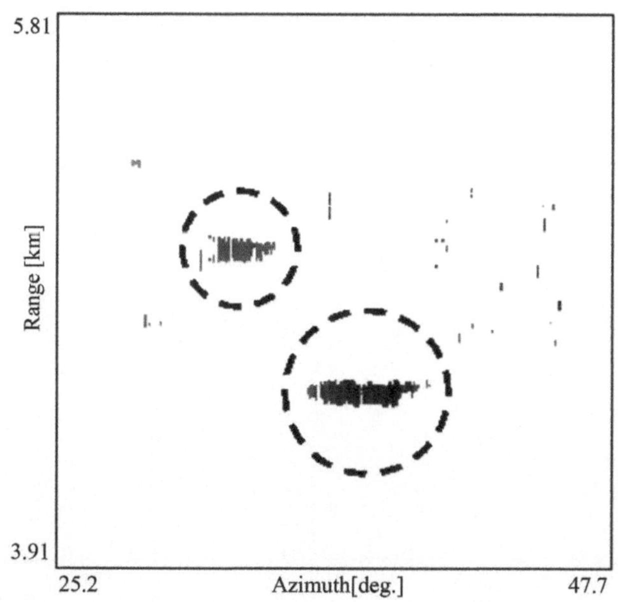

Figure 9. Log-normal/CFAR processed image ($N = 128$, $P_N = 10^{-3}$).

urement value of false alarm probability P_N is 1.3×10^{-2}. It is almost equal to the theory value. While, in **Figure 9**, part of targets are disappeared and targets are not completely detected. the measurement value of P_N is 4.8×10^{-3}. It is larger than the theory value. Because, calculating threshold levels on a computer, we did not consider the correlation between radar signals. In **Figures 8** and **9**, the vertical lines representing radar interference are not disappeared. Because the radar interference is artificial signal and does not obey the log-normal distribution.

To compare with the log-normal/CFAR processed image, the conventional LOG/CFAR processed image is shown in **Figure 10**. This is processed for $N = 128$, $P_N = 10^{-2}$. In **Figure 10**, white and black are reversed as in **Figures 8** and **9**. Clutter is suppressed but the target of SARUBIA MARU in upper left is disappeared and cannot be detected. Though weather clutter does not obey the Rayleigh distribution, it is processed by the LOG/CFAR circuit. Therefore, the threshold level T_h is too high, a target signal cannot exceed T_h and target is disappeared.

5. Conclusions

We observed weather clutter data containing targets of ships using an S-band radar with a frequency 3.05GHz, a beamwidth 1.8°, and a pulsewidth 0.5 μs. We investigated the weather clutter amplitudes by using the AIC and four probability distribution models. It is found that weather clutter amplitudes obey the log-normal, Weibull, and log-Weibull distributions with the shape parameters of 0.308 to 0.470, 4.42 to 4.51, and 15.91 to 16.44, respectively, for small data within the beam width of an antenna, and many small data containing targets obey the log-normal distribution. Therefore, we propose a new log-normal CFAR circuit modified a CA LOG/CFAR circuit. We apply this circuit to the real observed data, it is found that weather clutter is suppressed with a maximum improvement of target-to-clutter ratio 51.58 dB.

In the future, we will observe sea and ground clutter, and apply this circuit to such clutter. Then we will examine whether this circuit has an effect on suppression of such clutter or not. We will plan to develop the circuit

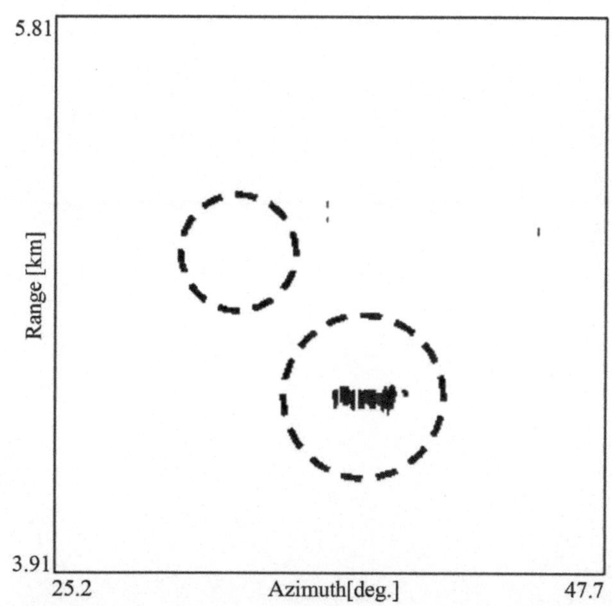

Figure 10. LOG/CFAR processed image (N = 128, P_N = 10^{-2}).

which can suppress two or three kinds of clutter at the same time.

REFERENCES

[1] D. K. Barton, "Radar Clutter," Artech House, Dedham, 1975.

[2] J. Croney, "Clutter on Radar Displays," *Wireless Engineering*, Vol. 33, 1956, pp. 83-96.

[3] M. Sekine, "Radar Signal Processing Techniques," The Institute of Electronics, Information and Communication Engineers, Tokyo, Japan, 1991 (in Japanese).

[4] S. Sayama and S. Ishii, "Amplitude Statistics of Ground Clutter from Town and Hill Observed by S-band Radar," *IEEJ Transactions Fundamentals and Materials*, Vol. 131, No. 11, 2011, pp. 916-923 (in Japanese).

[5] S. Sayama and S. Ishii, "Amplitude Statistics of Sea Clutter by MDL Principle," *IEEJ Transactions Fundamentals and Materials*, Vol. 132, No. 10, 2012, pp. 886-892 (in Japanese).

[6] S. Sayama, S. Ishii and M. Sekine, "Amplitude Statistics of Weather Clutter Observed by L-Band Radar," *IEEJ Transactions Fundamentals and Materials*, Vol. 126, No. 6, 2006, pp. 438-442.

[7] S. Sayama and M. Sekine, "Observations and Suppression of Sea Ice Clutter by Means of Millimeter Wave Radar," *IEICE Transactions on Communications (Japanese Edition)*, Vol. J85-B, No. 7, 2002, pp. 1104-1111 (in Japanese).

[8] M. Sekine and Y. Mao, "Weibull Radar Clutter," Peter Peregrinus Ltd., London, 1990.

[9] E. Jakeman and P. N. Pusey, "A Model for Non-Rayleigh Sea Echo," *IEEE Transactions on Antennas and Propagation*, Vol. AP-24, No. 6, 1976, pp. 806-814.

[10] J. K. Jao, "Amplitude Distribution of Composite Terrain Radar Clutter and the K-Distribution," *IEEE Transactions on Antennas and Propagation*, Vol. AP-32, No. 10, 1984, pp. 1049-1062.

[11] S. Watts, "Radar Detection Prediction in Sea Clutter Using the Compound K-Distribution Model," *IEE Proceedings*, Vol. 132, No. 7, 1985, pp. 613-620.

[12] S. Watts, "Radar Detection Prediction in K-Distributed Sea Clutter and Thermal Noise," *IEEE Transactions on Aerospace and Electronic Systems*, Vol. AES-23, No. 1, 1987, pp. 40-45.

[13] H. Akaike, "Information Theory and an Extension of the Maximum Likelihood Principle," In: B. N. Petrov and F. Csaki, Eds., *2nd International Symposium on Information Theory*, Akademiai Kiado, 1973, pp. 267-281.

[14] W. Weibull, "A Statistical Theory of Strength of Materials," *IVB-Handl.*, 1939, No. 151.

[15] M. Sekine, T. Musha, Y. Tomita, T. Hagisawa, T. Irabu and E. Kiuchi, "Log-Weibull Distributed Sea Clutter," *IEE Proceedings*, Vol. 127, No. 3, 1980, pp. 225-228.

[16] S. Kullback and R. A. Leibler, "On Information and Sufficiency," *Annals of Mathematical Statistics*, Vol. 22, No. 1, 1953, pp. 79-86.

[17] Y. Sakamoto, M. Ishiguro and G. Kitagawa, "Information Statistics," Kyoritsu Shuppan Co., Ltd., Tokyo, 1983 (in Japanese).

[18] M. Sekine and N. Oura, "Introduction to Electrical and Electronic Measurements," Shoko-do, Tokyo, 2006 (in Japanese).

26

Blind Modulation Recognition in Wireless MC-CDMA Systems Using a Support Vector Machine Classifier

Mohamed El-Hady Magdy Keshk[1], El-Sayed Elrabie[2], Fathi El-Sayed Abd El-Samie[2], Mohammed Abd El-Naby[2]

[1]Egyptian Space Program, National Authority for Remote Sensing and Space Science, Cairo, Egypt; [2]Faculty of Electronic Engineering, Menoufia University, Menouf, Egypt.

ABSTRACT

Automatic Digital Modulation Recognition (ADMR) is becoming an interesting problem with various civil and military applications. In this paper, an ADMR algorithm in Multi-Carrier Code Division Multiple Access (MC-CDMA) systems using Discrete Transforms (DTs) and Mel-Frequency Cepstral Coefficients (MFCCs) is proposed. This algorithm uses various DT techniques such as the Discrete Wavelet Transform (DWT), Discrete Cosine Transform (DCT) and Discrete Sine Transform (DST) with MFCCs to extract features from the modulated signal and a Support Vector Machine (SVM) to classify the modulation orders. The proposed algorithm avoids over fitting and local optimal problems that appear in Artificial Neural Networks (ANNs). Simulation results shows the classifier is capable of recognizing the modulation scheme with high accuracy up to 90% - 100% using DWT, DCT and DST for some modulation schemes over a wide Signal-to-Noise Ratio (SNR) range in the presence of Additive White Gaussian Noise (AWGN) and Rayleigh fading channel, particularly at a low Signal-to-Noise ratios (SNRs).

Keywords: ADMR; MC-CDMA; MFCC; DT; SVM; ANN

1. Introduction

ADMR is an intermediate step between signal detection and demodulation. It is a rapidly evolving area in various digital signaling systems currently developed or planned for various civilian and military communication applications [1]. Recent developments in software defined radio are based on perfect recognition of the modulation scheme of the received signal. Automatic recognition plays an important role in both military and civilian applications due to its capability of placing several receivers in one universal receiver. ADMR techniques usually can be categorized in two main principles, the first is based on the Decision-Theoretic Approach (DTA) and the second is based on the Pattern Recognition Approach (PRA).The DTA is a probabilistic solution based on a priori knowledge of probability functions and certain hypotheses [2], and the PRA consists of two subsystems. The first one is the feature extraction subsystem and the second one is the classifier.

Feature extraction extracts useful small amount of data from the signal. The classifier uses the useful information extracted from the feature extraction subsystem to classify the signal. In this paper, we use the pattern recognition approach.

In the last years, MC-CDMA has become the dominant transmission technique for both wired and wireless communication systems. Besides its advantages to combat multipath propagation with simple receiver structures, it furthermore enables a flexible adaption of modulation schemes on a subcarrier basis or group of subcarriers [3,4]. MC-CDMA is praised as a modulation solution that merges the insights due to Shannon (particularly those related to CDMA) with insights due to Fourier (particularly those explaining why OFDM has advantages in a dispersive channel).

There are many studies in ADMR field with different methods about the way the features can be extracted. In [5], the authors presented a kernel-based modulated signal classification method for a Cognitive Radio (CR) system, which cast into machine learning approaches. This method extracts both statistical and spectral features from the received signals. The Kernel-based Generalized Discriminant Analysis (KGDA) projects these features to a new feature space with C−1-dimensions such that the separability of the classes increases.

In [6], multi-class classification of analog and digital signals in CR was presented using SVMs. The author implemented a simulated model of an SVM signal classifier trained to recognize seven distinct modulation schemes; five digital (BPSK, QPSK, GMSK, 16QAM and 64QAM) and two analog (FM and AM). The results show that the SVM correctly classified signals with success rate of 85% - 98%.

In [7], an equalizer was used to reduce the channel effects. A selected combination of the fourth, sixth and eighth order moments and cumulants of the received signal was used to better represent the digital signal. A multi-class classifier based on SVMs was used to determine the types of the received signals. Simulation results showed that this method has a success rate in the presence of noise and fading.

In [8], an automatic digital modulation identifier in dispersive media was proposed. Prior to this identifier, the undesired effects of channel are mitigated with an equalizer. Higher order cumulants and moments (up to eight) were used as features. Simulation results showed that this identifier is able to identify different types of modulations (e.g. QAM64, V.29, and ASK8) with high accuracy even at low SNRs.

In [9], an approach to signal classification combining spectral correlation analysis and SVMs was introduced. Four spectral coherence characteristic parameters were chosen via spectral correlation analysis. By utilizing a nonlinear SVM, a significant amount of calculation was performed offline, thus the computational complexity was reduced. Simulations indicated that the overall success rate is above 92.8% when the SNR is equal to 4 dB.

In [10], a total of 7 statistical signal features were extracted and used to classify 9 modulation signals. The authors investigated the performance of the two types of SVM classifiers. Classification results up to 97% were obtained on an AWGN channel.

In this paper, we propose an algorithm capable of identifying digital modulation techniques in MC-CDMA system, and after identification it is capable of recovering transmitted data. This algorithm uses different discrete transform techniques (DWT, DCT and DST) for features extraction, and then applies MFCCs to obtain the useful features. The proposed classifier is a multi-class SVM which can classify more than two classes. This classifier has the capability of recognizing the M-ary Phase Shift Keying (M-PSK) signals and the order of the identified modulation. The performance of the proposed algorithm is examined based on Correct Recognition Probability (CRP). AWGN and multi-path fading (Rayleigh) channels are considered when developing the mathematical model and through most of the results.

The remainder of the paper is organized as follows. Section 2 shows the proposed ADMR technique for the MC-CDMA System. Section 3 defines the mathematical model of the proposed algorithm and presents the cepstral analysis methods. Section 4 shows feature extraction from discrete transforms. Section 5 is dedicated to the structure of SVM. The results, algorithm performance analysis, and a comparative study for various orders of digital modulation types are presented in Section 6. Conclusions of this research work are presented in Section 7.

2. Proposed ADMR Techniquein MC-CDMA System

In a multipath fading environment, not only a direct transmission signal but also many reflected signals arrive at the receiver at the same time, which causes Inter-Symbol Interference (ISI) leading to a significant degradation in the network performance. In broadband wireless communications, high bit rate transmission is required for high quality communications. The most important objective of Software Defined Radio (SDR) systems is to take care of the ISI resulting from high data rates. To overcome this problem, it is necessary to use parallel transmission, in which high speed data is converted to slow parallel data in several channels.

System Configuration

In **Figure 1**, we illustrate the sequence of processes of ADMR in the training mode and recognition mode.

Features extraction involves two stages, the first one is the discrete transforms (DCT, DST, DWT) and the second one is MFCCs extraction. Here, we use 13 features only, because these features are unique from a modulation technique to another.

In **Figure 2**, the transmitted high speed data (random number) is first converted into parallel data sub-channels to overcome the problem caused by multi path fading environment. then, the transmitted data of sub-channel is

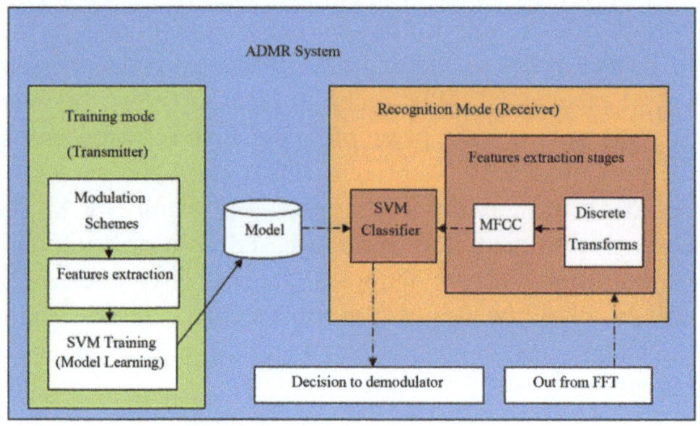

Figure 1. Proposed ADMR technique.

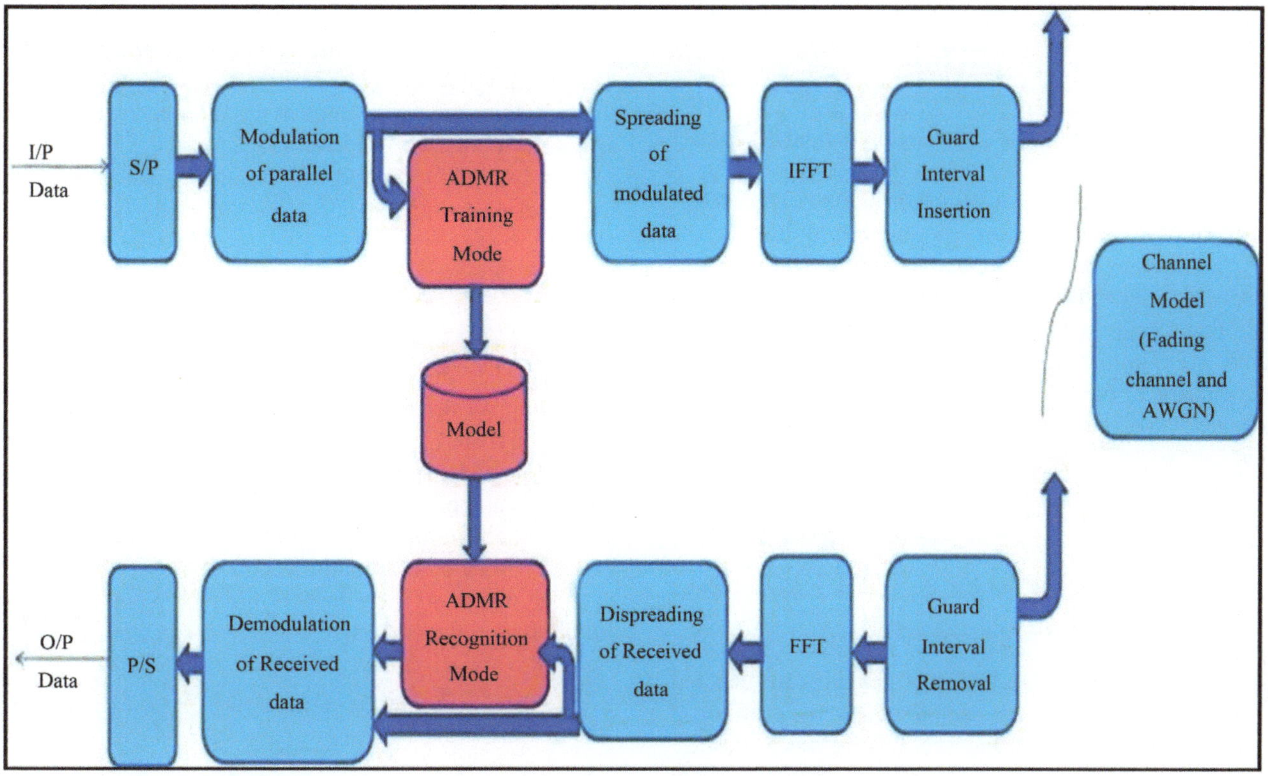

Figure 2. ADMR in MC-CDMA system.

modulated by the whole types of modulation techniques which mentioned before (BPSK, QPSK, 8PSK, 16PSK, 64PSK). In training mode, features of modulated data is extracted and fed to SVM

The classification process consists of two main modes; training mode and recognition mode. In the training mode, the SVM is trained to classify input features into unique classes (modeling), in our study we have 5 classes. Once a proper set of feature vectors is obtained from the previous stage, the SVM develop a model for each modulated signal features for building the reference models into the system database.Then the modulated signals are fed to the spreading process (hadamard code is used), then the Inverse Fast Fourier Transform (IFFT) is applied for the multiplexing processes to distinguish between the subchannels and to generate MC-CDMA signal. This MC-CDMA signal is fed to the guard interval insertion block which adds the cyclic guard interval before every symbol to reduce the problems resulting from ISI, and then the signal is transmitted.

In the recognition mode, unknown signal features are extracted and passed to the system. The system tests these features bycomputing a matching score between the unknown received signal feature vector and the reference models stored in the system database, and the output ofthis matching score is usedto classify the received signal into one of the reference model vectors (classify received signal order), *i.e.* the system is capable of taking a decision about the signal identity. This decision is made based on the result of the matching technique used.

3. Cepestral Analysis

The idea of cepstral analysis is to separate the spectral components (characteristics) of the signal so that signal dependent information can be obtained. Mathematically, the cepstrum is computed by taking the fast Fourier transform (FFT) of the signal, taking the log of the magnitude of the spectrum and then taking the IFFT.

$$\text{Cepstrum}(\text{frame}) = \text{FFT}^{-1}\left(\log\left(\left|\text{FFT}(\text{frame})\right|\right)\right) \quad (1)$$

For the calculation of the MFCCs of a modulated signal, the signal is firstly framed and windowed, the DFT is then taken, and the magnitude of the resulting spectrum is warped by the Mel-scale. The log of this spectrum is then taken and the DCT is applied. The DCT replaces the IDFT stage in practice for I simplicity. This is illustrated in **Figure 3**.The steps of extraction of the MFCCs are summarized in the following subsections.

3.1. Framing and Windowing

The modulated signal is a slowly time-varying signal. In a signal identification system, the modulated signal is partitioned into short-time segments called frames. To make the frame parameters vary smoothly, there is normally a 50% overlap between each two adjacent frames. Win-

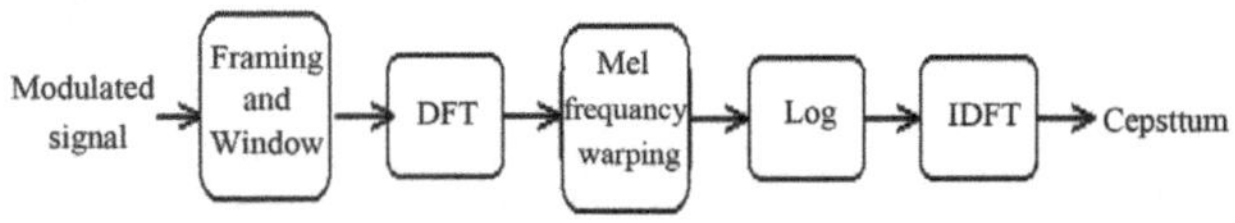

Figure 3. Extraction of MFCCs from a modulated signal.

dowing is performed on each frame with one of the popular signal processing windows like the Hamming window [11]. Windowing is often applied to increase the continuity between adjacent frames and smooth out the end points such that abrupt changes between ends of successive frames are minimized.

The shape of the Hamming window is similar to that of a cosine wave. The following equation defines the window:

$$w(n) = 0.54 - 0.46\cos\left(\frac{2n\pi}{N-1}\right) \qquad (2)$$

for $n = 0,1,2,\cdots,N-1$; where N is the length of the window.

As a frame is multiplied by a window, most of the data at the edges of the frame becomes insignificant causing loss of information. An approach to tackle this problem is to allow overlapping in the sections between frames, which allows adjacent frames to include portions of data in the current frame. This means that the edges of the current frame are included as the centre data of adjacent frames. Typically, around 50% of overlapping is sufficient to embrace the lost information.

3.2. The DFT

Fourier analysis provides a way of analysing the spectral properties of a given signal in the frequency domain. The Fourier transform converts a discrete signal $s(n)$ from time domain into frequency domain with the equation [12]:

$$S(k) = \sum_{n=0}^{N-1} s(n)e^{-j2\pi nk/N},\ 0 \le k \le N-1 \qquad (3)$$

where $n = 0,1,\cdots,N-1$, and N is the number of samples in the signal $s(n)$. k represents the discrete frequency index, and j is equal to $\sqrt{-1}$. The result of the DFT is a complex-valued sequence of length N. The IDFT is defined as:

$$s(n) = \frac{1}{N}\sum_{k=0}^{N-1} S(k)e^{j2\pi nk/N},\ 0 \le n \le N-1 \qquad (4)$$

3.3. The Mel Filter Bank

The power signal is then applied to this bank of filters called Mel filter bank to determine the frequency content across each filter. The Mel frequency filter bank is a series of triangular band pass filters. The filter bank is based on a non-linear frequency scale called the Mel scale which has been discussed earlier. The filters are overlapped in such a way that the lower boundary of one filter is situated at the centre frequency of the previous filter and the upper boundary is situated at the centre frequency of the next filter. The maximum response of a filter, that is, the top vertex of the triangular filter, is located at the filter centre frequency and is normalized to unity.

In the MFCCs method, the main advantage is that it uses Mel-frequency scaling, The Mel scale is defined as [12]:

$$\mathrm{Mel}(f) = 2595\log\left(1 + \frac{f}{700}\right) \qquad (5)$$

where Mel is the Mel-frequency scale and f is the frequency on the linear frequency scale.

The final step is to compute the MFCCs. For this purpose the log Mel spectrum is converted back to time. Then, the DCT is applied to it to get MFCCs. The DCT is applied to the filter outputs and the first few coefficients are grouped together as a feature vector

If the output of the mth Mel filter is $\tilde{S}(m)$, then the MFCCs are given as [13]:

$$c_g = \sqrt{\frac{2}{N}}\sum_{m=1}^{N_f}\log\left(\tilde{S}(m)\right)\cos\left(\frac{g\pi}{N_f}(m-0.5)\right) \qquad (6)$$

where $g = 0,1,\cdots,G-1$, G is the number MFCCs, N_f is the number of Mel filters and c_g is the gth MFCC.

The number of the resulting MFCCs is chosen between 12 and 20, since most of the signal information is represented by the first few coefficients. The 0th coefficient represents the mean value of the input signal.

4. Feature Extraction from Discrete Transforms

Discrete transforms can be used for extraction of robust MFCCs in modulation identification systems. The DWT, the DCT, and the DST have been investigated in the literature for this purpose [14]. **Figure 4** illustrates the utilization of discrete transforms in modulation identification systems.

4.1. The DWT

The wavelet transform is computed separately for differ-

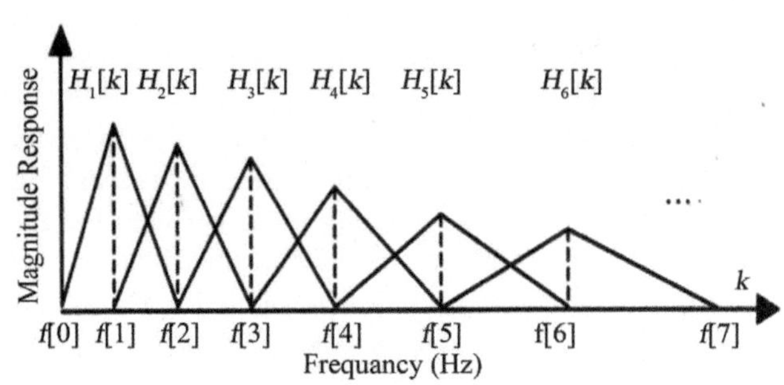

Figure 4. Mel-spaced filter bank.

ent segments of the time-domain signal at different frequencies. Wavelet transform allows Multi-Resolution Analysis (MRA), where it has time-frequency localization. This property allows locality both in time and frequency domains. MRA is designed to give good time resolution and poor frequency resolution at high frequencies and good frequency resolution and poor time resolution at low frequencies. The locality of transformation of a signal is important in two ways for pattern recognition. Firstly, different parts of the signal may convey different amounts of information. Secondly, when the signal corrupted by local noise in time and/or in frequency, the noise affects only a few coefficients if the coefficients represent local information in time and frequency.

The idea of DWT is to represent a signal as a series of approximations (low pass version) to the signal and details (high pass version) at different resolutions.

4.2. The DCT

The DCT is a 1-D transform with an excellent energy compaction property. For a modulated signal $x(n)$, the DCT is represented by [15]:

$$X(k) = \alpha(k)\sum_{n=0}^{N-1} x(n)\cos\left(\frac{\pi(2n+1)k}{2N}\right), \quad k = 0,1,2,\cdots,N-1 \tag{7}$$

where

$$\alpha(0) = \sqrt{\frac{1}{N}}, \quad \alpha(k) = \sqrt{\frac{2}{N}}$$

IDCT is given by:

$$x(n) = \sum_{K=0}^{N-1} \alpha(k)X(k)\cos\left(\frac{\pi(2n+1)}{2N}\right), \quad n = 0,1,2,\cdots,N-1 \tag{8}$$

The features are extracted from $X(k)$ and used for modulation identification.

4.3. The DST

The DST is another triangular transform with common properties with the DCT. The mathematical representation of the DST is given by [15]:

$$X(k) = \sum_{n=0}^{N-1} x(n)\sin\left(\frac{\pi}{N+1}(n-1)(k+1)\right) \quad k = 0,\cdots,N-1 \tag{9}$$

The features are extracted from $X(k)$ and used for modulation identification.

5. Classification

SVM

Support Vector Machine (SVM) is an empirical modeling algorithm and is the state-of-the-art for the existing classification methods. The SVM is basically a two-class classifier based on the ideas of "large margin" and "mapping data into a higher dimensional space", and the kernel functions in the SVM.

The first objective of the SVM classification is the maximization of the margin between the two nearest data points belonging to two separate classes. The second objective is to constraint that all data points belong to the right class. It is a two-class solution which can use multi-dimension features.

The two objectives of the Support Vector Classifier (SVC) problem are then incorporated into an optimization problem. SVC classifies the points from two linearly separable sets in two classes by solving a quadratic optimization problem in order to find the optimal separating hyper plane between these two classes. This hyper plane maximizes the distance from the convex hulls of each class. These techniques can be extended to the nonlinear cases by embedding the data in a nonlinear space using kernel functions. The robustness of SVC originates from the strong fundamentals of statistical learning theory.

SVC can be applied to separable and non-separable data points. In the non-separable case, the algorithm adds one more design parameter. This parameter is the weight of the error caused by the points present in the wrong class region. In MC, this issue occurs in the low SNR cases. Another degree of freedom in the SVC is the kernel function used. SVM must firstly transform the similarity space to a manageable space. This is done by defining a "kernel" which is an inner product to convert points in the input space to points in the feature space as shown in **Figure 5**, In MC using SVM, we used quadratic kernels. Since SVM is basically a binary classifier, received signal.

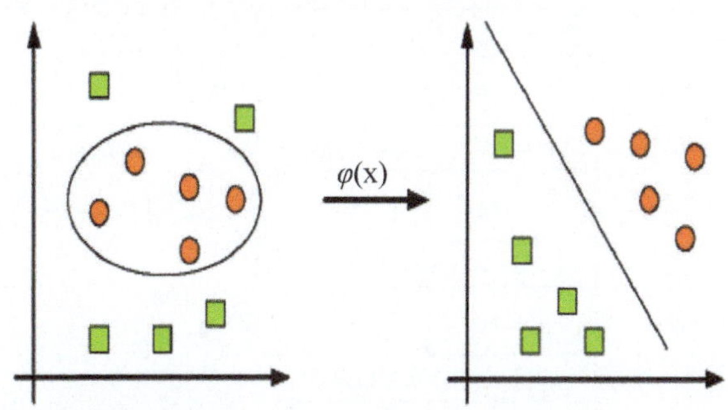

Input Space →Features Space

Figure 5. The idea of SVM.

it is not straightforward to apply it to multi-class classification problems. The most typical method for the multi-class problem is to classify one class from the other classes (refer 1-v-r), another typical method is to combine all possible two-class (pair wise) classifiers (refer 1-v-I) [10]. It's known as 1-v-I type SVM is superior to 1-v-r with respect to its learning time, but execution time for classification of 1-v-I is much worse than 1-v-r [16].

The third method is the DDAG SVM proposed in [17]. Its training phase is the same as the 1-v-I type but its testing time is less than the 1-v-I type. In **Figure 6** we illustrate the method we used in this paper to classify 5 modulation techniques, first we classify one class from the others and if the received signal features (which belong to one class of 5 classes) does not belong to the single class and belongs to the other class, we remove the single class and take one class from the other classes and classify it from the rest of other classes and so on until we correctly classify the received signal.

6. Results and Discussion

The proposed algorithm was verified and validated for various orders of digitally modulated signals (BPSK, QPSK, 8PSK, 16PSK and 64PSK) in AWGN and multipath fading channels. All the simulation steps for MC-CDMA system including the digitally modulated signals, the feature extraction, training of the SVM and performance evaluation were developed using MATLAB. The proposed classifier has shown an excellent performance over AWGN and fading channels even at low SNR.

In the training phase, 5 modulated signals of different orders are fed to the feature extraction stage (Discrete Transform and MFCCs) to generate the feature vectors, which are used to form the system database and these features are used to train the SVM. In the testing phase, each one of these modulated signals is received. Similar features to those used in the training are extracted from those received modulated signals and used for matching.

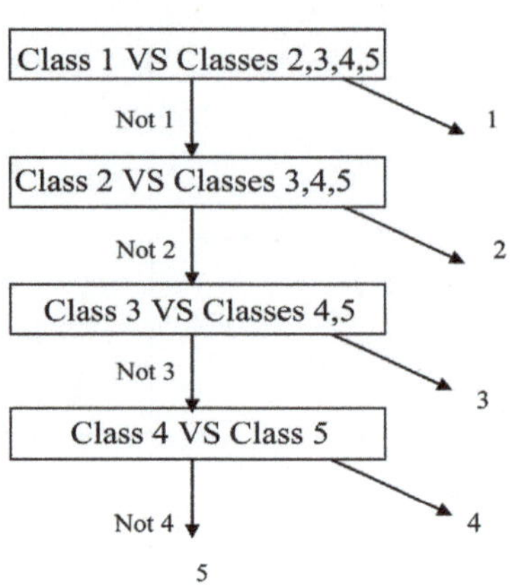

Figure 6. Method to classify 5 classes of modulation techniques.

The features used in all experiments are 13 or 26 MFCCs to form a feature vector of the modulated signal, depending on which method we choose. There are seven methods for extracting features adopted in the experiments. In the first method, the MFCCs (13) are extracted from the modulated signals, only. In the second one, the features (13 MFCCs) are extracted from the DWT of the modulated signals. In the third method, the features (26 MFCCs) are extracted from both the original modulated signals and the DWT of these signals and concatenated together. In the fourth method, the features (13 MFCCs) are extracted from the DCT of the modulated signals. In the fifth method, the features (26 MFCCs) are extracted from both the original modulated signals and the DCT of these signals and concatenated together. In the sixth method, the features (13 MFCCs) are extracted from the DST of the modulated signals. In the seventh method, the features (26 MFCCs) are extracted from both the original modulated signals and the DST of these signals and concatenated together. The recognition rate is used as the performance evaluation metric in all experiments. It is defined as the ratio of the number of success identifications to the total number of identification trials

The discrete transforms and cepstral coefficient were calculated, and the features set are extracted. Then, preprocessing these features subset of each modulation order is performed as a preparation of SVM training. In training mode, we use quadratic kernel function as a learning function. The performance of the classifier was examined for 200 realizations for each modulation order, and the results are presented using the Correct Recognition Probability (CRP).

In **Figure 7**, we found that at the low SNR (SNR= −10 dB), the recognition rate for all modulation techniques is among 5 to 97 percent and BPSK recorded the highest percent and 16PSK recorded the lowest one. Then, as the

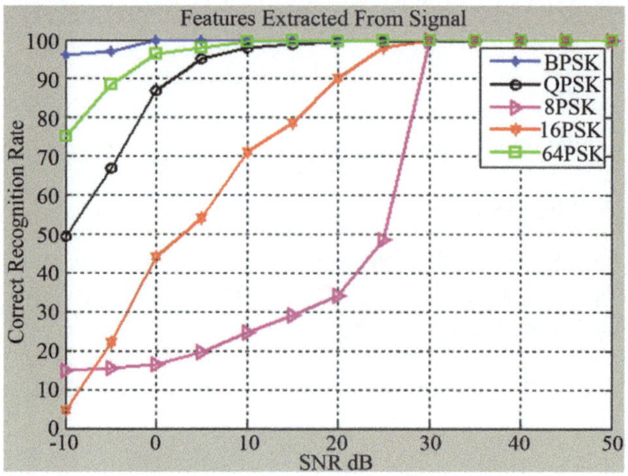

Figure 7. Recognition rate at different SNRs for various modulation schemes when features extracted from the signal.

SNR increases the modulation schemes BPSK, 64PSK QPSK, 16PSK and 8PSK, reach to 100 percent at SNR = 0, 10, 20, 30 and 30, respectively.

In **Figure 8**, we found that at the low SNR (SNR= −10 dB) the recognition rate for all modulation techniques is among 20 to 100 percent and BPSK recorded the highest percent and 16PSK recorded the lowest one. Then as the SNR increases the modulation schemes BPSK, QPSK, 8PSK, 16PSK and 64PSK are reached to 100 percent at SNR −10, 0, 25, 25 and 30 dB respectively.

In **Figure 9**, we found that at the low SNR (SNR= −10 dB), the recognition rate for all modulation techniques is among 18 to 99 percent. 64PSK recorded the highest percent and 8PSK recorded the lowest one. As the SNR increases, the modulation schemes 64PSK, BPSK, QPSK 8PSK and 16PSK reach 100 percent at SNR −5, 0, 15, 20, 30 dB respectively.

In **Figure 10**, we found that at the low SNR (SNR= −10 dB), the recognition rate for all modulation techniques is among 0 to 100 percent and QPSK recorded the highest percent. 8PSK, 16PSK, 64PSK recorded the lowest one. Then as the SNR increases the modulation schemes QPSK, BPSK, 8PSK, 164PSK and 64PSK reach 100 percent at SNR = −10, 0, 10, 10 and 10 dB, respectively.

In **Figure 11**, we found that at the low SNR (SNR = −10 dB) the recognition rate for all modulation techniques is among 0 to 100 percent. BPSK recorded the highest percent and 16PSK recorded the lowest one. Then, as the SNR increases, the modulation schemes BPSK, QPSK, 8PSK, 64PSK and 16PSK are reach 100 percent at SNR −10, −10, −5, 10 and 20 dB, respectively.

In **Figure 12**, we found that at the low SNR (SNR = −10 dB), the recognition rate for all modulation techniques is among 13 to 99 percent. QPSK recorded the highest percent and 16PSK recorded the lowest one. Then, as the SNR increases the modulation schemes QPSK, 64PSK, BPSK, 8PSK and 16PSK reach 100 percent at SNR = −5,

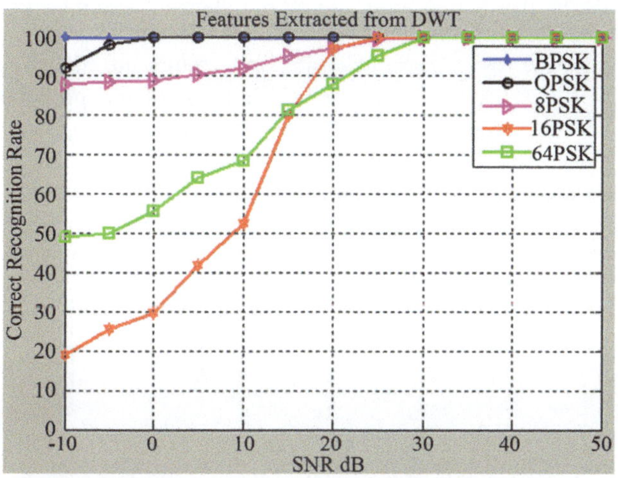

Figure 8. Recognition rate at different SNR for various modulation schemes, when features extracted from DWT.

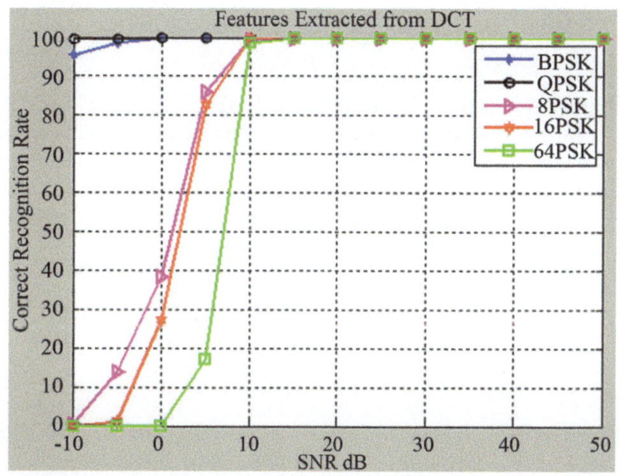

Figure 10. Recognition rate at different SNRs for various modulation schemes, when features extracted from the DCT of signal.

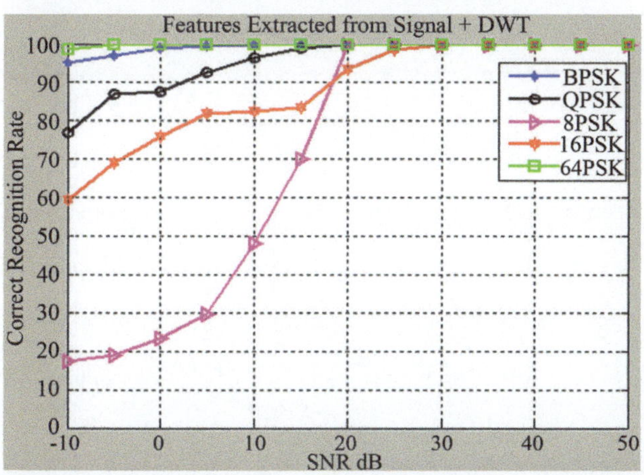

Figure 9. Recognition rate at different SNR for various modulation schemes, when features extracted from signal plus DWT of signal.

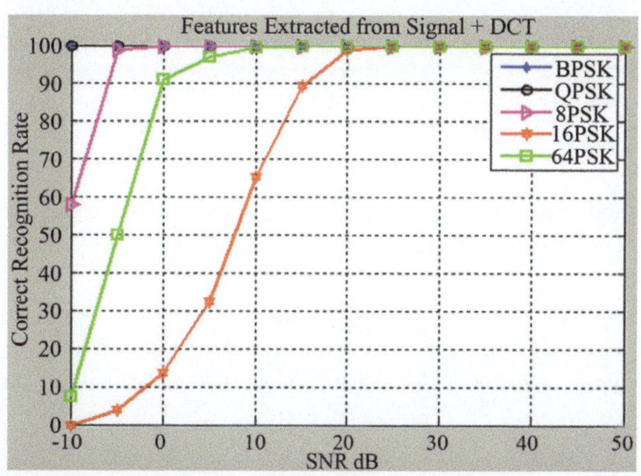

Figure 11. Recognition rate at different SNRs for various modulation schemes, when features extracted from signal plus DCT of signal.

−5, 0, 0 and 0 dB, respectively.

In **Figure 13**, we found that at the low SNR (SNR= −10 dB), the recognition rate for all modulation techniques is among 0 to 100 percent. BPSK, QPSK, and 8PSK recorded the highest percent and 16PSK recorded the lowest one. Then, as the SNR increases the modulation schemes BPSK, QPSK, 8PSK, 64PSK and 16PSK reach 100 percent at SNR −10, −10, −10, −5 and 10 dB, respectively.

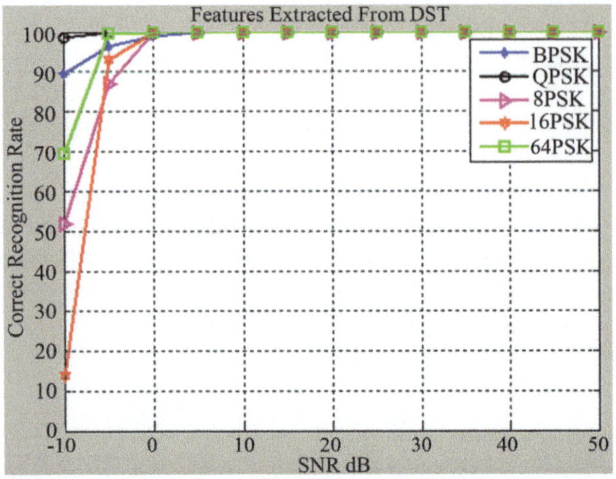

Figure 12. Recognition rate at different SNRs for various modulation schemes, when features are extracted from DST of signal.

So from the above results and by comparing them with the results of previous studies, we find that in [5], the CRPs of BPSK and QPSK when using 1024 samples at SNR = −10 dB are90.44 and 91.25 percent, respectively and at SNR = 25 dB are 99.71, and 99.90 percent, respectively. In [6], the CRPs of BPSK and QPSK at SNR = 0 dB are 93 and 89 percent, respectively, and at SNR = 30 dB are 99 and 100 percent, respectively. In [7], the CRP of BPSK, QPSK and 8PSK at SNR = 2 dB are 91, 84 and 78 percent, respectively, and at SNR = 8 are 100, 96 and 93 percent, respectively, and at SNR = 17 are 100, 100 and 100 percent, respectively. In [8], the CRPs of BPSK, QPSK and 8PSK at SNR = 2 dB are 91, 84 and 78 percent, respectively, and at SNR = 8 dB are 100, 96 and 93 percent, respectively, and at SNR = 17 dB are 100, 100 and 100 percent, respectively. In [9], the CRPs of BPSK and QPSK at SNR = 0 dB are 86.3 and 85.1 percent respectively, and at SNR = 4 dB are 93 and 92.8 percent, respectively. Comparison results are shown in **Table 1**. All these results are in favor of our proposed approach.

Table 1. Comparison between the proposed approach and other approaches.

50	45	40	35	30	25	20	15	10	8	5	4	2	0	−5	−10	SNR (dB) / Modulation	Algorithm
-	-	-	-	100	100	100	100	100	-	100	-	-	97.95	99.9	99.82	BPSK	[5]
-	-	-	-	100	-	-	-	-	-	-	-	-	98	-	-		[6]
-	-	-	-	-	-	-	100	-	100	-	-	91	-	-	-		[7]
-	-	-	-	-	-	-	100	-	100	-	-	91	-	-	-		[8]
-	-	-	-	-	-	100	100	-	98.2	-	93	-	86.3	-	-		[9]
100	100	100	100	100	100	100	100	100	100	100	100	100	100	100	100		Our Proposed
-	-	-	-	100	100	100	100	100	-	99.98	-	-	99.98	99.9	99.95	QPSK	[5]
-	-	-	-	100	-	-	-	-	-	-	-	-	97	-	-		[6]
-	-	-	-	-	-	-	100	-	96	-	-	84	-	-	-		[7]
-	-	-	-	-	-	-	100	-	96	-	-	84	-	-	-		[8]
-	-	-	-	-	-	100	99.9	-	97.3	-	92.8	-	85.1	-	-		[9]
100	100	100	100	100	100	100	100	100	100	100	100	100	100	100	100		Our Proposed
-	-	-	-	-	-	-	-	-	-	-	-	-	-	-	-	8PSK	[5]
-	-	-	-	-	-	-	-	-	-	-	-	-	-	-	-		[6]
-	-	-	-	-	-	-	100	-	93	-	-	78	-	-	-		[7]
-	-	-	-	-	-	-	100	-	93	-	-	78	-	-	-		[8]
-	-	-	-	-	-	-	-	-	-	-	-	-	-	-	-		[9]
100	100	100	100	100	100	100	100	100	100	100	100	100	100	100	100		Our Proposed
-	-	-	-	-	-	-	-	-	-	-	-	-	-	-	-	16PSK	[5]
-	-	-	-	-	-	-	-	-	-	-	-	-	-	-	-		[6]
-	-	-	-	-	-	-	-	-	-	-	-	-	-	-	-		[7]
-	-	-	-	-	-	-	-	-	-	-	-	-	-	-	-		[8]
-	-	-	-	-	-	-	-	-	-	-	-	-	-	-	-		[9]
100	100	100	100	100	100	100	100	100	100	100	100	97	81	24	0		Our Proposed
-	-	-	-	-	-	-	-	-	-	-	-	-	-	-	-	64PSK	[5]
-	-	-	-	-	-	-	-	-	-	-	-	-	-	-	-		[6]
-	-	-	-	-	-	-	-	-	-	-	-	-	-	-	-		[7]
-	-	-	-	-	-	-	-	-	-	-	-	-	-	-	-		[8]
-	-	-	-	-	-	-	-	-	-	-	-	-	-	-	-		[9]
100	100	100	100	100	100	100	100	100	100	100	100	100	100	100	98		Our Proposed

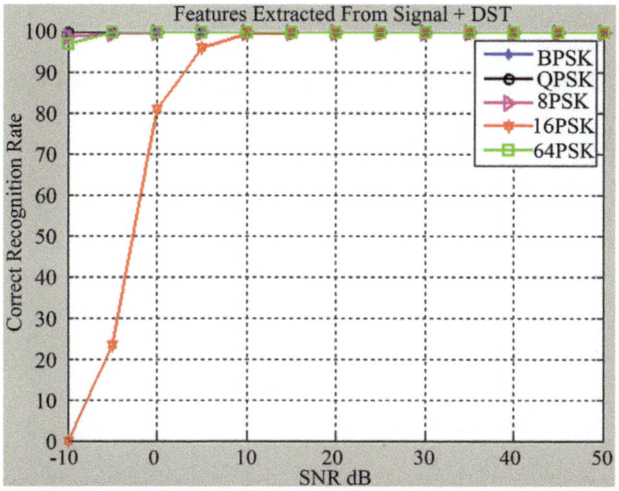

Figure 13. Recognition rate at different SNRs for various modulation schemes, when features are extracted from signal plus DST of signal.

7. Conclusion

This paper presented an efficient ADMR approachusing discrete transforms and cepstral analysis for MC-CDMA, and it can be used in any communication system this Simulation results have shown a great success of the proposed approach with various modulation techniques which opens the door for further research on automatic modulation classification with cepstral features.

REFERENCES

[1] M. W. Aslam, Z. Zhu and A. K. Nandi, "Automatic Digital Modulation Classification Using Genetic Programming with K-Nearest Neighbor," *The* 2010 *Military Communications Conference—Unclassified Program Waveforms and Signal Processing Track.*

[2] K. Hassan, I. Dayoub, W. Hamouda and M. Berbineau, "Automatic Modulation Recognition Using Wavelet Transform and Neural Networks in Wireless Systems," *EURASIP Journal on Advances in Signal Processing*, 2010.

[3] N. Yee, J. P. Linnartz and G. Fettweis, "Multi-Carrier CDMA in Indoor Wireless Radio Networks," *Proceedings PIMRC* '93, Yokohama, 8-11 September 1993, pp. 109-113.

[4] G. Fettweis, "Multi-Carrier Code Division Multiple Access (MC-CDMA): Basic Idea," Internal Report, Technekron Communication Systems.

[5] T. Zarey, J. Aboueiyy and K. N. Plataniotisyy, "KGDA for Signal Classification in Cognitive Radio."

[6] M. Petrova, P. M¨ah¨onen and A. Osuna, "Multi-Class Classification of Analog and Digital Signals in Cognitive Radios Using Support Vector Machines," Institute for Networked Systems, RWTH Aachen University Kackertstrasse 9, Aachen, Germany

[7] A. Ebrahimzadeh and G. Ardeshir, "A New Signal Type Classifier for Fading Environments," *Journal of Computing and Information Technology*, Vol. 15, No. 3, 2007, pp. 257-266.

[8] A. Ebrahimzadeh and S. A. Seyedin, "Automatic Digital Modulation Identification in Dispersive Channels," *Proceedings of the* 5*th WSEAS International Conference on Telecommunications and Informatics*, Istanbul, 27-29 May 2006, pp. 409-414.

[9] H. Hu, J. Song and Y. Wang, "Signal Classification Based on Spectral Correlation Analysis and SVM in Cognitive Radio," 22*nd International Conference on Advanced Information Networking and Applications*, Okinawa, 25-28 March 2008.

[10] C.-S. Park, W. Jang and S. P. Nah, "Automatic Modulation Recognition Using Support Vector Machine in Software Radio Applications," EW Lab, Agency for Defense Development, Korea.

[11] O. A. Dobre, A. Abdi, Y. Bar-Ness and W. Su, "Survey of Automatic Modulation Classification Techniques Classical Approaches and New Trends," Sarnoff Symposium, Princeton, 2005.

[12] T. Kinnunen, "Spectral Features for Automatic Test Independent Speaker Recognition," Licentiate's Thesis, Department of Computer Science, University of Joensuu, Finland, 2003.

[13] T. Matsui and S. Furui, "Comparison of Text-Independent Speaker Recognition Methods Using VQ-Distortion and Discrete/Continuous HMMs," *IEEE Transactions on Speech and Audio Processing*, Vol. 2, No. 3, 1994, pp. 456-459.

[14] B. S. Andrew and C. Lindgren, "Speech Recognition Using Features Extracted from Phase Space Reconstructions," 2003.

[15] J. S. Walker, "A Primer on Wavelets and Their Scientific Applications," CRC Press LLC, 1999.

[16] Z. Wu, X. Wang, Z. Gao and G. H. Ren, "Automatic Digital Modulation Recognition Based on Support Vector Machines."

[17] J. C. Platt and N. Cristianini, "Large Margin DAGs for Multiclass Classification," In: S. A. Solla, T. K. Leen and K.-R. M¨uller, Eds., MIT Press, Cambridge, 2000, pp. 547-553.

Performance Analysis of Decode and Forward Cooperative Relaying over the Generalized-K Channel

Ali Dziri[1], Michel Terre[1], Nidal Nasser[2]

[1]CEDRIC-LAETITIA, Conservatoire National des Arts et Metiers, Paris, France; [2]College of Engineering, Alfaisal University, Riyadh, KSA.

ABSTRACT

In this paper, we analyze the performance of cooperative diversity using adaptive Decode and Forward (DF) relaying over independent but non-necessarily identical flat composite fading channels which include multipath fading and shadowing simultaneously. We have considered the transmission of M-QAM modulated signals over Generalized-K channel model which is very versatile and accurately approximates many of the commonly used channel models as Nakagami-m, Log-Normal distributions and the mixture of the two distributions for the composite fading. Using an approximation of the Generalized-K pdf by a Gamma pdf, we derive analytical expressions for the outage probability and the Average Symbol Error Probability (ASEP) of the proposed scenario and analyze their dependence on the channel parameters. We assume in our context that the relay decides independently whether or not to forward the signal to the destination based on the received signal quality. Simulations are provided for 16-QAM modulation for both the outage and the ASEP. The obtained results are discussed and proved to be in good agreement with our theoretical analysis.

Keywords: Cooperative; Relay; Generalized-K Distribution; Composite Fading

1. Introduction

Multiple-Input Multiple-Output (MIMO) communication schemes have been widely studied to achieve diversity advantages, increased data rate, and interference suppression. However, one of the disadvantages of MIMO systems is the relatively large size needed of a wireless terminal. This is due to the requirement of an antenna spacing equal to half the wavelength to achieve sufficiently uncorrelated signals at the antenna connectors. Recently, it has been shown that cooperative diversity or user cooperation offers spatial diversity when multiple users cooperate to build a virtual antenna array [1,2], benefiting on MIMO advantages. Such a system aims to increase the channel capacity and to improve the link reliability of the wireless communications [2]. To this end, various relaying strategies have been investigated [3,4]. We distinguish mainly two techniques: Amplify and Forward (AF) (also known as analog relaying) and Decode and Forward (DF) (also known as digital relaying). In the former strategy, the relay node amplifies the received signal and then retransmits it. Thus, one disadvantage to be mentioned is the increased noise. In the latter strategy, the relay detects and possibly decodes the source message and then regenerates and retransmits it. This technique, is more reliable but it requires more processing compared to AF relaying [5]. The focus of this work is on DF relaying strategy.

To evaluate the system performance, we need to consider the propagation channel characteristics. In our context we consider the wireless channel which poses a severe challenge for reliable high-speed communication due to reflection, diffraction, scattering and absorption. To capture and model all these effects, many scientific research were carried out and they were classified mainly into three fading channel types: path-loss, multipath-fading (also known as small-scale fading) and shadowing (also known as large-scale fading).

Multipath fading is relatively fast and it is, therefore, responsible for the short-term signal variations. Depending on the radio propagation environment and the underlying communication scenario, there is a range of statistical multipath fading models available in the literature, such as Rayleigh, Rician, or Nakagami-m distributions [6]. The large-scale fading (shadowing) is due to objects obstructing the propagation path as trees and buildings, giving rise to random variations of the received power at a given distance. These fluctuations are experienced on local-mean powers. They are often modeled using a Log-normal distribution [7].

The combination of multipath fading and shadowing effects is usually known as composite fading. A versatile distribution to model the composite fading is the Generalized-K model which demonstrated a superior fit to the measured data over a wide range of physical channel conditions [8,9]. Thus, it has attracted a great deal of research interest recently. In [10], the authors have derived the outage probability, the average channel capacity and the ASEP for different modulated signals over the Generalized-K distribution. In [11], they have evaluated the spatial diversity performance over the Generalized-K channel.

Despite the multiplying number of scientific works on the cooperative diversity techniques mentioned above, most of them are limited to the Rayleigh, Rician, Log-normal and Nakagami-m channels. In [12,13], the authors studied the outage and the average error probability of adaptive DF cooperative networks with identical and non-identical Rayleigh fading channels. In [14,15] the authors derived the closed form for the outage probability of an adaptive DF over i.i.d Nakagami-m fading channels. In [16,17], the authors derived the closed form for the outage probability, the ASEP and the average channel capacity based on the MGF expression of the Nakagami-m distribution. In [18], AF cooperative diversity networks have been investigated under Generalized-K distribution. However, for a DF relaying over shadowed-Nakagami channels modeled by Generalized-K distribution, performance analysis has never done according to our best knowledge. Recently, we have evaluated the outage probability [19], and the ASEP for M-QAM modulated signals [20], of a DF relaying technique over a Generalized-K channel. However, a complete performance analysis for different propagation environments conditions is necessary. Thus, a more detailed theoretical analysis will be presented by considering the DF relaying over the Generalized-K channel. The performance metrics under consideration are the outage probability and the ASEP. Different propagation environments will be investigated. The remainder of this paper is organized as follows. In Section 2, we introduce the system model of the DF relaying over the Generalized-K fading model. We have introduced an approximation of the Generalized-K pdf by a Gamma pdf which serves for the outage and the ASEP analysis. The closed form derivations for the outage probability and the ASEP of the non cooperative system (direct link) and the cooperative system are presented in Sections 3 and 4, respectively. Simulation results are given in Sections 5 and 6 concludes this paper.

2. System and Channel Models

2.1. Relay System Model

We consider a dual-hop cooperative wireless system consisting of $J+2$ wireless nodes: one source node S, a set of J relay nodes, $R_1, R_2, \cdots, R_J$, and a destination node D. We assume that each node has a single antenna and all R_j operate in half-duplex mode. We consider a time-division channel allocation scheme with $J+1$ time slots. In the first time slot, the source broadcasts its message to all the nodes. During the following J time slots, the source stops transmitting and the relays of the decoding set will forward the source information to the destination in a scheduled order. Q denotes the set of indexes of relays which succeed in decoding correctly the source message for a given required SNR threshold, *i.e.*, each relay decides whether to forward the source information or not by comparing the received signal power with a decision SNR threshold. The cardinality of Q is denoted $\bar{J}$. In the considered scenario, we assume that the receivers at the relays and destination have exact channel state information but no transmitter channel state information is available at the source or relays [14].

Let us denote h_0, $\tilde{h}_j$ and h_j as the channel gains corresponding to the direct link, links between the source to the $j\text{-}th$ relay and the $j\text{-}th$ relay to the destination, respectively. Moreover, $\tilde{h}_j$ and h_0, h_j are supposed independent but not necessarily identical Generalized-K distributions with parameters $\tilde{m}_m^j$, $\tilde{m}_s^j$ for $j=1,\cdots,J$ and m_m^j, m_s^j for $j=0,1,\cdots,J$, respectively. Let $\tilde{\gamma}_j$ and γ_j denote the instantaneous SNR received at the the $j\text{-}th$ relay and the destination from the $j\text{-}th$ relay, respectively. Thus,

$$\tilde{\gamma}_j = \left|\tilde{h}_j\right|^2 \frac{Es}{N0} \text{ and } \gamma_j = \left|h_j\right|^2 \frac{Es}{N0}$$

where Es is the transmit energy signal and $N0$ represents the single sided power spectral density of the AWGN. The corresponding average SNR are denoted $\bar{\tilde{\gamma}}_j$ and $\bar{\gamma}_j$. Finally, we denote $\gamma_D = \sum_{j=0}^{\bar{J}} \gamma_j$ as the contribution of all the relays belonging to the set Q and the direct link using the MRC (Maximum Ratio Combiner) technique.

2.2. The Generalized-K Fading Channel Model

The small scale multipath fading is usually modeled using Nakagami-m distribution. The envelope of the transmitted signal over the Nakagami fading, X, has a probability density function $p_X(x)$ given as follows [21]

$$p_X(x) = \frac{2}{\Gamma(m_m)}\left(\frac{m_m}{\Omega}\right)^{m_m} x^{2m_m-1} e^{-\frac{m_m}{\Omega}x^2}, x > 0, \quad (1)$$

where m_m, $(0.5 \le m_m < +\infty)$, is the fading severity parameter called also the Nakagami parameter, Ω, $(\Omega = X^2)$, is the average received power and $\Gamma(.)$ is the Gamma function.

However, for example in terrestrial and satellite land-mobile systems, the link quality is also affected by slow variation of the mean signal level due to the shadowing from terrain, buildings, and trees. Consequently, multi-path fading are superimposed on shadowing. Hence, the pdf of the envelope of the received signal $p_X(x)$ will be conditioned on the average power Ω leading to:

$$p_{X|\Omega}(x|y) = \frac{2}{\Gamma(m_m)}\left(\frac{m_m}{y}\right)^{m_m} x^{2m_m-1} e^{-\frac{m_m}{y}x^2}, x > 0, \quad (2)$$

Consequently, in combined multipath and shadowing fading, (1) can be rewritten as follows

$$p_X(x) = \int_0^{+\infty} p_{X|\Omega}(x|y)\, p_\Omega(y)\mathrm{d}y, x > 0, \quad (3)$$

where $p_\Omega(y)$ is the pdf of the average power which is usually modeled by the Lognormal distribution [21]. To mitigate difficulties to derive closed form of some metric performance measure, using the moment matching method, it has been proposed to use the Gamma distribution to model the average power random variations given as follows

$$p_\Omega(y) = \frac{1}{\Gamma(m_s)}\left(\frac{m_s}{\Omega_0}\right)^{m_s} y^{m_s-1} e^{-\frac{m_s}{\Omega_0}y}, y \ge 0, \quad (4)$$

where $m_s > 0$ reflects the shadowing severity and Ω_0 is the local mean power.

Consequently, the pdf of the received signal envelope due to the composite fading (shadowing and multipath) leads to the Generalized-K distribution which is given by

$$p_X(x) = \frac{4\beta^{\frac{m_m+m_s}{2}}}{\Gamma(m_m)\Gamma(m_s)} x^{m_m+m_s-1} K_{m_s-m_m}\left(2\sqrt{\beta}x\right), \quad (5)$$
$$x \ge 0, \quad m_m \ge 0.5, \quad m_s > 0,$$

where $K_{m_s-m_m}$ is the modified Bessel function of the second kind and order $m_s - m_m$ and $\beta = m_m m_s / \Omega_0$.

In the cooperative system, as mentioned in [2], to capture the path-loss effects, we use

$$\Omega_0 = E\left(|h_0|^2\right) \sim \left(\frac{1}{d_{S,D}}\right)^\alpha, \quad \Omega_{0,j}^{S,R_j} = E\left(|\tilde{h}_j|^2\right) \sim \left(\frac{1}{d_{S,R_j}}\right)^\alpha,$$

$\Omega_{0,j}^{R_j,D} = E\left(|h_j|^2\right) \sim \left(\frac{1}{d_{R_j,D}}\right)^\alpha$ where $d_{i,j}$ is the distance between the terminals i and j, and α is the path-loss exponent.

Over a Generalized-K fading channel the pdf of the output instantaneous SNR γ follows a Generalized-K distribution given by [11]

$$p_\gamma(\gamma) = \frac{2\xi^{\frac{m_m+m_s}{2}}}{\Gamma(m_m)\Gamma(m_s)} \gamma^{\frac{m_m+m_s}{2}-1} K_{m_s-m_m}\left(2\sqrt{\xi\gamma}\right), \quad (6)$$
$$\gamma \ge 0, \quad m_m \ge 0.5, \quad m_s > 0,$$

where $\xi = \frac{m_m m_s}{\bar{\gamma}}$ and $\bar{\gamma}$ is the average SNR over the considered channel.

In the next section, we deal with the outage probability and the ASEP closed form derivations. To this end, in the cooperative case, we deal with the cdf of the sum of $\bar{J}+1$ Generalized-K random variables. As pointed in [22,23], the computation of the cdf expression is not straightforward and requires the use of approximations and asymptotic expansions or the use of the numerical inversion of the characteristic function. An alternative approach, to avoid the analytical difficulties, is to consider an approximation of the Generalized-K pdf, given by (6), using the moment matching method by a more tractable pdf [22], as shown in the next subsection.

2.3. The Approximation of a Generalized-K pdf

The main goal of this subsection is to give the analytical and main results of the approximation of the Generalized-K pdf by a suitable Gamma pdf. This approximation has been developed in two steps [22]:

- Approximation of the Generalized-K pdf by a Gamma pdf by matching the two first positive moments.
- Introduction of an adjustment factor for the shape and the scale parameters of the approximating Gamma pdf.

If we consider the distribution given by (5), the *n-th* moment of the Generalized-K distribution can be derived as [22]

$$E\left(X^n\right) = \frac{\Gamma(m_m+n)\Gamma(m_s+n)}{\Gamma(m_m)\Gamma(m_s)}\left(\frac{\Omega_0}{m_m m_s}\right)^n. \quad (7)$$

The *n-th* moment of the approximating Gamma distribution can be expressed as

$$E\left(X^n\right) = \frac{\Gamma(k+n)}{\Gamma(k)}\theta^n, \quad (8)$$

where θ and k denote the scale and shape parameters of the approximating Gamma distribution, respectively.

Matching the two first positive moments of the Generalized-K distribution to those of the approximating Gamma distribution results in

$$\theta = AF \cdot \Omega_0, \quad \text{and} \quad k = 1/AF \quad (9)$$

where $AF = \frac{1}{m_m m_s} + \frac{1}{m_m} + \frac{1}{m_s}$ is the amount of fading in the composite fading channel and $0 < AF < AF_{\max}$, $AF_{\max}$ is determined by the smallest physical values of m_m and m_s. However, this approximation results in a poor fitting in the lower and upper tail regions as compared to the original pdf. In order to bypass this inaccuracy, an adjustable form has been introduced for the scale and shape parameters of the region-wise approximating Gamma pdf for the two first moments. These parameters are given as follows

$$\tilde{\theta} = (AF - \varepsilon)\Omega_0, \quad \tilde{k} = \frac{1}{AF - \varepsilon}, \quad \tilde{\theta} > 0, \tilde{k} > 0, \quad (10)$$

where the adjustment parameter ε is chosen so that the pdf (cdf) of the approximating Gamma distribution becomes as close as desired to the pdf (cdf) of the Generalized-K distribution in a certain region of interest. We note that $-AF < \varepsilon < AF$. Since, the relevant practical range of AF is from zero (for non-fading channels) to 8 (for severe multipath fading and shadowing conditions corresponding to $m_m = 0.5$ and $m_s = 0.5$), the relevant range of the adjustment factor is $-8 < \varepsilon < 8$.

2.4. Approximation of the pdf of the Sum of N Independent Generalized-K Random Variables

As aforementioned in Subsection 2.3, the pdf of the sum of N Generalized-K random variables can be approximated by a Gamma pdf using the adjustable form.

$$\tilde{\theta}_N = \frac{\sum_{i=1}^{N} K_i \Omega_{0,i}^2 + 2\sum_{i=1}^{N}\sum_{j=i+1}^{N}\left(\Omega_{0,i}\Omega_{0,j}\right)}{\sum_{i=1}^{N}\Omega_{0,i}} - \sum_{i=1}^{N}\left(\Omega_{0,i} + \frac{\varepsilon_i}{N}\right), \quad \tilde{\theta}_N > 0, \quad (11)$$

$$\tilde{k}_N = \frac{\sum_{i=1}^{N}\Omega_{0,i}}{\tilde{\theta}_N}, \quad \tilde{k}_N > 0, \quad (12)$$

where $K_i = 1 + \frac{1}{m_m^i} + \frac{1}{m_s^i} + \frac{1}{m_m^i m_s^i}$. m_m^i and m_s^i are the shape parameters of the *i-th* Generalized-K random variable.

We note that for the i.i.d case, the expressions in (11) and (12) simplify to

$$\tilde{\theta}_N = (AF - \varepsilon)\Omega_0, \quad \tilde{\theta}_N > 0, \quad (13)$$

$$\tilde{k}_N = \frac{N}{AF - \varepsilon}, \quad \tilde{k}_N > 0, \quad (14)$$

An illustration of the choice of the optimal value of the adjustment factor is given in Section 5.

3. Performance of the Non Cooperative System (Direct Link)

In the following, we consider the outage probability and the ASEP as performance system metrics.

3.1. Outage Probability

The outage probability represents the probability of unsatisfactory signal reception. It can be defined as the probability that the received SNR is less than a specified SNR threshold value at the destination, denoted γ_D^{th} [10]. We consider the direct link composed by the source and the destination node communicating over the composite fading channel. The outage probability denoted P_{out} is given as [10]

$$P_{out} = P\left(\gamma_0 < \gamma_D^{th}\right) = \int_0^{\gamma_D^{th}} p_{\gamma_0}(\gamma)\,\mathrm{d}\gamma. \quad (15)$$

We note that $p_{\gamma_0}(\gamma)$ is given by (6). According to [10], we can derive P_{out} as follows

$$P_{out} = \pi \csc\left(\pi\left(m_s^0 - m_m^0\right)\right) \cdot \left[\frac{\left(\xi\gamma_D^{th}\right)^{m_m^0} {}_1F_2\left(m_m^0; 1 + m_m^0 - m_s^0; m_m^0 + 1; \xi\gamma_D^{th}\right)}{\Gamma\left(m_s^0\right)\Gamma\left(1 + m_m^0 - m_s^0\right)\Gamma\left(1 + m_m^0\right)} - \frac{\left(\xi\gamma_D^{th}\right)^{m_s^0} {}_1F_2\left(m_s^0; 1 + m_s^0 - m_m^0; m_s^0 + 1; \xi\gamma_D^{th}\right)}{\Gamma\left(m_m^0\right)\Gamma\left(1 + m_s^0 - m_m^0\right)\Gamma\left(1 + m_s^0\right)}\right], \quad (16)$$

where $\xi = \frac{m_m^0 m_s^0}{\bar{\gamma}_0}$, $\csc(\cdot)$ is the cosecant function, and ${}_pF_q(\cdot;\cdot;\cdot;\cdot)$ is the generalized hyper-geometric function [24].

3.2. Average Symbol Error Probability

The ASEP, P_{se}, for modulated signals over the fading channel h_0 can be obtained by averaging its conditional symbol error probability, $P_{se}(\gamma)$, under a given SNR γ in AWGN over the pdf of the received SNR $p_{\gamma_0}(\gamma)$

$$P_{se} = \int_0^{+\infty} P_{se}(\gamma)\, p_{\gamma_0}(\gamma)\,\mathrm{d}\gamma. \quad (17)$$

In this paper, we consider M-QAM modulated signals over Generalized-K fading channel. Thus, $P_{se}(\gamma)$ can be expressed as $P_{se}(\gamma) = A\,erfc\left(\sqrt{B\gamma}\right)$ [6], where $A = 2 - 2/\sqrt{M}$, $B = 3/\left(2(M-1)\right)$ and $erfc(x) = \frac{2}{\pi}\int_x^{+\infty} e^{-t^2}\,\mathrm{d}t$. By considering the same notations above, the closed form of the ASEP for the direct link is given by [10].

$$P_{se} = \frac{A\left(\frac{\xi}{B}\right)^{\frac{m_m^0+m_s^0}{2}}}{\sqrt{\pi}\Gamma\left(m_m^0\right)\Gamma\left(m_s^0\right)} \times G_{2,3}^{2,2}\left[\frac{\xi}{B}\left|\begin{matrix}1-\left(m_m^0+m_s^0\right)/2, -\left(m_m^0+m_s^0-1\right)/2 \\ \left(m_s^0-m_m^0\right)/2, \left(m_m^0-m_s^0\right)/2, -\left(m_m^0+m_s^0\right)/2\end{matrix}\right.\right] \quad (18)$$

where $G[\cdot]$, is the Meijer's G-function [24].

4. Performance of the Cooperative System

4.1. Outage Probability

In the cooperative case, the outage probability P_{out} is obtained by the contribution of all the relay nodes belonging to the set Q and the direct link using the MRC (Maximum Ratio Combining) technique. Thus, by using the law of total probability we obtain

$$\begin{aligned} P_{out} &= \sum_{Q} P\left(\gamma_D < \gamma_D^{th} \middle| Q\right) P(Q) \\ &= \sum_{\bar{J}=0}^{J} P\left(\gamma_D < \gamma_D^{th} \middle\| Q\right| = \bar{J}\right) P\left(|Q| = \bar{J}\right) \end{aligned} \quad (19)$$

where $P\left(|Q|=\bar{J}\right)$ is the probability that $\bar{J}$ nodes cooperate in relaying information to the destination.

$P\left(\gamma_D < \gamma_D^{th} \| Q| = \bar{J}\right)$ is the conditional outage probability in the case of $|Q| = \bar{J}$. In the following, we shall derive the expression of $P\left(|Q|=\bar{J}\right)$ and $P\left(\gamma_D < \gamma_D^{th} \| Q| = \bar{J}\right)$, respectively.

a) Calculation of $P\left(|Q|=\bar{J}\right)$

The probability that the j-th relay-node is in the decoding set Q, is defined as the probability that its SNR, $\tilde{\gamma}_j$, is larger than the threshold SNR at the j-th relay, $\tilde{\gamma}_j^{th}$. It is given as follows

$$A_j = P\left(\tilde{\gamma}_j > \tilde{\gamma}_j^{th}\right), \quad j \in [1, J] \quad (20)$$

For simplicity, we choose the same threshold for all the relays, *i.e,* $\tilde{\gamma}_j^{th} = \tilde{\gamma}^{th}$.

Consequently, A_j is given as follows

$$A_j = \int_{\tilde{\gamma}^{th}}^{+\infty} p_{\tilde{\gamma}_j}(\gamma)\,\mathrm{d}\gamma = 1 - \int_0^{\tilde{\gamma}^{th}} p_{\tilde{\gamma}_j}(\gamma)\,\mathrm{d}\gamma, \quad (21)$$

where $p_{\tilde{\gamma}_j}(\gamma)$ is the probability density of the SNR at the j-th relay. It is given by (6) and its cdf function is given by (16) by setting $\xi_j = \frac{\tilde{m}_m^j \tilde{m}_s^j}{\bar{\tilde{\gamma}}_j}$. Thus, A_j is deduced easily.

Let $\Phi(\bar{J})$ denotes a combination that $\bar{J}$ of J nodes cooperate in relaying information to the destination and $P_{\Phi(\bar{J})}$ as the corresponding probability. Assuming that the relay channels are independent, $P_{\Phi(\bar{J})}$ can be expressed as

$$P_{\Phi(\bar{J})} = \prod_{j=1}^{\bar{J}} A_j \prod_{j'=1, j' \notin \Phi(\bar{J})}^{J} \left(1 - A_{j'}\right). \quad (22)$$

Consequently,

$$P\left(|Q| = \bar{J}\right) = \sum_{\Phi(\bar{J})} P_{\Phi(\bar{J})}. \quad (23)$$

b) Calculation of $P\left(\gamma_D < \gamma_D^{th} \| Q| = \bar{J}\right)$

$P\left(\gamma_D < \gamma_D^{th} \| Q| = \bar{J}\right)$ can be seen as the cdf of $\gamma_D = \sum_{j=0}^{\bar{J}} \gamma_j$, evaluated at γ_D^{th}, *i.e.* the sum of the $\bar{J}+1$. Generalized-K random variables correspond to the $\bar{J}+1$ active links. As mentioned above in Section 2.4, to avoid analytical complexity, the computation of the cdf expression of γ_D can be approximated by the Gamma pdf with the scale and shape parameters given by (11) and (12), respectively. Consequently, $\gamma_D \sim \Gamma\left(\tilde{\theta}_{\bar{J}+1}, \tilde{k}_{\bar{J}+1}\right)$ and it is expressed as

$$p_{\gamma_D}(\gamma) = \frac{1}{\Gamma\left(\tilde{k}_{\bar{J}+1}\right)} \left(\frac{\tilde{k}_{\bar{J}+1}}{\bar{\gamma}_{\bar{J}+1}}\right)^{\tilde{k}_{\bar{J}+1}} \gamma^{\tilde{k}_{\bar{J}+1}-1} \mathrm{e}^{-\gamma \frac{\tilde{k}_{\bar{J}+1}}{\bar{\gamma}_{\bar{J}+1}}} \quad (24)$$

where $\bar{\gamma}_{\bar{J}+1} = \tilde{\theta}_{\bar{J}+1} \frac{Es}{N0}$.

Consequently $P\left(\gamma_D < \gamma_D^{th} \| Q| = \bar{J}\right)$ is given as follows

$$\begin{aligned} &P\left(\gamma_D < \gamma_D^{th} \| Q| = \bar{J}\right) \\ &= \int_0^{\gamma_D^{th}} p_{\gamma_D}(\gamma)\,\mathrm{d}\gamma = \frac{\gamma_{inc}\left(\tilde{k}_{\bar{J}+1}, \frac{\tilde{k}_{\bar{J}+1}}{\tilde{\theta}_{\bar{J}+1}} \gamma_D^{th}\right)}{\Gamma\left(\tilde{k}_{\bar{J}+1}\right)}, \end{aligned} \quad (25)$$

where $\gamma_{inc}(\cdot,\cdot)$ is the incomplete Gamma function. Finally, from (21)-(23) and (25) we derive the outage probability at the destination for the cooperative system as

$$P_{out} = \sum_{\bar{J}=0}^{J} \sum_{\Phi(\bar{J})} \prod_{j=1}^{\bar{J}} A_j \prod_{j'=1, j' \notin \Phi(\bar{J})}^{J} \left(1 - A_{j'}\right) \frac{\gamma_{inc}\left(\tilde{k}_{\bar{J}+1}, \frac{\tilde{k}_{\bar{J}+1}}{\tilde{\theta}_{\bar{J}+1}} \gamma_{th}^D\right)}{\Gamma\left(\tilde{k}_{\bar{J}+1}\right)} \quad (26)$$

4.2. Average Symbol Error Probability

In the cooperative case, the ASEP P_{se} is obtained by the contribution of all the relay nodes and the direct link. At the MRC output, by considering the same notation as above, P_{se} can be expressed as follows

$$P_{se} = \sum_{Q} P_{se}(Q) P(Q) = \sum_{\bar{J}=0}^{J} P_{se}\left(|Q| = \bar{J}\right) P\left(|Q| = \bar{J}\right) \quad (27)$$

$P\left(|Q|=\bar{J}\right)$ has been calculated in the last subsection given by (23). $P_{se}\left(|Q|=\bar{J}\right)$ is the conditional symbol error probability corresponding to the cooperation of the $\bar{J}$ branches and the direct link (source-destination). In the following, we shall derive its expression.

a) Calculation of $P_{se}\left(|Q|=\bar{J}\right)$

$P_{se}\left(|Q|=\bar{J}\right)$ can be seen as the probability of occurring an error when $\bar{J}$ relays are active. As mentioned in the last section, it can be expressed as

$$P_{se}\left(|Q|=\bar{J}\right)=\int_0^{+\infty}P_{se}\left(\gamma_D\right)p_{\gamma_D}\left(\gamma\right)\mathrm{d}\gamma, \quad (28)$$

where $p_{\gamma_D}(\gamma)$ is given by (24) and $P_{se}\left(|Q|=\bar{J}\right)$ is expressed as

$$P_{se}\left(|Q|=\bar{J}\right)=A\int_0^{+\infty}erfc\left(\sqrt{B\gamma}\right)p_{\gamma_D}\left(\gamma\right)\mathrm{d}\gamma. \quad (29)$$

To evaluate this integral, we use the relationship $Q(x)=\frac{1}{2}erfc\left(\frac{x}{\sqrt{2}}\right)$ which leads to a new integral form as follows

$$P_{se}\left(|Q|=\bar{J}\right)=2A\int_0^{+\infty}Q\left(a\sqrt{\gamma}\right)p_{\gamma_D}\left(\gamma\right)\mathrm{d}\gamma, \quad (30)$$

where $a=\sqrt{2B}$. Using ([6], (5.1), (5.18)), this integral is deduced as follows.

For integer values of $\tilde{k}_{\bar{J}+1}$, $P_{se}\left(|Q|=\bar{J}\right)$ can be written as

$$P_{se}\left(|Q|=\bar{J}\right)$$
$$=A\left[1-\sqrt{\frac{B\bar{\gamma}_{\bar{J}+1}}{\tilde{k}_{\bar{J}+1}+B\bar{\gamma}_{\bar{J}+1}}}\times\sum_{k=0}^{\tilde{k}_{\bar{J}+1}}\binom{2k}{k}\left(\frac{1}{4}-\frac{B\bar{\gamma}_{\bar{J}+1}}{4\tilde{k}_{\bar{J}+1}+4B\bar{\gamma}_{\bar{J}+1}}\right)^k\right]. \quad (31)$$

For non-integer values of $\tilde{k}_{\bar{J}+1}$, $P_{se}\left(|Q|=\bar{J}\right)$ can be written as

$$P_{se}\left(|Q|=\bar{J}\right)=A\cdot C\frac{\sqrt{B\bar{\gamma}_{\bar{J}+1}/\tilde{k}_{\bar{J}+1}}}{\left(1+B\bar{\gamma}_{\bar{J}+1}/\tilde{k}_{\bar{J}+1}\right)^{\tilde{k}_{\bar{J}+1}+\frac{1}{2}}}$$
$$\times{}_2F_1\left(1,\tilde{k}_{\bar{J}+1}+\frac{1}{2},\tilde{k}_{\bar{J}+1}+1,\frac{1}{1+\frac{B\bar{\gamma}_{\bar{J}+1}}{\tilde{k}_{\bar{J}+1}}}\right). \quad (32)$$

where $C=\Gamma\left(\tilde{k}_{\bar{J}+1}+1/2\right)/\left(\Gamma\left(\tilde{k}_{\bar{J}+1}+1\right)\sqrt{\pi}\right)$.

Thus, using (23), (27), (31) or (32), the ASEP P_{se} can be deduced easily.

5. Simulation Results

In this section, we present Monte-Carlo simulations to illustrate the previous theoretical results for both the outage probability and the ASEP.

5.1. Outage Probability

To be more accurate in our illustrations, we have simulated the shadowed Nakagami channel by taking into account the Nakagami-m distribution and the Log-normal distribution. For the non cooperative system, **Figure 1** depicts the outage probability versus the threshold SNR, γ_D^{th}, for a transmit SNR of $Es/N0=12\,\text{dB}$. The Threshold SNR at the relays is chosen $\tilde{\gamma}^{th}=\gamma_D^{th}$. We have considered three different environments with increasing propagation hardness:

- Generalized-K channel with $m_m=1$ and $m_s=50$: for high values of m_s, greater than 50, we consider that there is no shadowing. It is equivalent to Rayleigh channel. The corresponding plots in **Figure 1**, show less outage probability compared to the others configurations. Moreover, they are close each other. We note the robustness of the channel simulation compared to the theoretical rayleigh performance. Besides, we deduce the accuracy of the approximated model compared to the exact formula;
- Generalized-K channel with $m_m=1$ and $m_s=2$;
- Generalized-K channel with $m_m=0.5$ and $m_s=2$.

For the two last configurations, we note similar performances between the Generalized-K cdf and the approximated Gamma cdf with an optimal value of $\varepsilon=1.6$ for $m_m=0.5$, and $\varepsilon=1$ for $m_m=1$. More precisely, the optimal value of ε is obtained for lower tail approximation, since we are interested in small threshold values. We note just a small difference between theoretical results and simulation which is due to the approximation of the Log-normal pdf distribution by a Gamma distribution to model the shadowing fading.

In some cases, it is important to illustrate the outage

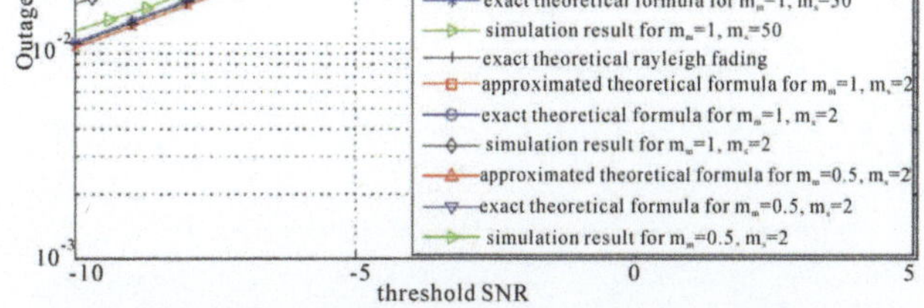

Figure 1. Outage probability versus the threshold SNR γ_D^{th} for the non cooperative system.

probability measure as function of the average SNR for a given threshold SNR. **Figure 2** shows this illustration for $\gamma_D^{th} = 5$ dB for the same parameters of the three configurations mentioned above.

In the cooperative communication scenario, we have omitted the propagation environment relative to Rayleigh case; we have considered the two others environments cases corresponding to $m_m = 1$, $m_s = 2$ and $m_m = 0.5$, $m_s = 2$. For both cases, we have considered identical Nakagami-m shadowed fading channels for the links of the first hop and those of the second dual hop.

We have considered a number of relay nodes, J = 1, 2, and 3. Similarly to the non cooperative system, we have illustrated:

- The outage probability versus the threshold SNR γ_D^{th} for a transmit SNR value of $Es/N0 = 12$ dB for both the first and the second hops. **Figures 3** and **4** show the obtained results.
- The outage probability versus $Es/N0$ for a given threshold SNR $\gamma_D^{th} = 12$ dB. **Figures 5** and **6** show the obtained results.

We note that, since we consider i.i.d channels, the optimal value of ε is the same for all links, as mentioned in the theoretical analysis. Thus, $\varepsilon = 1$ for $m_m = 1$, $m_s = 2$ and $\varepsilon = 1.6$ for $m_m = 0.5$, $m_s = 2$. We observe that theoretical results match well simulation results. The little difference is again due to the approximation of the Log-normal distribution by a Gamma distribution. As expected, the use of more relays provides better outage performance.

5.2. Average Symbol Error Probability

In this subsection, we have illustrated the results relative

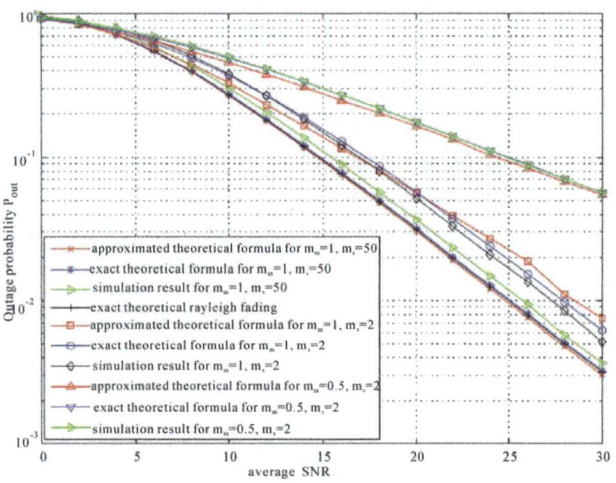

Figure 2. Outage probability versus $Es/N0$ for the non cooperative system.

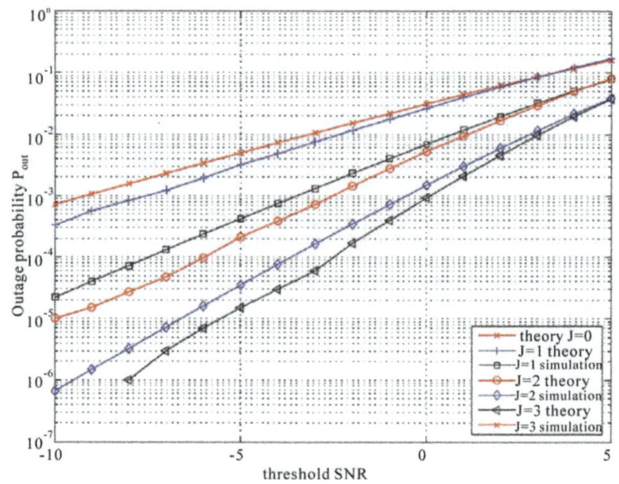

Figure 4. Outage probability versus the threshold SNR γ_D^{th} for the cooperative system for $m_m = 1$, $m_s = 2$.

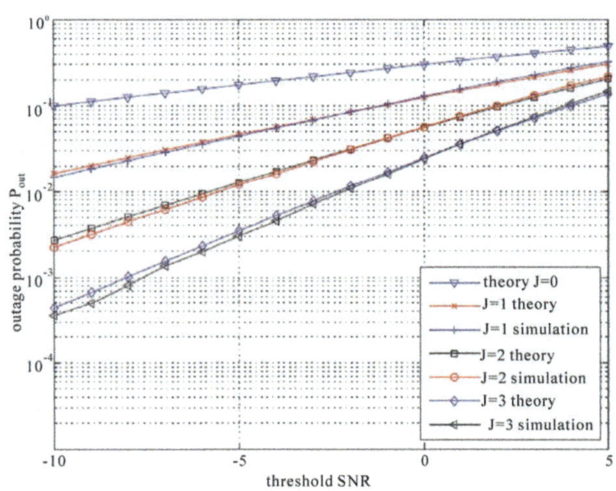

Figure 3. Outage probability versus the threshold SNR γ_D^{th} for the cooperative system for $m_m = 0.5$, $m_s = 2$.

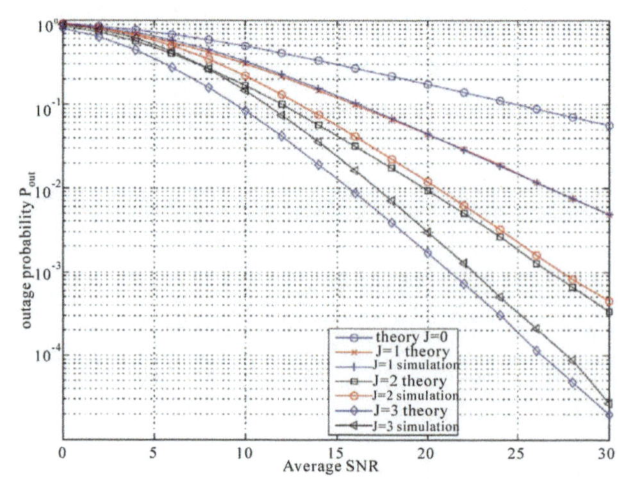

Figure 5. Outage probability versus $Es/N0$ for the cooperative system for $m_m = 0.5$ and $m_s = 2$.

to the ASEP for 16-QAM modulated signals. We plot the performance curves in terms of the analytical and simulation ASEP versus $Es/N0$ in dB. Similarly to the outage analysis, we consider two scenarii: the non co-operative and the cooperative systems. **Figure 7** shows the obtained results for the first scenario.

We note a similar performance between the theoretical Rayleigh curve, the approximated results and the simulation results.

For the second scenario, we have considered identical and independent channels with the following parameters $m_s = 2$ and $m_m = 1$. We have considered a threshold SNR $\gamma_D^{th} = \tilde{\gamma}^{th} = 15$ dB. To capture the path-loss effects, we have considered the following distances $d_{S,R} = 0.7$, $d_{R_j,D} = 0.8$ and $d_{S,D} = 1$. We have chosen a number of relay nodes, J = 1, 2, and 3. Obtained results are depicted in **Figure 8**. Optimal values of ε are calculated for each $Es/N0$ value. They are given by **Table 1**.

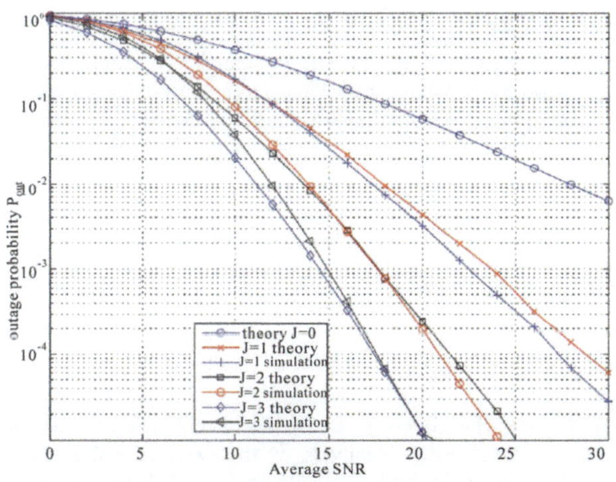

Figure 6. Outage probability versus $Es/N0$ for the co-operative system for $m_m = 1$ and $m_s = 2$.

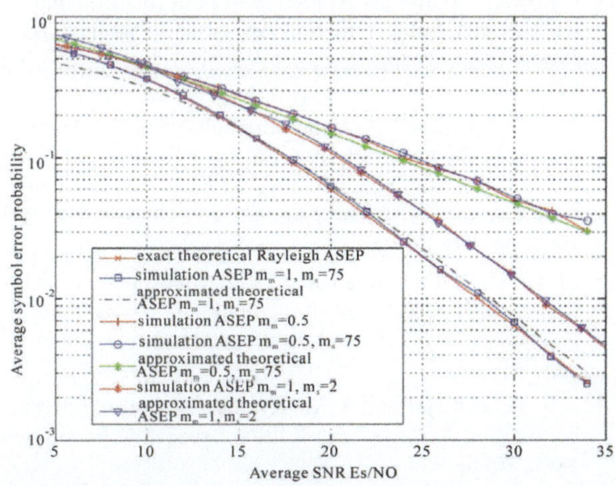

Figure 7. Average symbol error probability versus $Es/N0$ for the non cooperative system.

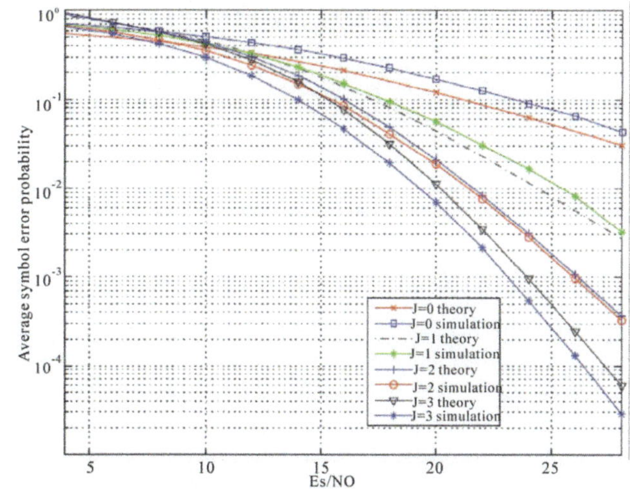

Figure 8. Average symbol error probability versus $Es/N0$ for $m_m = 1$ and $m_s = 2$.

Table 1. ε_{opt} versus $Es/N0$.

$Es/N0$	4	6	8	10	12	14	16
ε_{opt}	0.67	0.69	0.7	0.71	0.72	0.73	0.74

$Es/N0$	18	20	22	24	26	28
ε_{opt}	0.74	0.75	0.75	0.76	0.76	0.76

We note a good match between the simulation and theoretical results. The small difference between theoretical results and simulation includes the combination of the two approximations: approximation of the Lognormal pdf distribution by a Gamma distribution and the approximation of the Generalized-K distribution by a Gamma distribution. As expected, the use of more relays provides better ASEP performance.

6. Conclusion

We have analyzed the outage probability and the ASEP of an adaptive DF relaying over the composite fading multipath/shadowing channels modeled by a Generalized-K distribution. The utilization of an approximation of the Generalized-K pdf by an adjustable form of a Gamma distribution based on the two first moments matching method has been presented and validated by simulation under severe fading conditions. Based on this approximation, we derived closed-form expressions for the outage probability and the ASEP. We have provided Monte-Carlo simulations which are in good agreement with our theoretical analysis. This solution is helpful to investigate the performance analysis of cooperative networks under different environment conditions.

REFERENCES

[1] A. Nosratinia, T. E. Hunter and A. Hedayat, "Cooperative Communication in Wireless Networks," *IEEE Communications Magazine*, Vol. 42, No. 10, 2004, pp. 74-80.

[2] J. N. Laneman, D. N. C. Tse and G. W. Wornell, "Cooperative Diversity in Wireless Networks: Efficient Protocols and Outage Behavior," *IEEE Transactions on Information Theory*, Vol. 50, No. 12, 2004, pp. 3062-3080.

[3] M. O. Hasna and M. S. Alouini, "Outage Probability of Multihop Transmission over Nakagami Fading Channels," *IEEE Communications Letters*, Vol. 7, No. 5, 2003, pp. 216-218.

[4] M. O. Hasna and M.-S. Alouini, "End-to-End Performance of Transmission Systems with Relays over Rayleigh-Fading Channels," *IEEE Transactions on Information Theory*, Vol. 1, No. 6, 2003, pp. 1126-1131.

[5] E. Zimmermann, P. Herhold and G. Fettweis, "On the Performance of Cooperative Relaying Protocols in Wireless Networks," *Special Issue on European Wireless*, Vol. 16, No. 1, 2005, pp. 5-16.

[6] M. K. Simon and M. S. Alouini, "Digital Communications over Fading Channels," 2nd Edition, Wiley-IEEE Press, Hoboken, 2004.

[7] A. J. Coulson, A. G. Williamson and R. G. Vaughan, "A Statistical Basis for Lognormal Shadowing Effects in Multipath Fading Channels," *IEEE Transactions on Communications*, Vol. 46, No. 4, 1998, pp. 494-502.

[8] A. Abdi and M. Kaveh, "K-Distribution: An Appropriate Substitute for Rayleigh-Lognormal Distribution in Fading-Shadowing Wireless Channels," *IEEE Electron Device Letters*, Vol. 34, No. 9, 1998, pp. 851-852.

[9] F. Hansen and F. Meno, "Mobile Fading-Rayleigh and Lognormal Superimposed," *IEEE Transactions on Vehicular Technology*, Vol. 26, No. 4, 1977, pp. 332-335.

[10] P. S. Bithas, N. C. Sagias, P. T. Mathiopoulos, G. K. Karagiannidis and A. A. Rontogiannis, "On the Performance Analysis of Digital Communications over Generalized-K Fading Channels," *IEEE Communications Letters*, Vol. 10, No. 5, 2006, pp. 353-355.

[11] P. S. Bithas, P. T. Mathiopoulos and S. A. Kotsopoulos, "Diversity Reception over Generalized-K (KG) Fading Channels," *IEEE Transactions on Wireless Communications*, Vol. 6, No. 12, 2007, pp. 4238-4243.

[12] N. C. Beaulieu and J. Hu, "A Closed-Form Expression for the Outage Probability of Decode-and-Forward Relaying Dissimilar Rayleigh Fading Channels," *IEEE Communications Letters*, Vol. 10, No. 12, 2006, pp. 813-815.

[13] J. Hu and N. C. Beaulieu, "Closed-Form Expressions for the Outage and Error Probabilities of Decode-and-Forward Relaying in Dissimilar Rayleigh Fading Channels," *IEEE International Conference on Communication*, Glasgow, 24-28 June 2007, pp. 5553-5557.

[14] H. A. Suraweera, P. Smith and J. Armstrong, "Outage Probability of Cooperative Relay Networks in Nakagami-m Fading Channels," *IEEE Communications Letters*, Vol. 10, No. 12, 2006, pp. 834-836.

[15] C. K. Datsikas, N. C. Sagias, F. I. Lazarakis, and G. S. Tombras, "Outage Analysis of Decode and Forward Relaying over Nakagfami-m in Fading Channels," *IEEE Signal Processing Letters*, Vol. 15, 2008, p. 4144.

[16] S. S. Ikki and M. H. Ahmed, "Performance Analysis of Multi-Branch Decode-and-Forward Cooperative Diversity Networks over Nakagami-m Fading Channels," *IEEE International Conference on Communications*, St. John's, 14-18 June 2009, pp. 1-6.

[17] S. Ikki and M. H. Ahmed, "Performance Analysis of Cooperative Diversity Wireless Networks over Nakagami-m Channels," *IEEE Communications Letters*, Vol. 11, No. 4, 2007, pp. 334-336.

[18] G. P. Efthymoglou, N. Bissias and V. A. Aalo, "On the Error Rate Analysis of Dual-Hop Amplify-and-Forward Relaying in Generalized-K Fading Channels," *Journal of Electrical and Computer Engineering*, 2010, pp. 1-5.

[19] A. Dziri, D. L. Ruyet, D. Roviras and M. Terre, "Outage Probability Analysis of the Decode and Forward Relaying over the Composite Fading Multipath/Shadowing Channels," 2010 *IEEE Eleventh International Workshop on Signal Processing Advances in Wireless Communications* (*SPAWC*), Paris, 20-23 June 2010, pp. 1-5.

[20] A. Dziri, D. Le Ruyet, D. Roviras and M. Terré, "Symbol Error Probability Analysis of the Decode and Forward Relaying over the Composite Fading Multipath/Shadowing Channels," *IEEE International Symposium on Signal Processing and Information Technology*, Luxor, 2010, pp. 171-176.

[21] I. Kostic, "Analytical Approach to Performance Analysis for Channel Subject to Shadowing and Fading," *IEEE Proceedings-Communications*, Vol. 152, No. 6, 2005, pp. 821-827.

[22] S. Al-Ahmadi and H. Yanikomeroglu, "On the Approximation of the Generalized-K pdf by a Gamma pdf Using the Moment Matching Method," *IEEE Wireless Communications and Networking Conference*, Ottawa, 5-8 April 2009, pp. 1-6.

[23] M. Gu and D. A. Abraham, "Using McDaniels Model to Represent Non-Rayleigh Reverberation," *IEEE Transactions on Oceanic Engineering*, Vol. 26, No. 3, 2001, pp. 348-357.

[24] I. S. Gradshteyn and I. M. Ryzhik, "Table of Integrals, Series, and Products," 6th Edition, Academic Press, New York, 2000.

Effect of Partial Ground Plane Removal on the Radiation Characteristics of a Microstrip Antenna

Hong-Min Lee[1], Wong-Sang Choi[2]

Department of Electronic Engineering, Kyonggi University, Suwon, South Korea.

ABSTRACT

This study presents a new, simple method for reducing the back-lobe radiation of a microstrip antenna (MSA) by a partially removed ground plane of the antenna. The effect of the partial ground plane removal in different configurations on the radiation characteristics of a MSA are investigated numerically. The partial ground plane removal reduces the back-lobe radiation of the MSA by suppressing the surface wave diffraction from the edges of the antenna ground plane. For further improving the front-to-back (F/B) ratio of the MSA, a new soft-surface configuration consisting of an array of stand-up split ring resonators (SRRs) are placed on a bare dielectric substrate near the two ground plane edges. Compared to the F/B ratio of a conventional MSA with a full ground plane of the same size, an improved F/B ratio of 9.7 dB has been achieved experimentally for our proposed MSA.

Keywords: Front-to-Back Ratio; Microstrip Antenna; Removed Ground Plane; Soft-Surface; Split Ring Resonator

1. Introduction

Microstrip antennas (MSAs) are used in modern communication systems due to its low cost, lightweight, and planar structure. One of the major concerns in practical MSA design is surface wave excitation. When an MSA is fabricated on a substrate, it shows significant performance degradation by surface waves. On a finite ground plane, surface waves propagate until they reach an edge where they reflected back and diffracted by the edges. Particularly when the patch antenna is printed on high dielectric substrates, its back radiation pattern increases owing to surface wave diffraction from the edges of the antenna ground plane. Numerous efforts have been made earlier to reduce the surface waves on an MSA. One approach is to construct an artificial periodic structure such as an electromagnetic band-gap (EBG) [1] or a conventional artificial soft/hard surface [2-5]. Both EBG and soft surfaces can be used to suppress surface wave propagation. The main difference is that soft surfaces exhibit band-gaps only one direction, but they offer the best performance in most applications of antennas. Another approach is to use a micro-machining technology [6], in which part of the substrate beneath the radiating patch is removed to realize a low effective dielectric constant of the substrate. This structure can be used to reduce the surface waves, it involves high fabrication cost. In recent years, methods for reducing the mutual coupling between antennas [7] and enhancing the front-to-back (F/B) ratio of the MSA [8,9] have been investigated. Although the structure reported in [8], can be used to reduce the surface waves, they require a considerable area to form a band-gap. In [9], the broadside gain of an MSA is increased by around 2 dBi by partially removing the substrate surrounding the patch element. However, this method cannot be used to reduce the back-lobe gain. We present a new method for reducing the back radiation of an MSA by employing a partially removed ground plane and soft-surface structures.

2. Effect of the Partial Ground Plane Removal on the Gains of MSA

To investigate the effect of the ground plane size on the radiation characteristics of an MSA, a conventional MSA with a full metallic ground plane that operates in the dominant mode (TM_{10}) at 2.5 GHz is designed.

The simulated peak broadside gain curve of the probe feed MSA with a different size of the ground plane (S mm × S mm) is plotted in **Figure 1**. This MSA with a size of square patch element (P × P = 18 mm × 18 mm) is designed on the Rogers R3210 substrate (having a relative dielectric constant of 10.2, thickness of 1.27 mm, and loss tangent of 0.0025), and simulations are carried out using CST Microwave Studio. It is observed from **Figure 1** that as the size of the ground plane increases from 40 mm to 140 mm, the broadside gain remains around 3.2 dBi, and it subsequently decreases to −1.0 dBi at 210 mm. We choose the length of the ground plane of the MSA as 120 mm.

Next, in order to investigate the effect of partial ground plane removal on the F/B ratio of the MSA, some metallic parts of the MSA ground plane with a width of G_L is removed, as shown in **Figure 2**. The simulated F/B ratio curves of the MSA with a different size of the partially removed ground plane are plotted in **Figure 3** and summarized in **Table 1**.

In the back-lobe region radiation pattern of the MSA, the E-plane edge diffraction of surface waves has a much larger contribution to the magnitude than the H-plane edge diffraction [10], and it decreases by reducing the length of the ground plane (H-plane removal). However, in the case of E-plane removal the back-lobe gain (at θ = 180°) was not improved. As a result, the F/B ratio remains around 14 dB. In the both E- and H-plane removal case, F/B ratio of the MSA was improved significantly as the width of the partially removed ground plane G_L increased. It is observed from **Figure 3** that as the width of the partially removed ground plane G_L increased from 0 to 20 mm, the F/B ratio of the MSA shows 23.1 dB. The F/B ratio of the MSA increases to a maximum at G_L = 22.5 mm, with the F/B ratio peak of 23.6 dB. We choose the partially removed ground plane size of the MSA as 75 mm × 75 mm. **Figure 4** shows the plot of the simulated normal electric field (E_z) distributions in the E-plane of the MSA with a full ground plane and a partial ground plane removal. In the case of a conventional MSA with a full ground plane, surface waves are diffracted at the end of the ground plane edges, as shown in **Figure 4(a)**.

As a result, the partial ground-plane removal method can reduce the back radiation of the MSA owing to the suppression of surface wave diffraction from the edges of the conventional MSA ground plane. On the other hand, in the case of a MSA with a partial ground plane removal, there are field minima at the ends of the removed ground plane, as shown in **Figure 4(b)**.

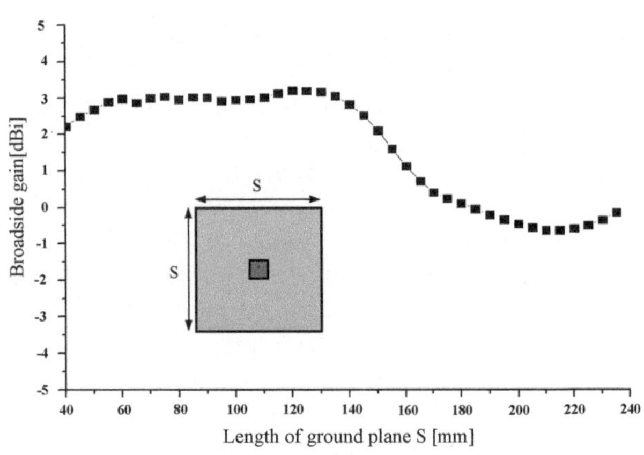

Figure 1. Simulated broadside gain versus the ground plane size S of a conventional MSA.

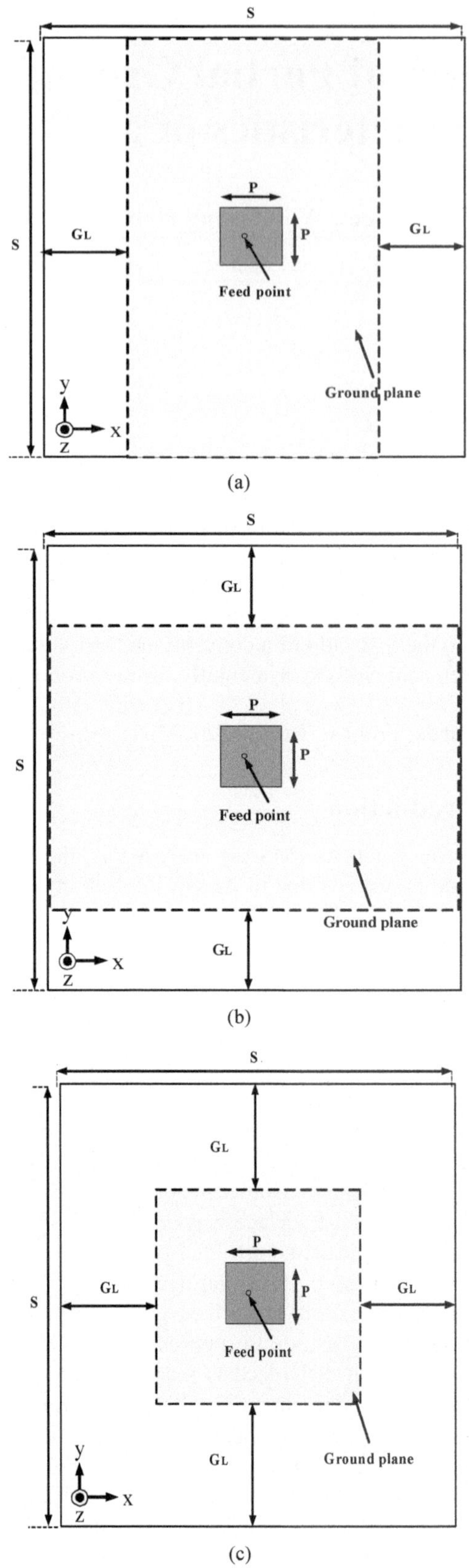

Figure 2. Geometry of MSA with different partial ground plane removal width G_L: (a) E-plane removal; (b) H-plane removal; (c) Both E- and H-plane removal.

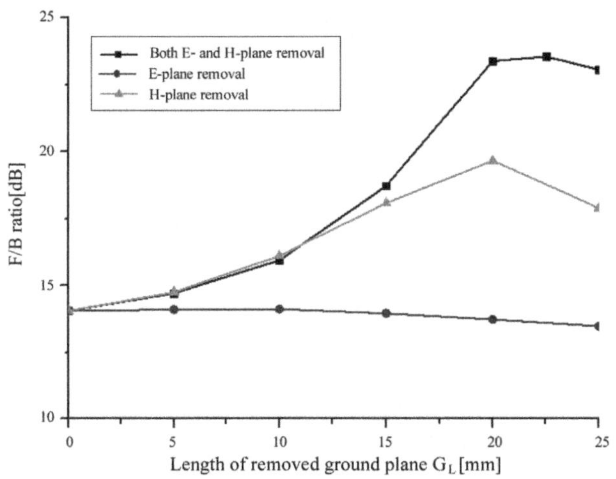

Figure 3. Simulated F/B ratio versus the length of the partially removed ground plane G_L.

Table 1. Comparison of simulated F/B ratio for the different partial ground plane removal width G_L.

Partial Removed Ground Plane Width G_L [mm]	F/B ratio [dB]		
	E-Plane Removal	H-Plane Removal	Both E- and H-Plane Removal
0	14.0	14.0	14.0
5	14.1	14.8	14.7
10	14.1	16.1	15.9
15	13.9	18.1	18.7
20	13.7	19.7	23.3
25	13.5	17.9	23.0

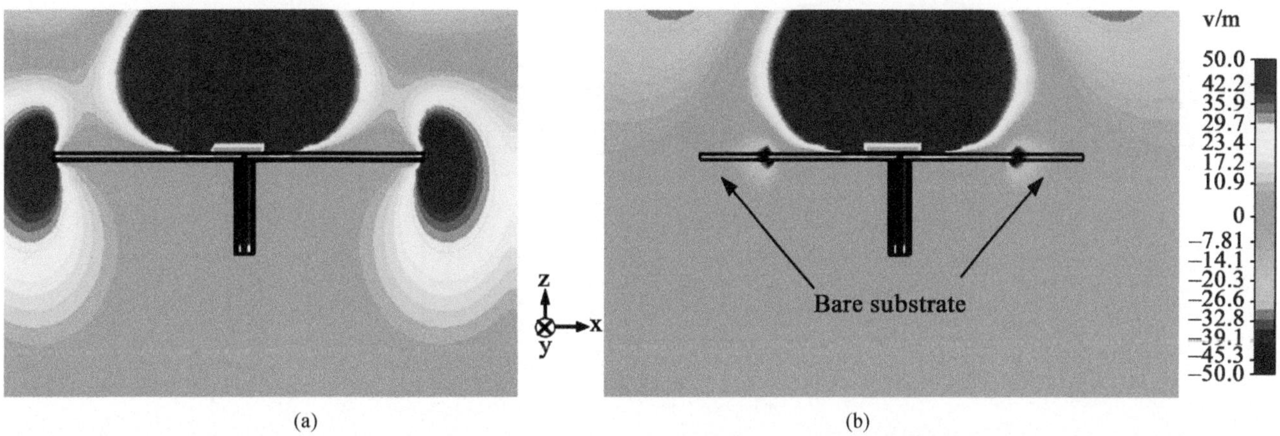

Figure 4. Simulated normal electric field (E_z) distributions of the MSA: (a) With a full ground plane; (b) With a partial ground plane removal.

For further improving the F/B ratio of the MSA with a simple soft-surface structure consist of split ring resonators (SRRs), we investigate the field excited near the MSA. **Figure 5** shows the plot of the simulated tangential magnetic field component (H_y) distributions within the substrate excited by the MSA. It is observed that the magnitudes of the tangential magnetic field component (H_y) are strongly distributed along the E-plane edge direction of the MSA and decay slowly with an increase in the distance along the y-axis. In contrast, the magnitudes of the tangential magnetic field component (H_y) along the H-plane edge direction of the MSA decay rapidly with an increase in the distance along the x-axis. The complex power density in the vicinity of the MSA on the ground plane can be expressed with the tangential magnetic field components (H_x and H_y), and the normal electric field component (E_z), according to equation as

$$\bar{S} = \frac{1}{2}\left(E_z H_x^* \hat{y} - E_z H_y^* \hat{x}\right) \quad (1)$$

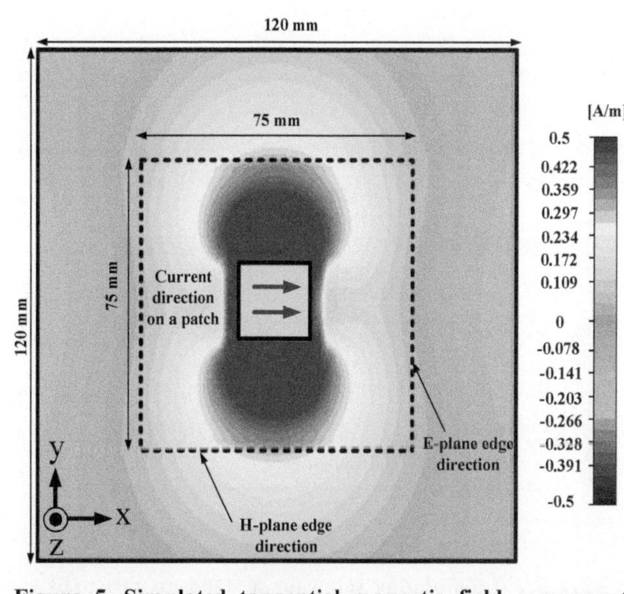

Figure 5. Simulated tangential magnetic field component (H_y) distributions on the MSA.

For reducing the ground plane edge diffraction along the two H-plane edges, which account for back radiation of the MSA, we prepared a soft-surface structure consisting of an array of stand-up SRRs. By placing a stand-up SRR near the side of the partially removed ground plane edge, where the magnitudes of the tangential magnetic field component (H_y) are strongly distributed, some parts of the complex power density can be reduced. A schematic of the simulation setup for an optimum stand-up SRR unit cell within a dielectric substrate and the simulated transmission coefficient are shown in **Figure 6**. As shown in **Figure 6**, the optimized physical dimensions (in millimeters) of the stand-up SRR are as follows: a = 13.8, w = 0.4, h = 1.27 and g = 0.7. A single SRR consisted of the metal strips and the two metallic via-holes, as shown in **Figures 6(a)** and **(b)**. The metal used for the metallic SRR patterns is a copper with a conductivity $\sigma = 5.8 \times 10^7$ S/m. It was placed within a Rogers R3210 substrate. Computer simulations for one SRR with appropriate boundary conditions are carried out. The perfect electric conductor boundary conditions are applied to the top and bottom walls of the waveguide. Perfect magnetic conductor boundary conditions are applied to the side walls of the waveguide.

A single SRR unit cell is placed inside a waveguide, and a vertically polarized transverse electromagnetic (TEM) wave is incident normally on the front side of port 1, as shown in **Figure 6(a)**.

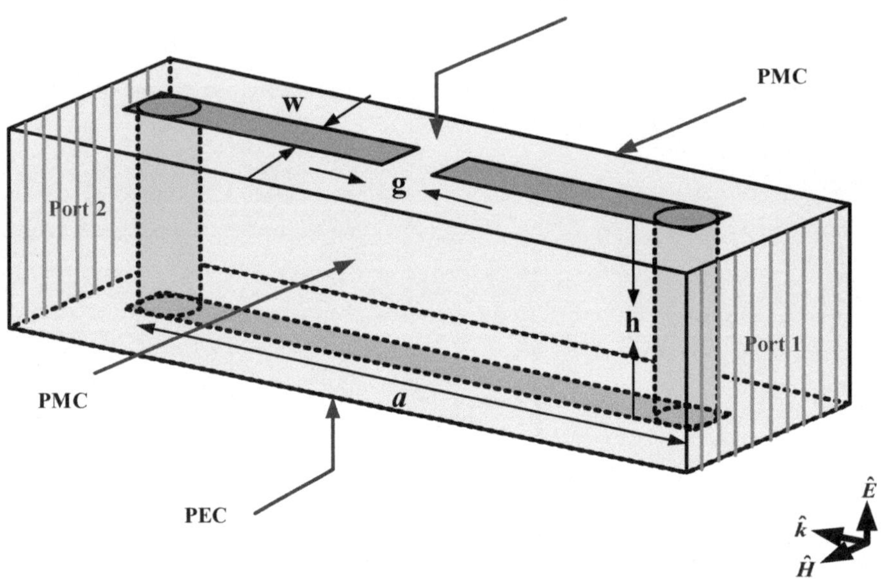

(a)

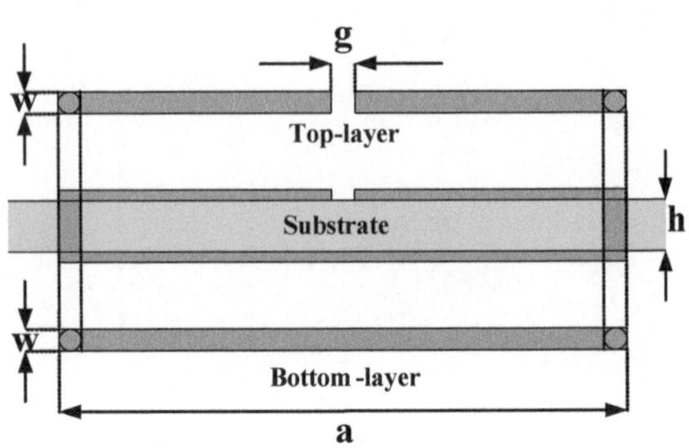

(b)

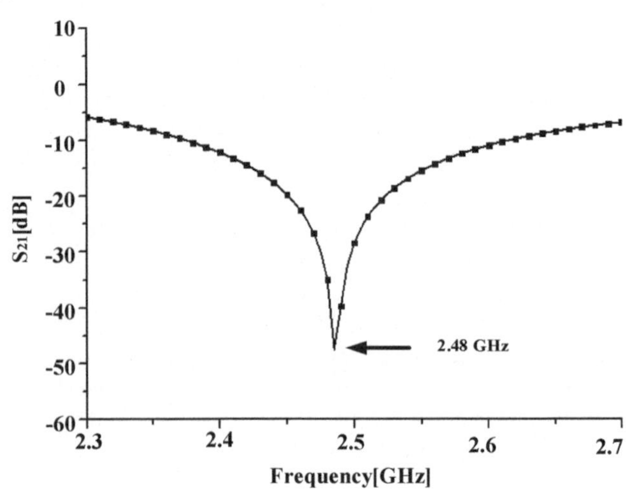

(c)

Figure 6. (a) Simulation setup for a stand-up SRR; (b) Cross-sectional view of a stand-up SRR in dielectric substrate; (c) Simulated transmission coefficient of a stand-up SRR.

It is well known that incident electromagnetic wave can resonantly couple to the LC resonance of an SRR through either the electric or magnetic field. This occurs either when the electric field is parallel to the side containing the SRR gap or when the magnetic field has a component normal to the plane of the SRR ring. When the incident wave is polarized as indicated in **Figure 6(a)**, the excitation is purely magnetic. The simulated transmission coefficient (S_{21}) of the SRR as a function of the frequency is shown in **Figure 6(c)**. We observe that the magnitudes of the transmission coefficients of a stand-up SRR are very low at the fundamental resonance frequency of 2.48 GHz, and the simulated stop-bandwidth (≤ -10 dB) is 180 MHz. **Figure 7** shows the geometry of the proposed MSA loaded with the soft surface which composed of the stand-up SRRs in partially removed ground plane.

The optimized physical dimensions (in millimeters) of the MSA and the stand-up SRR are as follows: S = 120, G = 75, P = 18, d = 15.1, a = 13.8, w = 0.4, h = 1.27 and g = 0.7. In total, six SRRs are placed 0.5 mm off the ground plane along the H-plane edge direction (x-axis) of the MSA. Initially, the effect of varying the number of periods N in terms of the F/B ratio of the MSA was studied. **Figure 8** shows the geometry of the soft surface made with two periods of SRRs (N = 2) near ground plane. It can be seen from **Table 2** that the back-lobe gain of the MSA is reduced as the number of period increases. However, the broadside gain of the MSA decreases significantly when the number of periods is three (N = 3). This is attributed to the fact that a portion of the diffracted waves radiated from the ground plane edges also contributes to space waves. As a result, two period of array structure with four SRRs was used in this design. The distance l between the ground plane edge and the arrayed SRRs is another determinant parameter. The effect of introducing the soft surface made with two periods of SRRs (N = 2) near ground plane was studied as a function of the distance l from the ground plane edge to the SRRs. **Figure 9** shows the simulated broadside and back-lobe gain of the MSA for the different values of distance l. An important fact related to F/B ratio of the MSA is that the decrease of the distance l does not necessary guarantee a reduction in the back radiation, as it

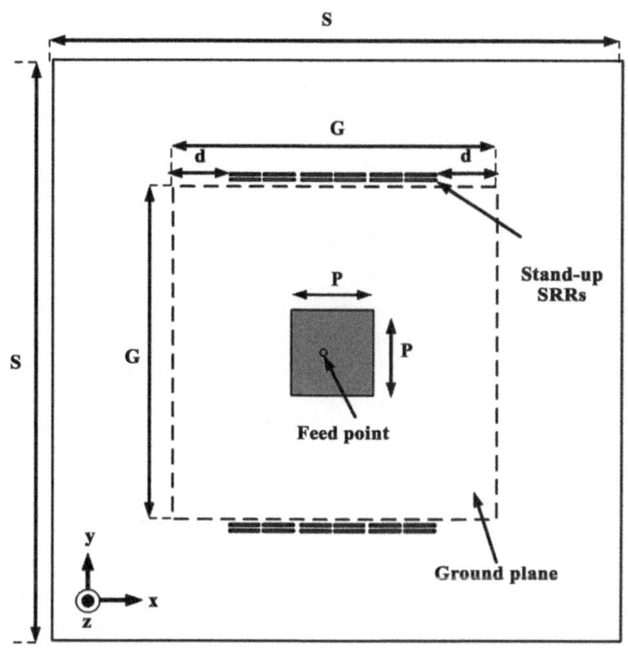

Figure 7. Geometry of a MSA loaded with the soft surface in partially removed ground plane.

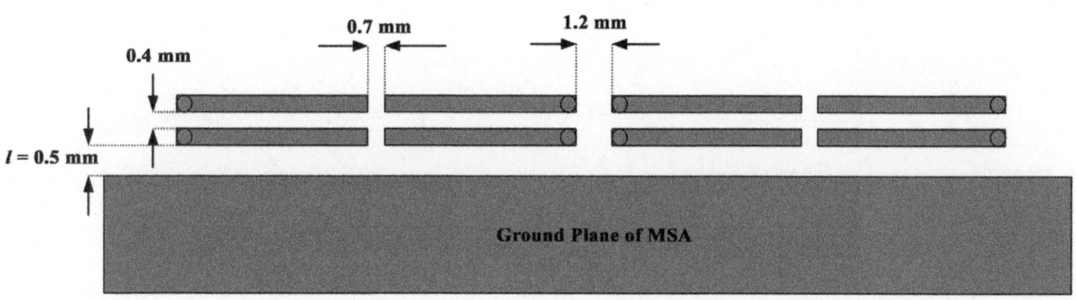

Figure 8. Geometry of the soft surface made with two periods of stand-up SRRs (N = 2) near the ground plane edge.

Table 2. Comparison of simulated gain and F/B ratio for the different number of period N.

Number of Period N	Antenna Gain [dBi]		F/B Ratio [dB]
	Broadside ($\theta = 0°$)	Back-lobe ($\theta = 180°$)	
1	2.4	–20.0	22.4
2	2.3	–21.8	24.1
3	1.6	–26.3	27.9

could be expected. From **Figure 9**, the maximum F/B ratio of 24.1 dB is obtained for the following case: $l = 0.5$ mm. **Figure 10** shows a comparison of the simulated input reflection coefficient (S_{11}) of the MSA with/without stand-up SRRs. The resonant frequency of the MSA decreases from 2.5 to 2.445 GHz owing to the loading effect of the array of stand-up SRRs.

For the sake of comparison, the simulated E- and H-plane radiation patterns of the different MSAs at 2.445 GHz are shown in **Figure 11**. It is observed that the back radiation is reduced when both partial ground-plane removal and a soft-surface structure consisting of the arrayed stand-up SRRs are used. The simulated gains at different angles (θ = 0°, 180°) and the F/B ratios of the MSAs are summarized in **Table 3**. Notably, the back radiation level of the proposed MSA shows improvements in both the E-plane (xz-plane) and the H-plane (yz-plane). The partially removed ground plane structure of the proposed MSA can suppress the surface wave propagation along the E-plane edge direction (y-axis) and the soft surface structure with stand-up SRRs reduces the ground-plane edge diffraction along the H-plane edge direction (x-axis) simultaneously. As a result the back radiation level of the proposed MSA can be reduced further. It is noted that a slight reduction in the broadside gain of the MSA is caused by the fact that a portion of the diffracted waves radiated from the ground plane edges contributes to both the forward waves and the backward waves. In comparison with the conventional MSA having a full ground plane of the same size, the broadside gain (at θ = 0°) of the proposed MSA reduced by 0.9 dBi. However, the back-lobe gain (at θ = 180°) was improved by around 10.7 dBi. As a result, the F/B ratio increased from 14.38 to 24.1 dB owing to partial ground-plane removal and the use of the soft-surface structure.

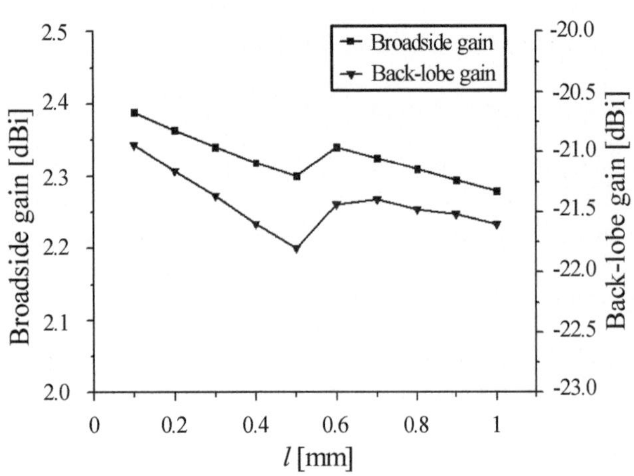

Figure 9. Simulated broadside and back-lobe gain of the MSA as a function of the distance *l* between the ground plane edge and the arrayed SRRs (*N* = 2).

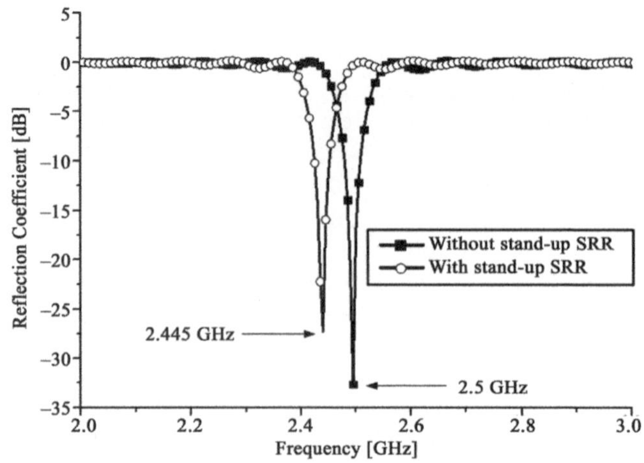

Figure 10. Simulated input reflection coefficient (S_{11}) of a MSA with/without stand-up SRRs.

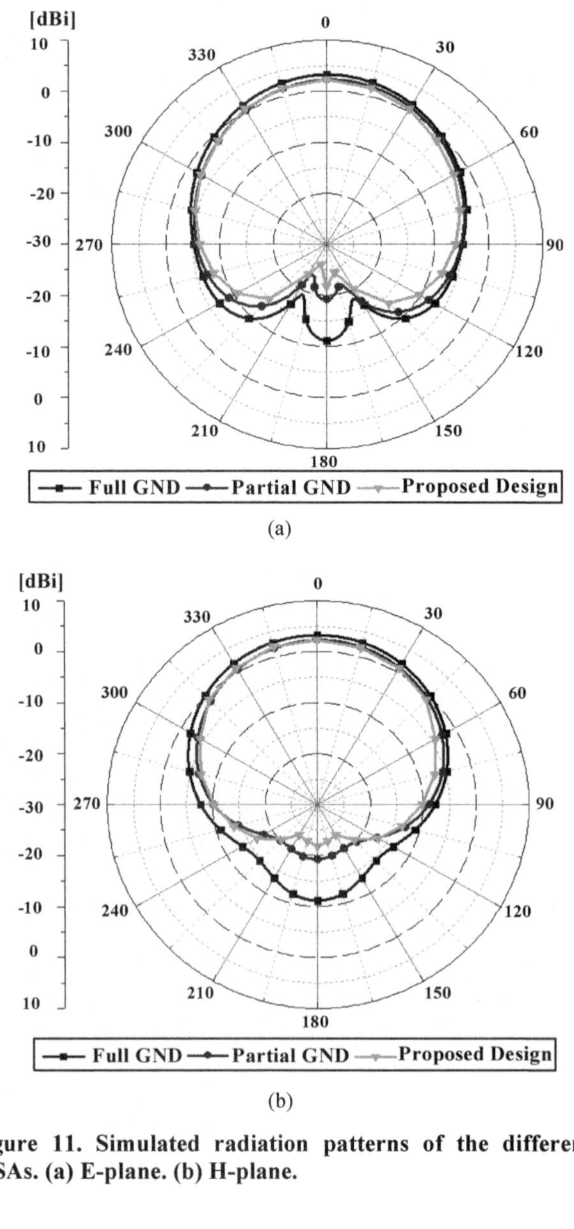

Figure 11. Simulated radiation patterns of the different MSAs. (a) E-plane. (b) H-plane.

3. Experimental Results

In order to verify the simulations, a prototype MSA loaded with the soft surface in a partially removed ground plane is fabricated on a Rogers R3210 substrate (having a relative dielectric constant of 10.2, a thickness of 1.27 mm, and a loss tangent of 0.0025); a 50 Ω SMA coaxial probe connecter is installed for feeding the MSA. The photograph of the fabricated antenna is shown in **Figure 12**. The total size of the MSA is 120 mm × 120 mm. **Figure 13** shows a comparison of the simulated and measured reflection coefficient of the proposed MSA. It can be noted that a slight reduction in the measured bandwidth ($S_{11} < -10$ dB) in comparison with that of the simulated result is attributed to the fabrication tolerances. However, the measured and simulated resonant frequencies of the MSA agree well with each other. The simulated and measured radiation patterns at a frequency of 2.445 GHz are shown in **Figure 14**.

In both the E-and the H-planes, the MSA shows a smooth symmetric omni-directional pattern with slight backward radiation. The measured broadside (at $\theta = 0°$) and back-lobe gains (at $\theta = 180°$) are around 2.8 and –21.4 dBi, respectively. As a result, an F/B ratio of 24.2 dB was achieved experimentally.

Table 3. Comparison of simulated gain and F/B ratio.

Antenna Type	Antenna Gain [dBi]		F/B Ratio [dB]
	Broadside ($\theta = 0°$)	Back-lobe ($\theta = 180°$)	
Conventional MSA with a Full Ground Plane	3.21	−11.17	14.38
MSA with a Partially Removed Ground Plane	2.33	−19.28	21.61
Proposed MSA	2.30	−21.80	24.10

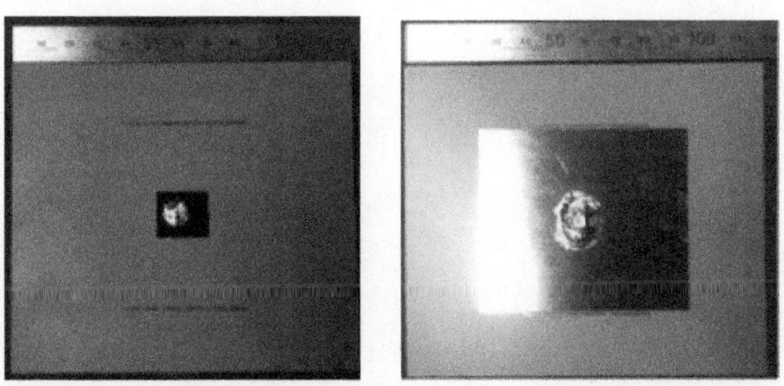

(a) (b)

Figure 12. Photographs of the fabricated antenna: (a) Top-view; (b) Bottom-view.

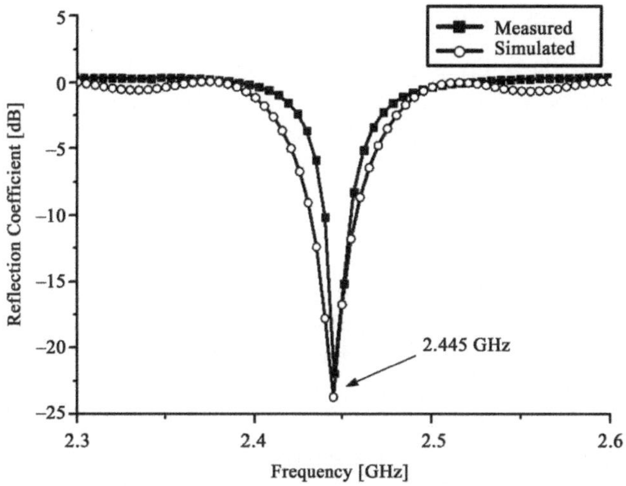

Figure 13. Simulated and measured reflection coefficient of the proposed MSA.

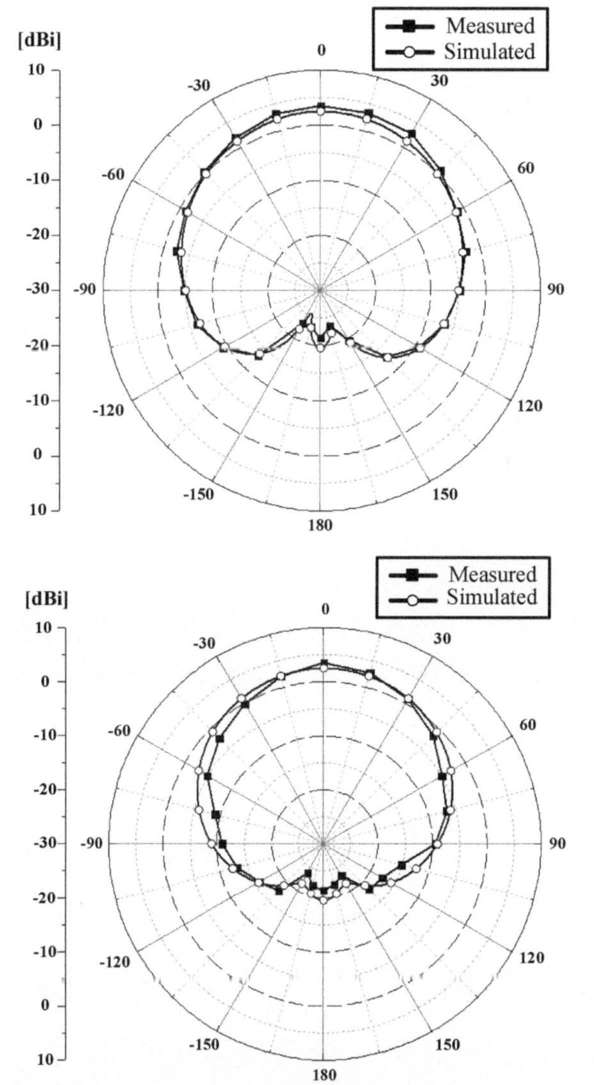

Figure 14. Simulated and measured radiation patterns of the different MSAs at 2.445 GHz.

4. Conclusion

We have investigated the effects of partial metallic ground plane removal on the radiation characteristics of a MSA. The partial ground plane removal method has been shown to improve the F/B ratio of the MSA by at least 9.7 dB as compared with that of a conventional MSA having a full ground plane of the same size. The structure of the proposed MSA is very simple and can be implemented with ease. When this structure is used between two MSAs, a mutual coupling reduction effect can also be generated. This structure can be applied for reducing back radiation in the MSA antenna design.

REFERENCES

[1] N. Llombart, A. Neto, G. Gerini and P. de Maagt, "Planar Circularly Symmetric EBG Structures for Reducing Surface Waves in Printed Antennas," *IEEE Transactions on Antennas Propagation*, Vol. 53, No. 10, 2005, pp. 3210-3218.

[2] H. Boutayeb and T. A. Denidni, "Gain Enhancement of a Microstrip Patch Antenna Using a Cylindrical Electro Magnetic Crystal Substrate," *IEEE Transactions on Antennas Propagation*, Vol. 55, No. 11, 2007, pp. 3140-3145.

[3] T. T. Thai, G. R. Dejean and M. M. Tentzeris, "Design and Development of a Novel Compact Soft-Surface Structure for the Front-to-Back Ratio Improvement and Size Reduction of a Microstrip Yagi Array Antenna," *IEEE Antennas and Wireless Propagation Letters*, Vol. 7, 2008, pp. 369-373.

[4] E. Rajo-Iglesias, Q. Quevedo-Teruel and L. Inclan-Sanchez, "Planar Soft Surfaces and Their Application to Mutual Coupling Reduction," *IEEE Transactions on Antennas Propagation*, Vol. 57, No. 12, 2009, pp. 3852-3859.

[5] E. Rajo-Iglesias, L. Inclan-Sanchez and O. Quevedo-Teruel, "Back Radiation Reduction in Patch Antennas using Planar Soft Surfaces," *Progress in Electromagnetics Research Letters*, Vol. 6, 2009, pp. 123-130.

[6] J. Papapolymerou, R. F. Drayton and L. P. B. Katehi, "Micromachined Patch Antennas," *IEEE Transactions on Antennas Propagation*, Vol. 46, No. 2, 1988, pp. 275-283.

[7] O. Quevedo-Teruel, L. Inclan-Sanchez and E. Rajo-Iglesias, "Soft Surfaces for Reducing Mutual Coupling between Loaded PIFA Antennas," *Progress in Electromagnetics Research Letters*, Vol. 9, 2010, pp. 91-94.

[8] I. Gallego-Gallego, O. Quevedo-Teruel, L. Inclan-Sanchez, E. Rajo-Iglesias and F. J. García-Vidal, "On the Use of Soft Surfaces to Reduce Back Radiation in Textile Microstrip Patch Antennas," *Proceedings of the 5th Eu-CAP*, 2011, pp. 534-537.

[9] S. B. Yeap and Z. N. Chen, "Microstrip Patch Antennas with Enhanced Gain by Partial Substrate Removal," *IEEE Transactions on Antennas Propagation*, Vol. 58, No. 9, 2010, pp. 2811-2816.

[10] J. Huang, "The Finite Ground Plane Effect on the Microstrip Antenna Radiation Patterns," *IEEE Transactions on Antennas Propagation*, Vol. 31, No. 4, 1983, pp. 649-653.

Very Compact Bandstop Filters Based on Miniaturized Complementary Metamaterial Resonators

Hicham Lalj[1,2], Hafid Griguer[1,2], M'hamed Drissi[1]

[1]European University of Bretagne, National Institute of Applied Science of Rennes, Institute of Electronic and Telecommunication of Rennes, Rennes, France; [2]Moroccan School of Engineering Science of Rabat, Rabat, Morocco.

ABSTRACT

In this paper, a design of very compact microstrip bandstop filters based on complementary split ring resonators (CSRRs) is proposed. Two techniques of metamaterial miniaturization are used to optimize the physical and electrical size of the CSRR. The bandstop filter is produced by an array of miniaturized loaded CSRRs etched on the center line of a microstrip. The size of the proposed filter, is as small as 0.58 cm^2, and its electrical length is very small with only 0.08 $\lambda 0$), compared to a conventional bandstop filter, a miniaturization of a factor 5 while the bandstop performance is maintained. A very good agreement obtained between the measurement and the simulation results.

Keywords: Metamaterials; Bandstop Filter; Complementary Split Ring Resonators (CSRRs)

1. Introduction

Many modern telecommunications systems, such as the embedded systems, systems of mobile phone and wireless communication, use filters; many conventional filter designs has reached their technologically defined limits. To satisfy the increasing demand for high performance, circuit integration and reduced size, alternative concepts have to be explored [1].

In recent years, left-handed Metamaterials have attracted considerable interest of scientists and engineers working in the field of microwave technology. These Metamaterials exhibit both a negative permittivity and permeability which result in a negative index of refraction, a property not available within any natural material [2-4]. From duality arguments, it has been shown that negative permittivity media can also be generated by means of a resonant element, namely the complementary split ring resonator (CSRR) introduced by Falcone *et al.* in 2004 [5]. These resonators can be considered as quasi-lumped elements and are, therefore, also very interesting for the miniaturization of planar microwave devices such as filters and diplexers, or for improving their performances.

In previous research of some of the authors, SRRs and CSRRs have been successfully applied to the design of microwave filters [6,7].

The aim of this paper is to apply a bandstop filter synthesis proposed in the literature [4] based on CSRRs, and two techniques of Metamaterials miniaturization [8-10], to design a very compact microwave bandstop filter.

In Section 2, we propose a design of conventional bandstop filter based on CSRRs etched on the center line of a microstrip. In Section 3, I will describe how to obtain a reduction of the filter size by the optimization of electrical and physical CSRR size. Finally, in Section 4, the obtained calculated results are discussed and compared with measurements.

2. Bandstop Filter Design

A conventional bandstop filter based on a 50 Ω line and the CSRR etched in the conductor strip [4] is designed in the microstrip technology. This filter is considered as reference of comparison with the realised miniaturized filters.

CSRR is dual counterparts of SRR. Therefore a dual electromagnetic behaviour for them is expected according to the duality theorem. The incident electric field needs to be polarized in the axial direction of the resonator. In this way, CSRRs are etched on center line of the microstrip technology.

This arrangement makes sure that the CSRRs are properly exited by the electric field applied parallel to the ring axis.

The CSRR topology and equivalent circuit model are illustrated in **Figure 1**. The CSRR unit cell was designed to operate around 5.70 GHz. The geometry of the cell is as follows: c = d = 0.3 mm, g = 0.6 mm and the global size is 7.4 mm × 3 mm.

The substrate used is a RT/Duroid having the following characteristics (relative permittivity εr = 2.2, loss tangent tg(φ) = 0.0001 and thickness h = 0.8 mm).

The resonator is simulated by using a commercially available 3D full-wave solver (Ansoft HFSS).

Figure 2 shows the [S] parameters simulated results. It shows a rejected frequency band around the designed frequency of the CSRR resonator explained by a transmission of about −25 dB.

Figure 3 shows a conventional bandstop filter design; it consists of 3 rings CSRRs. Our simulation results show that three rings could provide significant improvement. Further increase of rings would have just an increase of reject level.

Figure 4 presents the conventional bandstop filter simulated results. It has a central frequency of 5.7 GHz and a rejection band as wide as 1.2 GHz with 20% fractional bandwidth. A deep rejection band (S21 = −40 dB) is obtained in the designed frequency with low return loss.

Figure 5 shows the simulated group delay. It presents a very small variation in the operating frequency (<1 ns).

The overall area for the filter is 0.67 cm^2, and the electrical length of device, excluding access lines, is 0.44 λ0.

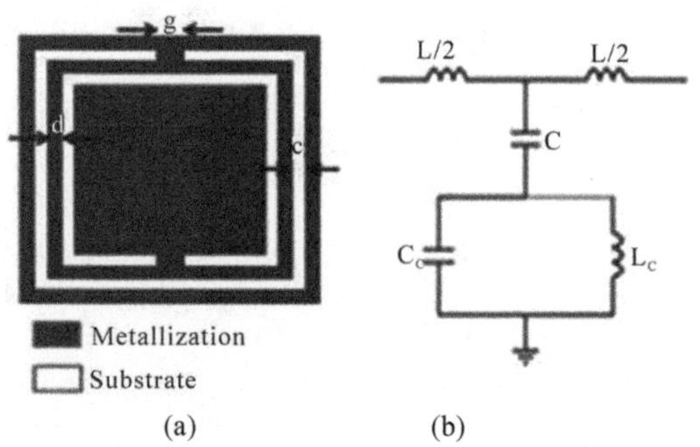

Figure 1. (a) CSRR particle with relevant dimensions; (b) CSRR microstrip line equivalent circuit model.

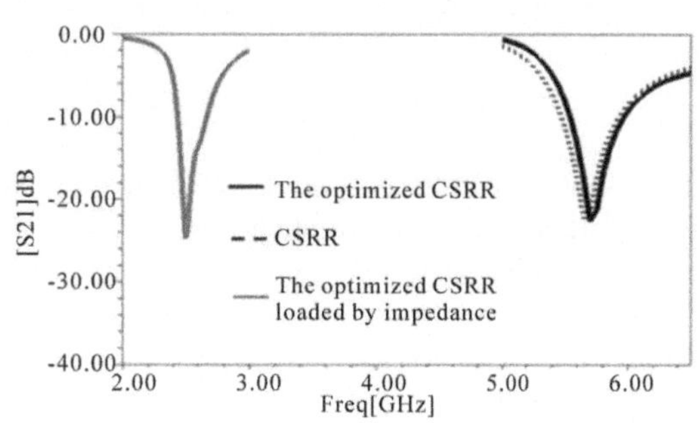

Figure 2. Simulated S21 parameters for different CSRRs structures.

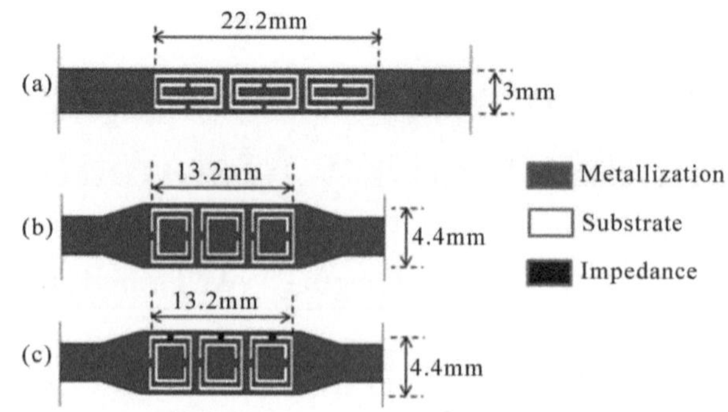

Figure 3. (a) Design of conventional bandstop filter based on CSRRs; (b) Design of proposed bandstop filter based on optimized CSRRs; (c) Design of bandstop filter based on optimized CSRRs loaded by an impedance.

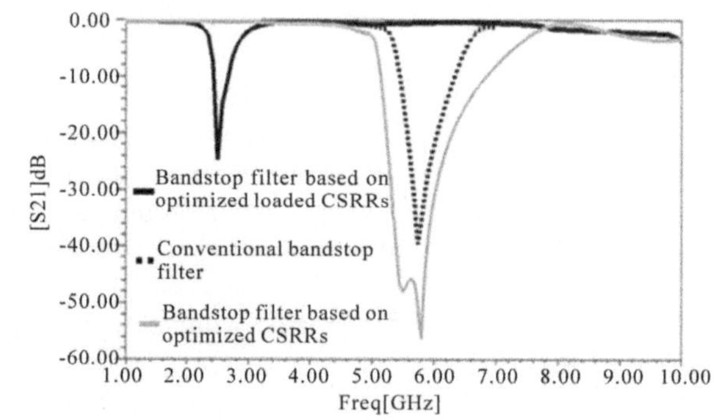

Figure 4. Simulated S21 parameters for different bandstop filters structures.

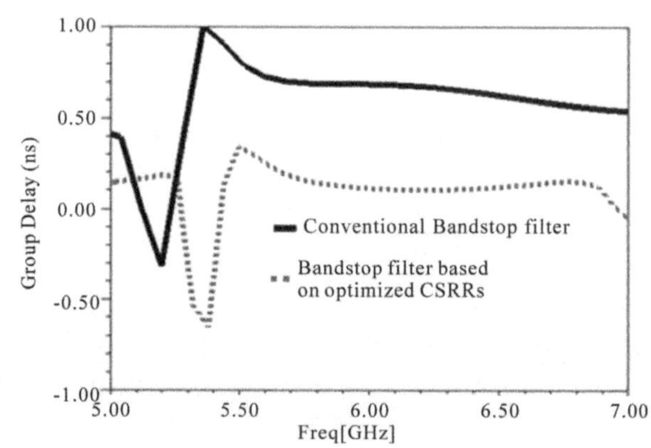

Figure 5. Simulated group delay for bandstop filters.

3. Bandstop Filter Miniaturization

To reduce the bandstop filter size whilst sustaining the filter bandstop performances, two techniques of metamaterial miniaturization are utilized. The first one is based on the dimensions optimization, and the second is based on the increasing of the equivalent distributed capacitance.

The first technique consists in reducing the physical size of the resonator, knowing that the CSRR resonant frequency depends on their dimensions [2,5].

The optimized geometry of the CSRR, while having the same metamaterial phenomenon around 5.7 GHz, is as follows: c = d = 0.2 mm, g = 0.4 mm and the global size is 4.4 mm × 4.4 mm. The substrate used is a RT/Duroid.

The **Figure 2** shows the optimized CSRR [S] simulated parameters. It presents a rejected frequency band around the designed frequency of the CSRR resonator explained by a transmission of about −25 dB.

The design of the proposed filter based on optimized CSRRs is presented in **Figure 3**.

The width of the center line is w = 3 mm (corresponding to the characteristic impedance of 50 Ω in the region without CSRR) and w = 4.4 mm (in the region with CSRR). To match this discontinuity of center line, tapered line is used.

Figure 4 shows the proposed bandstop filter [S] parameters simulated results. It displays a bandstop central frequency of 5.7 GHz with a 32% fractional bandwidth. A deep rejection band (S21 = −60 dB) is obtained in the designed frequency with low return loss. **Figure 5** shows the simulated group delay. It presents a very small variation in the operating frequency (<1 ns). The overall area for the conventional bandstop filter is 0.67 cm^2, whereas the proposed metamaterial filter consumes an area of 0.58 cm^2, and the electrical length of device, excluding access lines, is 0.26 $\lambda 0$, thus, less than the half of the conventional filter electrical length (**Table 1**), while the bandstop performance is maintained.

Thereafter we load the optimized CSRR cell by impedance to vary the distributed equivalence capacitance of the CSRR, and reduce their electrical size.

The CSRR itself can be described as an L_C resonant; the resonant frequency is described by the following expression [11]:

$$f = \frac{1}{2\pi\sqrt{L_C C_C}} \quad (1)$$

If the equivalence capacitance of the CSRR is increased, the resonant frequency shifts towards the low frequencies.

Figure 2 shows the optimized CSRR loaded by impedance [S] parameters simulated results. It shows a rejected frequency band around 2.5 GHz. In this case, we obtained a global size of the CSRR resonator around $\lambda g/20$, showing a 50% reduction in size.

Figure 3 shows the bandstop filter based on optimized CSRR loaded by impedance design.

Table 1. Reduction factor of different bandstop filters.

Type of filter	Conventional bandstop filter	Bandstop filter based on optimized CSRRs	Bandstop filter based on optimized loaded CSRRs
Reduction factor	1	2	5

The simulated results of this filter are shown in **Figure 4**. The latter shows a bandstop central frequency of 2.5 GHz with 23% fractional bandwidth. A deep rejection band (S21 = −25 dB) is obtained with low return loss.

Figure 6 shows the simulated group delay. It presents a very small variation in the operating frequency (<1 ns).

The electrical length of optimal filter is very small with only 0.08 $\lambda 0$ thus, less than a fifth of the conventional filter electrical length (**Table 1**), while the bandstop performance is maintained.

4. Measured Results and Discussions

The filter based on the optimized CSRR without impedance was fabricated and shown in **Figure 7**. AGILENT TECHNOLOGIE-N5230A vector network analyzer was used to measure the response of the proposed filter and the results are shown in **Figure 8**.

A very good agreement is obtained between simulated and measured results. The measurement shows a bandstop central frequency of 5.75 GHz with 35% fractional bandwidth. A deep rejection band (S21 = −50 dB) is obtained in the vicinity of the designed frequency with low return losses.

5. Conclusions

In this paper, a very compact planar left-handed metama-

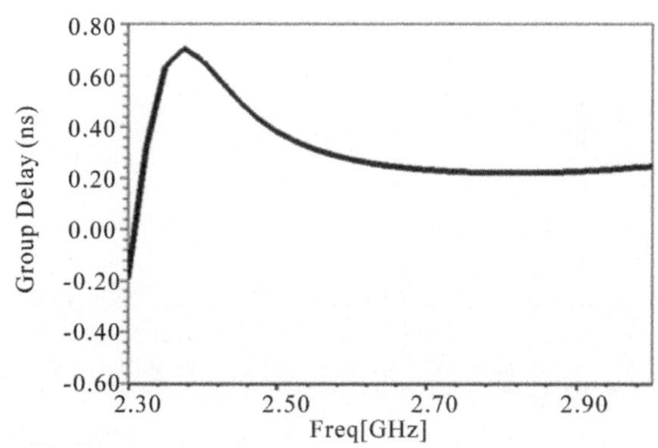

Figure 6. Simulated group delay for bandstop filter based on optimized loaded CSRRs.

(a) (b)

Figure 7. Photograph of the fabricated device: (a) Top plane; (b) Bottom plane.

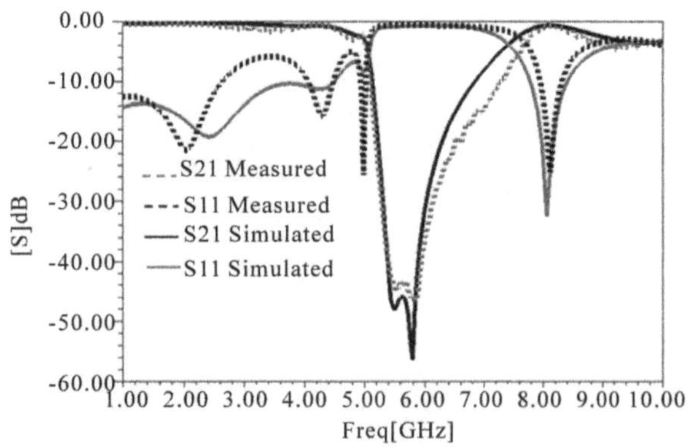

Figure 8. Simulated and Measured S parameters of the bandstop filter.

terial bandstop filter, which includes a 50 Ω line and an array of miniaturized CSRR has been demonstrated.

The combination of these two elements yields a resonator design with extremely small electrical outline dimensions. The electrical length of filter is very small with only 0.08 $\lambda 0$, compared to conventional bandstop filter, a miniaturization of a factor 5 while the bandstop performance is maintained. The bandstop filters based on this concept can be very promising for the applications where miniaturization and compatibility with planar millimeter wave technology are the key issues.

REFERENCES

[1] J.-S. G. Hong and M. J. Lancaster, "Microstrip Filters for RF/Microwave Applications," Wiley, New York, 2001.

[2] J. B. Pendry, A. J. Holden, D. J. Robbins and W. J. Stewart, "Magnetism from Conductors and Enhanced Nonlinear Phenomena," *IEEE Transactions on Microwave Theory and Techniques*, Vol. 47, No. 11, 1999, pp. 2075-2084.

[3] R. Marqués, F. Martin and M. Sorolla, "Metamaterials with Negative Parameters: Theory, Design and Microwave Applications," John Wiley & Sons, Inc., Upper Saddle River, 2007.

[4] R. W. Ziolkowski and N. Engheta, "Metamaterial Special Issue Introduction," *IEEE Transaction on Antennas and Propagation*, Vol. 51, No. 10, 2003, pp. 2546-2549.

[5] F. Falcone, T. Lopetegi, J. D. Baena, R. Marqués, F. Martín and M. Sorolla, "Effective Negative Epsilon Stop-Band Microstrip Lines Based on Complementary Split Ring Resonators," *IEEE Microwave and Wireless Components Letters*, Vol. 14, 2004, pp. 280-282.

[6] J. Garcia-Garcia, F. Martin, F. Falcone, J. Bonache, J. D. Baena, I. Gill, E. Amat, T. Lopetegi, M. A. G. Laso, J. A. M. Iturmendi, M. Sorolla and R. Marques, "Microwave Filters with Improved Stopband Based on Sub-Wavelength Resonators," *Microwave and Optical Technology Letters*, Vol. 46, 2005, pp. 283-286.

[7] J. D. Baena, J. Bonache, F. Martin, R. Marqués Sillero, F. Falcone, T. Lopetegi, M. A. Glaso, J. Garcia-Garcia, I. Gil, M. F. Portillo and M. Sorolla, "Equivalent-Circuit Models for Split-Ring Resonators and Complementary Split-Ring Resonators Coupled to Planar Transmission Lines," *IEEE Transactions on Microwave Theory and Techniques*, Vol. 53, No. 4, 2005, pp. 1451-1461.

[8] K. Aydin, I. Bulu, K. Guven, M. Kafesaki, C. M. Soukoulis and E. Ozbay, "Investigation of Magnetic Resonances for Different Split-Ring Resonator Parameters and Designs," *New Journal of Physics*, Vol. 7, No. 1, 2005, p. 168.

[9] R. Marques, F. Medina and R. Rafii-El-Idrissi, "Role of Bianisotropy in Negative Permeability and Left-Handed Metamaterials," *Physical Review B*, Vol. 65, No. 14, 2002, Article ID: 144440.

[10] H. Griguer, H. Lalj and M. Drissi, "SRR Miniaturization for Patch Antenna Stacked with Metamaterial Substrate," *Fourth International Congress on Advanced Electromagnetic Materials in Microwaves and Optics*, Karlsruhe, 13-16 September 2010.

[11] F. Falcone, T. Lopetegi, J. D. Baena, R. Marques, F. Martin and M. Sorolla, "Effective Negative ε Stop-Band Microstrip Lines Based on Complementary Split-Ring Resonators," *IEEE Microwave and Wireless Components Letters*, Vol. 14, 2004, pp. 280-282.

Permissions

The contributors of this book come from diverse backgrounds, making this book a truly international effort. This book will bring forth new frontiers with its revolutionizing research information and detailed analysis of the nascent developments around the world.

We would like to thank all the contributing authors for lending their expertise to make the book truly unique. They have played a crucial role in the development of this book. Without their invaluable contributions this book wouldn't have been possible. They have made vital efforts to compile up to date information on the varied aspects of this subject to make this book a valuable addition to the collection of many professionals and students.

This book was conceptualized with the vision of imparting up-to-date information and advanced data in this field. To ensure the same, a matchless editorial board was set up. Every individual on the board went through rigorous rounds of assessment to prove their worth. After which they invested a large part of their time researching and compiling the most relevant data for our readers. Conferences and sessions were held from time to time between the editorial board and the contributing authors to present the data in the most comprehensible form. The editorial team has worked tirelessly to provide valuable and valid information to help people across the globe.

Every chapter published in this book has been scrutinized by our experts. Their significance has been extensively debated. The topics covered herein carry significant findings which will fuel the growth of the discipline. They may even be implemented as practical applications or may be referred to as a beginning point for another development.

The editorial board has been involved in producing this book since its inception. They have spent rigorous hours researching and exploring the diverse topics which have resulted in the successful publishing of this book. They have passed on their knowledge of decades through this book. To expedite this challenging task, the publisher supported the team at every step. A small team of assistant editors was also appointed to further simplify the editing procedure and attain best results for the readers.

Our editorial team has been hand-picked from every corner of the world. Their multi-ethnicity adds dynamic inputs to the discussions which result in innovative outcomes. These outcomes are then further discussed with the researchers and contributors who give their valuable feedback and opinion regarding the same. The feedback is then collaborated with the researches and they are edited in a comprehensive manner to aid the understanding of the subject.

Apart from the editorial board, the designing team has also invested a significant amount of their time in understanding the subject and creating the most relevant covers. They scrutinized every image to scout for the most suitable representation of the subject and create an appropriate cover for the book.

The publishing team has been involved in this book since its early stages. They were actively engaged in every process, be it collecting the data, connecting with the contributors or procuring relevant information. The team has been an ardent support to the editorial, designing and production team. Their endless efforts to recruit the best for this project, has resulted in the accomplishment of this book. They are a veteran in the field of academics and their pool of knowledge is as vast as their experience in printing. Their expertise and guidance has proved useful at every step. Their uncompromising quality standards have made this book an exceptional effort. Their encouragement from time to time has been an inspiration for everyone.

The publisher and the editorial board hope that this book will prove to be a valuable piece of knowledge for researchers, students, practitioners and scholars across the globe.

List of Contributors

Shravan Sriram, Naveenkumar Vijayakumar, P. Aditya Kumar, Akash S. Shetty, V. P. Prasshanth and K. A. Narayanankutty
Department of Electronics and Communication Engineering, Amrita School of Engineering, Amrita Vishwa Vidyapeetham University, Coimbatore, India

Jean-Marie Floc'h and Ahmad El Sayed Ahmad
Unité Mixte de Recherche 6164 (UMR), Institut d'Electronique et de Télécommunications de Rennes (IETR), Rennes, France

Slimane Benmahmoud
Telecommunications and Digital Signal Processing Laboratory, Djillali Liabes University, Sidi Bel-Abbes, Algeria
Department of Electronics Engineering, M'sila University, M'sila, Algeria

Ali Djebbari
Telecommunications and Digital Signal Processing Laboratory, Djillali Liabes University, Sidi Bel-Abbes, Algeria

Ahmed A. Hamad
Department of Electrical Engineering, University of Babylon, Babel, Iraq

Iftekhar Alam and Vikas Srivastva
Wireless Communication and Computing, IIIT Allahabad, India

Arun Prakash and Rajeev Tripathi
ECED, MNNIT, Allahabad, India

A. K. Shankhwar
Electronics Engineering, HBTI, Kanpur, India

Jan Schorer and Jens Bornemann
Department of Electrical and Computer Engineering, University of Victoria, Saanich, Canada

Abdelhakim Elouadih and Ahmed Oulad-Said
Department of Electrical and Telecommunications Engineering, Royal Air Academy (ERA), Marrakesh, Morocco

Moha Mrabet Hassani
Department of Physics, Semlalia University of Sciences (FSSM), Marrakesh, Morocco

Aliakbar Hosseinzadeh, Mirhamed Mirmozafari, Mohammad Vatankhah Varnoosfaderani, Changiz Ghobadi and Javad Nourinia
Department of Electronics and Communication Engineering, Urmia University, Urmia, Iran

Meng-Ting Hsu, Ying-Hsiang Huang and Cheng-Chuan Chung
Microwave Communication and Radio Frequency Integrated Circuit Lab, Department and Institute of Electronic Engineering, National Yunlin University of Science and Technology, Douliou, Chinese Taipei

Reham A. El-Mayet and Salwa H. Elramly
Electronic and Communication Department, Ain Shams University, Cairo, Egypt

Hesham M. El-Badawy
National Telecommunication Institute, Cairo, Egypt

K. Vandhana, P. V. S. Sowmya, P. Roshni, K. Divya and S. Ashwin
Department of Electronics and Communication Engineering, Amrita Vishwa Vidyapeetham, Coimbatore, India

Bouna Fall
Université Lille Nord de France, Lille, France
Institut Français des Sciences et Technologies des Transports de L'aménagement et des Réseaux (IFSTTAR), Laboratoire Electronique, Ondes et Signaux Pour les Transports (LEOST), Villeneuve d'Ascq, France
Département Opto-Acousto-Electronique (DOAE), Université de Valenciennes et du Hainaut-Cambrésis (UVHC), Institut D'électronique, de Microélectronique et de Nanotechnologie (IEMN), Valenciennes, France
Centre National de la Recherche Scientifique (CNRS), Unité Mixte de Recherche (UMR), Villeneuve d'Ascq, France

Atika Rivenq
Université Lille Nord de France, Lille, France
Département Opto-Acousto-Electronique (DOAE), Université de Valenciennes et du Hainaut-Cambrésis (UVHC), Institut D'électronique, de Microélectronique et de Nanotechnologie (IEMN), Valenciennes, France
Centre National de la Recherche Scientifique (CNRS), Unité Mixte de Recherche (UMR), Villeneuve d'Ascq, France

Fouzia Elbahhar and Marc Heddebaut
Université Lille Nord de France, Lille, France
Institut Français des Sciences et Technologies des Transports de L'aménagement et des Réseaux (IFSTTAR), Laboratoire Electronique, Ondes et Signaux Pour les Transports (LEOST), Villeneuve d'Ascq, France

G. A. Bidkar and P. V. Hungund
Department of PG Studies and Research in Applied Electronics, Gulbarga University, Gulbarga, India

R. M. Vani
University Science Instrumentation Centre, Gulbarga University, Gulbarga, India

Hsin-Chia Yang, Cheng-Huang Tsao, Mu-Chun Wang and Sungching Chi
Department of Electronics Engineering, Ming Hsin University of Science and Technology, Hsin-Chu City, Chinese Taipei

Tsin-Yuan Chang and Jui-Ming Tsai
Department of Electrical Engineering, National Tsing-Hua University, Hsin-Chu City, Chinese Taipei

Ajay Shukla, Radheshyam Gamad and Rohan Raikwar
Department of Electronics & Instrumentation Engineering, S. G. S. Institute of Technology and Science, Indore, India

Hao-Yung Lo and Ting-Wei Lin
Department of Electrical Engineering, National Tsing Hua University, Hsinchu City, Chinese Taipei

Boudali Ouarzazi, Atika Menhaj-Rivenq and Iyad Dayoub
Department OAE, Institut d'Electronique de Microelectronique et de Nanotechnologie, Université de Valenciennes et du Hainaut- Cambrésis, Valenciennes, France

Marion Berbineau
Université Lille Nord de France, Villeneuve d'Ascq, France

A. Okassa M'foubat, C. Tatkeu and F. Elbahhar
Université Lille Nord de France, Lille, France

Hannu Olkkonen, Simo Ahtiainen and Kari Jarvinen
Department of Applied Physics, University of Eastern Finland, Kuopio, Finland

Juuso T. Olkkonen
VTT Technical Research Centre of Finland, Espoo, Finland

Lei Wang, Zhizhong Ding, Rui Huang and Qian Li
Department of Communication Engineering, Hefei University of Technology, Hefei, China

Hesheng Cheng
Department of Computer Science and Technology, Hefei Normal University, Hefei, China

Abdelati Reha
Department of Telecommunications Engineering, Higher Institute of Applied Engineering (IGA), Marrakesh, Morocco

Ahmed Oulad Said
Department of Electrical and Telecommunications Engineering, Royal Air Academy (ERA), Marrakesh, Morocco

Ayoola Akinloye Akinniranye
MTN, Nigeria

Samson Adenle Oyetunji
Federal University of Technology Akure, Akure, Nigeria

Mahammad A. Safwat and Hesham M. El-Badawy
Network Planning Department, National Telecommunication Institute, Cairo, Egypt

Ahmad Yehya
Department of Electrical Engineering, Alazhar University, Cairo, Egypt

Hosni El-Motaafy
Department of Electronics and Computer Engineering, HIT, Egypt

Shuji Sayama
Department of Communications Engineering, National Defense Academy, Yokosuka, Japan

Seishiro Ishii
Department of Information Networks, Sendai National College of Technology, Sendai, Japan

Mohamed El-Hady Magdy Keshk
Egyptian Space Program, National Authority for Remote Sensing and Space Science, Cairo, Egypt

El-Sayed Elrabie, Fathi El-Sayed Abd El-Samie and Mohammed Abd El-Naby
Faculty of Electronic Engineering, Menoufia University, Menouf, Egypt

Ali Dziri and Michel Terre
CEDRIC-LAETITIA, Conservatoire National des Arts et Metiers, Paris, France

Nidal Nasser
College of Engineering, Alfaisal University, Riyadh, KSA

Hong-Min Lee
Department of Electronic Engineering, Kyonggi University, Suwon, South Korea

Wong-Sang Choi

Hicham Lalj and Hafid Griguer
European University of Bretagne, National Institute of Applied Science of Rennes, Institute of Electronic and Telecommunication of Rennes, Rennes, France
Moroccan School of Engineering Science of Rabat, Rabat, Morocco

M'hamed Drissi
European University of Bretagne, National Institute of Applied Science of Rennes, Institute of Electronic and Telecommunication of Rennes, Rennes, France